Douglas Montgomery's Introduction to
Statistical Quality Control
A JMP® Companion

Brenda S. Ramírez
José G. Ramírez

sas.com/books

The correct bibliographic citation for this manual is as follows: Ramirez, Brenda S., M.S., and Jose G., Ramirez, Ph.D. 2018. *Douglas Montgomery's Introduction to Statistical Quality Control: A JMP® Companion*. Cary, NC: SAS Institute Inc.

Douglas Montgomery's Introduction to Statistical Quality Control: A JMP® Companion

Copyright © 2018, SAS Institute Inc., Cary, NC, USA

978-1-63526-022-9 (Hard copy)
978-1-63526-825-6 (Web PDF)
978-1-63526-823-2 (epub)
978-1-63526-824-9 (mobi)

Contents

Foreword

Statistical Process Control or SPC has been called one of the greatest technological innovations of the 20th century. I think this is because the techniques have a sound intuitive basis, are straightforward mathematically, and have broad applicability to a wide range of industrial and business environments, including but not limited to manufacturing, process development, product design, supply chain operations, financial operations, health care, logistics and distribution, and many other transactional and service operations. The application of SPC along with other techniques for quality and business improvement have led to significantly improved quality and reliability of many products and services and contributed in an important way to business success and economic development.

Two other innovations have also played a key role in the successful deployment of SPC and other quality improvement tools. These are the use of deployment frameworks, the most successful of which in my view is Six Sigma, and computer software. Because SPC can be very data-intensive appropriate software is essential to any successful application, and JMP is an outstanding package. It has all of the fundamental and advanced techniques that are necessary to a successful SPC implementation.

The authors have done an excellent job of demonstrating how the key ideas of SPC in my book, both basic Shewhart control charts, and more advanced techniques, can be implemented in JMP. The software package has a logical design and the authors provide detailed step-by-step help along with screen shots and output from JMP to guide the reader to successful use of the technology. In many places they also provide additional insights about the methodology or extensions of some of the basic ideas that are extremely useful to the practitioner. The authors have an extensive background in the application of these methods across a variety of industrial and business settings, and this comes through clearly in the writing. Some of their own innovations such as measures of process stability are included and thoroughly illustrated in the book.

I highly recommend this book. It is well-written, and provides clear, authoritative guidance on the implementation of SPC through the JMP software package. Even if you are an experienced JMP user you will find the book a rewarding and useful reference. For new users, the book is an invaluable aid that will quickly facilitate your successful use of the SPC toolkit.

Douglas C. Montgomery
Regents' Professor of Industrial Engineering
ASU Foundation Professor of Engineering
Ira A. Fulton School of Engineering
Arizona State University, Tempe, AZ

About This Book

Why Statistical Quality Control?

What comes to mind when you think of statistical quality control (SQC)? The Encyclopedia Britannica defines this phrase as "the use of statistical methods in the monitoring and maintaining of the quality of products and services." This definition is in line with our initial exposure to SQC during our college years, in classes like Statistical Process Control. These ideas continued to take shape when we studied for the American Society for Quality Certified Quality Engineering exam, which had us memorize numerous facts about different statistical quality tools. But it was not until we started using these tools and techniques in a real-world manufacturing environment that we truly understood their impact on improving products and processes.

Thirty years and several industries later, we have become great stewards of SQC techniques, and their use and application have become second nature. Therefore, when we were asked to author a companion book to Prof. Montgomery's *Introduction to Statistical Quality Control* (ISQC), we enthusiastically agreed. Like many, we were introduced to his work through his many books. They are among our favorites because they are very readable, practical, and relevant, not only to the industries that we have worked in but also to the engineers and scientists with whom we often interact. This is no coincidence since Professor Montgomery holds BS, MS, and PhD degrees, all in engineering, and has spent many years both as a professor of Industrial Engineering and Statistics at Arizona State University and as a practitioner collaborating with people in industry.

The synergy between engineering, science, and statistics is always found in Prof. Montgomery's teachings. Take ISQC, for example. This book provides applications for many of the common SPC techniques using data sources from well-known manufacturing and business processes. For example, the book educates the reader about XBar and Range charts using dimensional measurements from a Hard-Bake process, C charts are applied to nonconformities on a printed circuit board, and we interpret the results of an attribute gauge capability analysis to understand the consistency of a manual underwriting process for mortgage loan applications. ISQC Chapter 10, "Other Univariate Statistical Process-Monitoring and Control Techniques," contains many useful monitoring techniques that are very effective in practice but may be overlooked or misunderstood. We encourage you to check out his discussions for how to adapt SPC charts for the following scenarios: short production runs, nonstationary and autocorrelated output, change-point models, profile monitoring, and multistream processes.

Following in Prof. Montgomery's footsteps, we have written a companion book that is geared toward the practitioner of SQC, one who is using these techniques to monitor and improve products and processes. One of our goals in writing this book is to share valuable lessons that we have learned from applying SQC techniques to solve problems in a variety of industries, including semiconductors, electronics, chemical, and biotechnology.

Finally, to fully answer the question of why SQC, we must turn our attention to JMP software. We have been avid JMP users for almost as long as we have been industrial statisticians and know the software well. JMP not only has powerful SQC tools that are easy to use, but it also has plenty of state-of-the-art analysis and visualization tools if the need arises. We have included more than 20 JMP SQC platforms in our book, with step-by-step instructions and tips and tricks

What Does This Book Cover?

As the title suggests, this book is a JMP companion to *Introduction to Statistical Quality Control, Seventh Edition* by Douglas C. Montgomery, which we refer to as ISQC throughout this book. However, the main emphasis of this book is on statistical process control and capability analysis. Therefore, we focus on the techniques provided in ISQC Part 3, "Basic Methods of Statistical Process Control and Capability Analysis," and ISQC Part 4, "Other Statistical Process Monitoring and Control Techniques." These include topics such as Statistical Process Control (SPC), Process Capability Analysis (PCA), Measurement System Analysis (MSA), and Advanced Statistical Process Control (SPC).

For ISQC Chapters 6, 7, 8, 9, 10, and 11, we systematically reproduce the examples and relevant output using JMP. We provide the reader with easy step-by-step instructions, screen captures, and tips and tricks to follow along with. This book is useful for the practitioner because we emphasize the interpretation of the output and provide practical advice for how to navigate common challenges when using these techniques, based on our many years of experience using SPC.

Some recent advances in JMP related to these topics are highlighted in this book. This includes a thorough review of the **Control Chart Builder** and **CUSUM Control Chart**, which are relatively new additions to the **Quality and Process** menu. We are also excited to include a chapter on the **Process Screening** platform, new to JMP version 13, which includes the Stability Ratio in B. Ramírez and G. Runger (2006), and JMP Process Performance Graph, based on the process performance dashboard of J. Ramírez. This information is used to identify the overall health of a process through a Process Health Assessment (PHA).

Is This Book for You?

The main audience for this book is you, the practitioner, who uses these valuable quality and productivity statistical techniques and for which JMP provides a state-of-the-art implementation of them. This book provides the reader with an overview of concepts and tools used to statistically monitor process output, determine the ability of a process to meet specification limits, understand measurement system variability, and assess and prioritize the overall health of many processes. These techniques are used to aid development and manufacturing activities in a variety of industries, including, but not limited to, the automotive, biotechnology, electronics, pharmaceutical, medical devices, chemical, military, and aerospace industries.

This book is also suitable for anyone using Prof. Montgomery's *Introduction to Statistical Quality Control* book to increase your knowledge of these techniques. This includes students taking a SQC course with ISQC as the textbook. In addition to emphasizing the key topic-related content of ISQC, we also provide additional analyses that offer insight to effectively implementing these important tools. Finally, for those who want to learn how to use JMP to more easily explore your data using tools associated with SPC, PCA, MSA, and Advanced SPC, this book is a must.

What Are the Prerequisites for This Book?

Although we provide an overview of each statistical quality tool introduced in this book, we refer the reader to Prof. Montgomery's *Introduction to Statistical Quality Control* for detailed discussions on theory and concepts. We also assume a familiarity with the basic functions of JMP, such as importing and manipulating data, navigating around the JMP menus and windows, and using the basic JMP tools. A summary of related JMP help and resources is provided in the subsequent section called JMP Software.

What Should You Know about the Examples?

For most of the examples presented in ISQC Parts 3 and 4, step-by-step instructions are provided for the reader to follow along, with lots of JMP screen captures. A discussion of the analysis results is also included, and the output is interpreted in the context of the analysis goals. Supplementary examples are provided in the Statistical Insights section in each chapter to illustrate additional JMP functionality not previously covered or to elaborate on important points. A summary of the examples used in this book is provided in Chapter 1.

Software Used to Develop the Book's Content

This book was written using JMP version 14. We have included more than 20 JMP SQC platforms in our book, which are primarily part of the **Quality and Process** menu. As mentioned previously, we are aware of two platforms discussed in this book that were added in the last several versions, Process Screening (version 13) and CUSUM Control Chart (version 14). A summary of the JMP platforms used in this book is shown in Chapter 1.

Example Data

The data used in this book is available at http://support.sas.com/jramirez or http://support.sas.com/bramirez. Users can download the JMP tables and follow along.

About the Authors

Brenda S. Ramírez, MS, is an industrial statistician with many years of experience working in the semiconductor, chemical, and biotechnology industries. In this role, Brenda partners with engineers and scientists to bring new products to market, sustain manufacturing operations, and guide process improvements through the union of science and statistics. She has spent her career using and promoting Statistical Quality Control techniques, such as SPC and Process Stability metrics. Brenda received an MS in applied statistics from Worcester Polytechnic Institute and an MS in industrial and management engineering from Rensselaer Polytechnic Institute. She is an avid user of SAS and JMP statistical software from SAS. Her book, *Analyzing and Interpreting Continuous Data Using JMP: A Step-by-Step Guide*, written with her husband José Ramírez, won the 2010 Award of Excellence in the Society for Technical Communications International Technical Publications Competition.

José G. Ramírez, PhD, is a statistical engineer with years of experience in the semiconductor, electronics and biotech industries. A JMP user for more than 25 years, he works closely with engineers and scientists to help them make sense of data, and through collaborative education, helps promote statistical thinking and JMP usage. He received a degree in mathematics from Universidad Simón Bolívar in Caracas, Venezuela, and both an MS in applied statistics and a PhD in statistics from the University of Wisconsin-Madison. He was one of the founding members of the Center for Quality and Productivity Improvement at the University of Wisconsin-Madison. At the 1998 international SAS users conference, Ramírez won the best contributed statistics paper, and in 2002 he received the SAS User Feedback Award. His book, *Analyzing and Interpreting Continuous Data Using JMP: A Step-by-Step Guide*, written with his wife Brenda Ramírez, won the 2010 Award of Excellence in the Society for Technical Communications International Technical Publications Competition.

Learn more about these authors by visiting their author pages, where you can download free book excerpts, access example code and data, read the latest reviews, get updates, and more:
http://support.sas.com/brenda-ramirez
http://support.sas.com/jose-ramirez

We Want to Hear from You

SAS Press books are written *by* SAS users *for* SAS users. We welcome your participation in their development and your feedback on SAS Press books that you are using. Please visit sas.com/books to do the following:

- Sign up to review a book
- Recommend a topic
- Request information on how to become a SAS Press author
- Provide feedback on a book

Do you have questions about a SAS Press book that you are reading? Contact the author through saspress@sas.com or https://support.sas.com/author_feedback.

SAS has many resources to help you find answers and expand your knowledge. If you need additional help, see our list of resources: sas.com/books.

Acknowledgments

Many engineers, scientists, statisticians, practitioners, and students have benefited from Professor Montgomery's prolific writings, in a wide range of topics, including engineering statistics, design of experiments, linear regression, time series, response surfaces, and statistical quality control. We have applied his teachings throughout our careers and have enjoyed our interactions with him over the years. We are fortunate to have been able to write a companion book to one of his most popular books, *Introduction to Statistical Quality Control*. This companion book is a way for us to express our appreciation to him.

We also want to express our gratitude to John Sall, who had the vision more than 30 years ago to create JMP, an intuitive, powerful, and innovative statistical software for engineers and scientists, to help them make sense out of their data. This book would have not been possible without him. We would also like to thank our many friends and collaborators at JMP, including the ones who took the time to review this book: Karen Copeland, Jianfeng Ding, Annie Dudley Zangi, Curt Hinrichs, Laura Lancaster, Peng Liu, Tonya Mauldin, Di Michelson, Sarah Seligman, Sue Walsh, Byron Wingerd, and Leo Wright.

Special thanks go to our editor and 'project manager', Catherine Connolly. Her critical eye for detail and editorial assistance were invaluable, as well as her suggestions to make the manuscript better. Thanks also to the SAS Press team for their dedication to bringing this book to print.

Gracias to our friend and superb graphic designer Víctor Fiol, who took a histogram, a control chart and a process performance graph and created a wonderful and meaningful cover.

Finally, with much love, we dedicate this book to our daughter Oriana, for her patience and for reminding us, "Mom, Dad you have to work on the book this weekend."

Chapter 1: Using This Book

Overview

This chapter illustrates how to use this book. The linkages between Prof. Douglas Montgomery's *Introduction to Statistical Quality Control* (ISQC) and this book are clearly defined to help the reader navigate the examples. Each JMP platform discussed in this book is described in context with the statistical tools used to solve problems. Finally, the chapter layout and the typographical conventions that are highlighted throughout each chapter are reviewed.

Chapter Contents

As mentioned in About This Book, this book was written for the 7th edition of *Introduction to Statistical Quality Control*. References in our book refer to specific ISQC chapters, chapter sections, examples, tables, figures and equations. We provide overviews and discussions for most of the examples presented in ISQC Part 3, Basic Methods of Statistical Process Control and Capability Analysis, and ISQC Part 4, Other Statistical Process-Monitoring and Control Techniques, and readers should be able to locate them in the corresponding ISQC chapter of future editions. In addition, this book was written using JMP version 14.

Table 1.1 Overview of Chapters and Links to ISQC, 7th Edition

JMP Companion Book Chapter Number and Title	Introduction to Statistical Quality Control (ISQC) Chapter Number and Title	Statistical Quality Control Techniques
CH 3. Control Charts for Variables	CH 6. Control Charts for Variables	XmR, XBar & Range, XBar & S, and 3-way control charts
CH 4. Control Charts for Attributes	CH 7. Control Charts for Attributes	P, NP, C, and U charts, Pareto Plot and Fishbone Diagram
CH 5. Process and Measurement System Capability Analysis	CH 8. Process and Measurement System Capability Analysis	Process Capability Indices (C_p, P_p, C_{pk}, P_{pk}), Gauge R&R and EMP, tolerance intervals
CH 6. Process Health Assessment	N/A	Stability Ratio, P_{pk}, and the Process Performance Dashboard
CH 7. Cumulative Sum and Exponentially Weighted Moving Average Control Charts	CH 9. Cumulative Sum and Exponentially Weighted Moving Average Control Charts	CUSUM control chart, EWMA control chart, and UWMA control chart
CH 8. Other Univariate Statistical Process Monitoring and Control Techniques	CH 10. Other Univariate Statistical Process-Monitoring and Control Techniques	Time Series Residuals control charts, Time Series Forecast charts, and the Cuscore chart
CH 9. Multivariate Process Monitoring and Control	CH 11. Multivariate Process Monitoring and Control	Hotelling's T^2 control chart, Multivariate Variability chart and Regression adjusted chart

Chapter 2: Overview of Statistical Quality Control Topics and JMP

This chapter introduces the topics covered in this book. We discuss both key concepts and JMP platforms associated with statistical quality control (SQC) topics. This includes several concepts associated with statistical process control (SPC), including techniques used with continuous and attribute data. The importance of assessing measurement (or gauge) error and how to carry out a measurement system analysis (MSA) for variable and attribute measurements are also discussed. Finally, the concept of a process health assessment, which combines process capability and process stability metrics, is introduced using the Process Performance graph.

This chapter also includes a high-level overview of JMP platforms used in this book. We provide screen captures that illustrate how to navigate to all of the JMP platforms used to support each of the main SQC topic areas: SPC, MSA and PHA. We also include a section on JMP table formats, where the most efficient JMP table structure to facilitate the different analysis types is discussed. In addition, we provide tips for how to facilitate the creation of multiple control charts from one JMP dialogue window.

Chapter 3: Control Charts for Variables

In this chapter, we discuss control charts for variable data, also known as Shewhart control charts. Dr. Walter Shewhart's XBar and Range ($\overline{\mathbf{X}}$ and R) control charts are the most well-known, and probably the most used, among practitioners and students alike. The XBar & R, XBar and Standard Deviation ($\overline{\mathbf{X}}$ and S), and Individual Measurement and Moving Range (XmR) control charts are highlighted in this chapter.

We provide step-by-step directions for how to reproduce Prof. Montgomery's ISQC Chapter 6 examples using JMP and offer detailed discussions of the analysis results. For example, we re-create the XBar & R chart for Flow Width measurements, XBar & S chart for Piston Ring data, XmR chart for Loan Processing Cost data, XmR chart for the log of Resistivity data, and a 3-way control chart for Vane Height data.

Several JMP platforms are highlighted in this chapter. The **Control Chart Builder** is the JMP platform that is used to create variable control charts in this chapter. This platform, which was introduced in version 10, is an easy and fun way to design process behavior charts and offers many powerful features. For those who are using earlier versions of JMP, the **Control Chart** platform, which is becoming a legacy platform, is also illustrated in this chapter.

In the Statistical Insights section, we show how to use the Operating Characteristics (OC) curve to evaluate the sensitivity of a control chart to detect a mean shift of interest. A phase control chart is created, which plots multiple control charts with unique control limits in the same panel, to compare the performance of a parameter during two different time periods. Finally, two examples are included to address departures from model assumption using lognormal probability limits to monitor a skewed parameter in the original units of the data and a variance components analysis to understand the sources of variation in hierarchical data.

Chapter 4: Control Charts for Attributes

In this chapter, we discuss control charts for attribute data. These types of charts are appropriate for data measured on a nominal or ordinal scale, such as, fraction good/bad or total defect counts. The P and NP control charts, for fraction nonconforming, and C and U control charts, for total defects, are highlighted in this chapter.

We provide step-by-step directions for how to reproduce Prof. Montgomery's ISQC Chapter 7 examples using JMP and offer detailed discussions of the analysis results. For example, we re-create a P chart for the fraction nonconforming of Orange Juice Cans, a P chart with variable control limits for the fraction nonconforming of Purchase Order data, C chart for nonconformities on printed circuit boards, U chart for Supply Chain Operations data and Textile Finishing Dyed Cloth data and a rare-events chart for Valve Failures.

Several JMP platforms are highlighted in this chapter. The **Control Chart Builder** is the JMP platform that is used to create variables control charts in this chapter. This platform, which was introduced in version 10 as an easy and fun way to design process behavior charts, has many powerful features. For those who are using earlier versions of JMP, the **Control Chart** platform, which is becoming a legacy platform, is also illustrated in this chapter.

In the Statistical Insights section, we show how to use the Operating Characteristics (OC) curve to evaluate the sensitivity of a control chart to detect a mean shift of interest. We also discuss the impact of overdispersion, which inflates the binomial or Poisson variance, on the performance of P/NP and U/C charts and show how to adjust these charts accordingly. Finally, we present an alternative approach for monitoring rare events, monitoring the rare event count directly with a G chart.

Chapter 5: Process and Measurement System Capability Analysis

In this chapter, we discuss process and measurement system capability analysis. Process Capability Analysis (PCA) is used to assess the ability of a process to meet stated requirements, such as, specification limits. Process capability indices, such as C_p and C_{pk}, which rely on short-term variability, and P_p and P_{pk}, estimated from long-term variability, are discussed. Measurement System Analysis (MSA) is used to understand the impact of measurement variability. In this chapter, we show how to carry out a Gauge R&R analysis for gauges that measure both continuous and attribute properties.

We provide step-by-step directions for how to reproduce Prof. Montgomery's ISQC Chapter 8 examples using JMP and offer detailed discussions of the analysis results. For example, we re-create the process capability analysis for Bursting Strengths of glass containers, Gauge R&R for Thermal Impedance data, and an attribute Gauge R&R for Loan Evaluation data. For some examples, data were simulated using the information provided. For example, we simulate data to address process centering for ISQC Example 8.3 and to investigate the impact of tolerance stack up in ISQC Example 8.8.

Several JMP platforms are highlighted in this chapter. The **Distribution** and **Control Chart Builder** are used to carry out process capability analysis. It should be noted that process capability analysis can also be done using the **Process Capability** and **Process Screening** platforms, which are covered in depth in Chapter 6. The **Measurement System Analysis** platform is used for Gauge R&R and Evaluating the Measurement Process (EMP) studies using continuous variables. The **Variability/Attribute Gauge Chart** performs a Gauge R&R for continuous and attribute measurement variables.

In the Statistical Insights section, we show how to calculate process capability indices for parameters for which the normal distribution is not a good approximation. We use non-normal probability distributions, such as the lognormal distribution, to calculate P_p and P_{pk}. We discuss the impact of sample size on the uncertainty of process capability indices. Finally, we show how the EMP methodology provides what we believe is a better way to analyze Gauge R&R data for continuous measurements.

Chapter 6: Process Health Assessment

In this chapter, we show how to conduct a process health assessment (PHA) using process capability and process stability concepts. We discuss the Stability Ratio test, introduced by B.

Ramírez and G. Runger (2006), to classify a parameter as stable and exhibiting common cause variation or as unstable and symptomatic of special cause variation. We use many examples to show how to interpret this statistic in the context of SPC. We then combine the process Stability Ratio with the process capability metric P_{pk} to determine the overall health of the process. Four process states are possible, including ideal (stable and capable), non-ideal (unstable and incapable), a potential yield issue (stable and incapable), or a potential process issue (unstable and capable). JMP's Process Performance Graph, based on the process performance dashboard of J. Ramírez (2018), is used visualize the process health for many parameters among the four quadrants.

There is not a 1-1 correspondence with this chapter and a chapter in Prof. Montgomery's ISQC book. The reader can refer to ISQC Chapter 6 and ISQC Chapter 8 for detailed discussions of process stability and process capability, respectively. At the end of ISQC Chapter 7, there is a discussion of actions taken to improve the process. The data used in this chapter was gathered from a few different sources and slightly altered in some places to illustrate the concepts.

The **Process Screening** platform, which was new to JMP version 13, takes center stage in this chapter. We provide an overview of this platform and step-by-step directions for how to navigate the various options. We also demonstrate the **Control Chart Builder** and the **Process Capability** platforms because both can be launched from within the **Process Screening** platform. The **Control Chart Builder** provides the control chart, process stability metrics (short- and long-term variation and the Stability Ratio), and process capability analysis for each individual parameter. We use the **Process Capability** platform to show a few additional graphs related to process capability.

In the Statistical Insights section, we show how to alter the size of the points (bubble) used in the Process Performance Graph using a third variable. The accumulated cost of goods for each process step is used to increase or decrease the size of each bubble used to plot the P_{pk} and Stability Ratio results for each parameter displayed in the Process Performance Graph. This information is used to help prioritize improvement efforts.

Chapter 7: Cumulative Sum and Exponentially Weighted Moving Average Control Charts

In this chapter, we discuss statistical process monitoring approaches that are commonly used when it is desirable to detect a smaller shift in the process mean. The Cumulative Sum, or CUSUM, control chart is discussed first. The CUSUM charting statistic is the cumulative sum of the deviations of each result from a specified target value. This control chart quickly produces a signal when there is a smaller shift in the process mean from the target, either above or below it. The CUSUM chart is our favorite among the smaller shift detection control charts. The exponentially weighted moving average (EWMA) control chart is also evaluated for smaller shifts. The EWMA charting statistic is a weighted average of the individual points, where the weights decrease in an exponential fashion with older data. Finally, a simple moving average of the individual points is used to create a Moving Average, or UWMA, control chart.

Examples from Prof. Montgomery's ISQC Chapter 9 are presented in this chapter. We provide step-by-step directions for how to reproduce these examples using JMP and offer detailed discussions of the analysis results. For example, we re-create the tabular CUSUM and graph using the random normal data in ISQC Table 9.1 and the EWMA and Moving Average control charts using the same data. We also show how to create a CUSUM chart to detect changes in variability using the Piston Ring data in ISQC Chapter 6.

Two JMP platforms are highlighted in this chapter. The **CUSUM Control chart** platform is part of the new generation of quality tools that is used to create CUSUM charts for shifts in the lower and/or upper directions. It uses a specified decision limit h, as compared with the traditional V-mask, to produce a signal. The **Control Chart** platform is used to create EWMA and UWMA control charts. Note that this platform can also create CUSUM control charts and, for completeness, one such example is included in this chapter.

In the Statistical Insights section, we show how to evaluate the performance of a CUSUM chart using h, k, and the average run length (ARL). We produce an ARL graph and discuss how it can be used to help our selection of h and k. We show two approaches to create a CUSUM chart to detect a shift in the process variation using a standardize quantity and a technique developed by J. Ramírez (1989).

Chapter 8: Other Univariate and Statistical Process Monitoring and Control Techniques

In ISQC Chapter 10, Montgomery provides a treasure trove of charting techniques for parameters with non-standard data structures. In this chapter, we focus on his discussion of process monitoring approaches for autocorrelated data using model-based approaches and Cuscore charts. We use a class of models called AutoRegressive Integrated Moving Average (ARIMA) to model the autocorrelation in the baseline time series data, and then monitor the output using an XmR control chart of the residuals. We also show how to use the time series forecast equation and prediction limits to monitor the parameter in its original scale.

The data for the examples in Prof. Montgomery's ISQC Chapter 10 were unavailable, so we are not able to reproduce the results for the relevant examples. Instead, we illustrate the techniques using a data set from the semiconductor industry (J. Ramírez 1998). We provide step-by-step directions for how to perform statistical process monitoring for autocorrelated data using JMP, and have detailed discussions of the analysis results. For example, we emulate the technique in ISQC Example 10.2, and show how to fit an IMA(1, 1) model to the data, and control chart the residuals. We also provide our version of the moving centerline EWMA control chart shown in ISQC Figure 10.17. While the Cuscore chart was discussed in ISQC Section 10.7, a specific example was not provided. Again, we used the data in Ramírez (1998) to illustrate this technique.

The **Time Series** platform in JMP is highlighted in this chapter. This platform is located under **Specialized Modeling**. This platform is used to fit an appropriate model and export the charting statistics. The control charts are produced in the **Control Chart Builder** or **Graph Builder** platforms. At this time, an automated tool is not available to create the Cuscore chart, so it is generated using the formula editor and the **Graph Builder**.

In the Statistical Insights section, we further discuss the Residuals control chart and provide additional insight into the signals observed in our baseline data. We fit several different ARIMA models to the baseline data and show how to evaluate and select the most appropriate one for the autocorrelation structure. We illustrate how to monitor future observations of the data using the forecast equation from the original fit. Finally, the Cuscore chart is used to detect changes in the time series model parameter with the new observations.

Chapter 9: Multivariate Process Monitoring and Control

In this chapter, we discuss process monitoring techniques that are used to monitor multiple parameters simultaneously. These techniques can be found in the literature under the banner Multivariate SPC. The Hotelling T^2 control chart is highlighted because it is one of the most common charts used for multivariate data. We also discuss a regression adjustment approach that is used to remove the effect of all other specified variables on the variable of interest and control charting the residuals. Finally, we use a multivariate approach to detect changes in the variance-covariance matrix of multiple parameters.

Examples from Prof. Montgomery's ISQC Chapter 11 are presented in this chapter. We provide step-by-step directions for how to reproduce these examples using JMP and offer detailed discussions of the analysis results. For example, we re-produce the Hotelling T^2 control chart for the Textile Fiber data, the Hotelling T^2 control chart for the Grit Composition data, and the regression adjustment chart for the Cascade Process data. We also show the multivariate analysis of the Chemical Process data.

Two JMP platforms are highlighted in this chapter, **Multivariate Control Chart** and **Multivariate**. The first platform is found under the **Quality and Process** menu and is used for SPC applications. The second platform is found under the **Multivariate Methods** menu, which also includes partial least squares, principal components, and discriminate analysis, for example. Although this platform is not specifically for statistical process monitoring, it is included in this chapter because it has more robust tools to investigate multivariate relationships among parameters.

In the Statistical Insights section, we explore an additional tool called the Partitioned T Square provided in JMP's multivariate SPC platform to help interpret and understand multivariate data. We also show how to save the multivariate limits developed in Phase I monitoring and apply them to new data obtained during Phase II monitoring.

Chapter Layout

To make it easier for the reader to locate information, we have used the same layout for Chapter 3 through Chapter 9, as shown in Table 1.2. Each chapter starts with a brief description of the SQC tools, a reference to the relevant ISQC chapter, and a list of the JMP platforms that are used. The concepts and techniques related to the SQC chapter topic are reviewed next, which includes a discussion of terminology, key concepts, underlying assumptions, and applications. In the next section, the JMP platforms featured in the chapter are listed and a table is provided that summarizes the important features that will be demonstrated using the examples. Since this is a companion book, a table is provided that summarizes the ISQC examples that will be reproduced in the chapter. This table includes information needed to follow along with the examples, including the ISQC example number, JMP table name, JMP platform, and key points.

The ISQC example numbers and JMP table names use a consistent nomenclature throughout this book. The ISQC example numbers are presented using ISQC nomenclature of Example followed by chapter number and example number; that is, "Example #.#". For example, Example 6.1, found on page 239 of ISQC, is the first example in ISQC Chapter 6, Control Charts for Variables. In some instances, we reproduce specific ISQC Figures or Tables that are found in a particular section. These are noted as "Table" or "Figure" with the corresponding ISQC number. For example, in Chapter 9 in this companion book, there is an entry for "Table 11.2 Grit Composition", which refers to an example discussed in Section 11.3.2 in ISQC Chapter 11, Multivariate Process Monitoring and Control, using data provided in ISQC Table 11.2.

Similarly, the JMP table names provided in the ISQC Examples section use nomenclature that references the chapter number in this companion book and the table number in ISQC; that is, Chapter X – ISQC Table Y.Z. For example, Chapter 4 – ISQC Table 7.10, found on page 325 of ISQC, refers to the JMP table name used in Chapter 4 of this companion book, which includes data from Table 7.10 of ISQC Chapter 7, Control Charts for Attributes. This nomenclature is not followed for data sources outside of ISQC. For example, the JMP data table Chapter 6 – Semiconductor 6.1, referenced in Chapter 6 of this companion book, has no specific reference back to ISQC.

The step-by-step instructions are provided next, which include lots of JMP screen captures, so the reader can follow along. A discussion of the analysis results is also included in this section, where important JMP features are highlighted, and the output is interpreted in the context of the analysis goals. Additional examples are provided in the Statistical Insights section in each chapter to illustrate additional JMP functionality not previously covered or to elaborate on important points. Note the list of references used in each chapter is provided in the end of this book.

Table 1.2 Chapter Sections and Descriptions

Section	Description
Overview	A brief description of the SQC tools discussed in the chapter and a reference to the relevant ISQC chapter are provided. The JMP platforms that are highlighted in the chapter are also included.
SQC Topic Review	This section provides a review of the SQC topics examined in the chapter. This review includes a discussion of terminology, key concepts, underlying assumptions, and applications. In some instances, additional references are provided.
JMP Platforms	The JMP platforms featured in the chapter are identified and discussed. The key features of each one are summarized in a table, which reflect the features that are discussed in the examples.
ISQC Examples	In this section, we reiterate the relevant ISQC chapter covered in the companion chapter. We also provide a table that summarizes the ISQC examples that are reproduced in the chapter, including the ISQC example number, JMP table name, JMP platform, and key points. The ISQC example number and JMP table name use a specific nomenclature, which is described in the main body of this section.
Step-by-step Directions and Discussions	For each example in the chapter, context for the analysis objectives and a description of the data is provided. Step-by-step instructions and screen shots for how to conduct the analysis using JMP are presented next. Finally, the results of the analysis are discussed. To make it easier to follow along, for longer examples, discussions of the analysis results are interspersed at appropriate points throughout the sequence of steps.
Statistical Insights	Additional examples of SQC concepts are provided in this section. The highlighted content reflects some of our key lessons learned from applying many different SQC techniques to solve problems in a variety of industries. For some examples in this section, additional output not provided in ISQC is included to illustrate JMP functionality or to elaborate on important points.

JMP Software and JMP Tables

In this section, we provide a list of the JMP platforms discussed in this book, additional JMP resources, and instructions for how to access the data used in the examples.

JMP Platforms

This book was written using JMP version 14. As mentioned previously, we are aware of two platforms discussed in this book that were added in the past several versions, **Process Screening** (version 13) and **CUSUM Control Chart** (version 14). The first level JMP menus and platforms covered in this book are shown in Figure 1.1.

Figure 1.1: First Level JMP Menus and Some Platforms Covered in this Book

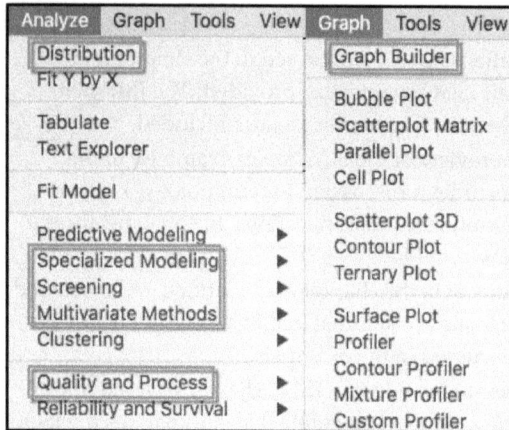

More than 20 JMP platforms are demonstrated in this book. Except for **Graph Builder**, all the platforms covered are located under the **Analyze** menu. This includes thirteen platforms to generate control charts, two platforms to carry out measurement system analysis, three platforms to assess process capability and process stability, two platforms to generate some of the seven basic quality control tools and four platforms are used as supplementary. A summary of the platforms used in each Chapter was described above and is further discussed in Chapter 2 of this book.

JMP Resources

The platforms listed in Table 1.3 are discussed in detail in this book. However, additional JMP resources are discussed in this section. The reader is encouraged to go to http://www.jmp.com/getstarted/ to learn more about the vast amount of resources available.

The **JMP Help** menu, which is accessed by clicking on **Help** in the main menu, is show in in Figure 1.2.

Figure 1.2 JMP Help and Books Options

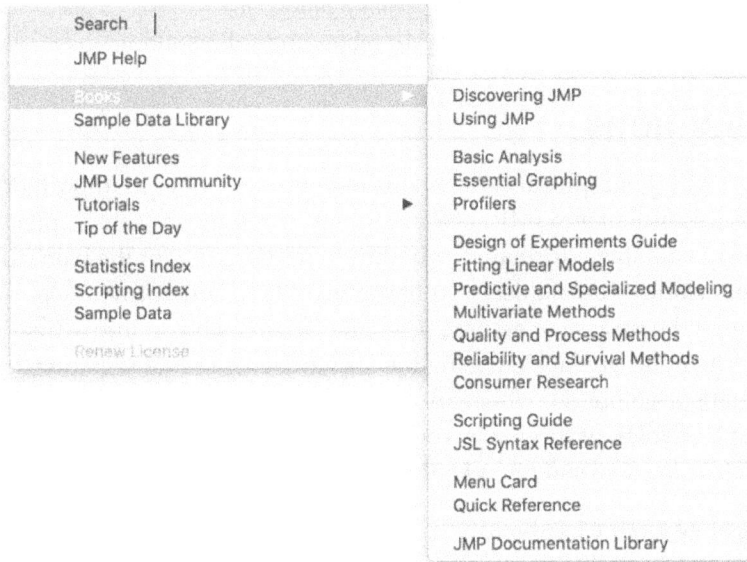

A few of these resources are summarized here:

- **Books ▶ Discovering JMP**: This is a helpful resource for those who are new to JMP. It teaches the user how to maneuver in JMP windows, how to prepare and manipulate data, how to visualize data, how to perform simple and more complex analyses, and how to save and share your work.

- **Books ▶ Quality and Process Methods**: This is a book that covers all the tools in the Quality and Process platform. Each chapter includes an overview of the tool and data sets from the Sample JMP library, with step-by-step instructions for how to create the desired charts or analysis, explanations of the output, and statistical details for the platform.

- **Books ▶ Quick Reference**: This is a quick-reference guide that provides keyboard shortcuts for Windows and Macintosh. It includes shortcuts for data table actions, working with files, window commands, editing files, and JMP tools, for example.

- **Sample Data**: This is an organized way to find sample JMP tables and scripts that are illustrated in the JMP books. The Sample Data Index is organized in several ways, according to the analysis type, type of data, teaching resources, and the Discovering JMP conference.

- **JMP User Community**: This option opens a web site with links to other JMP users through discussions, blogs, file exchanges, and users groups. Links to on-demand web casts and videos and the Discovery Summit are also included here.

JMP Tables Used in This Book

The data used in this book is available at http://support.sas.com/jramirez or http://support.sas.com/bramirez. Users can download the JMP tables and follow along.

Typographical Conventions

We use several different types of typographical conventions to point the reader to important information regarding statistical content and JMP software usage. Throughout each chapter, we include Statistics Notes to relay valuable information we want to reinforce about the particular example being discussed. For example, the Statistics Note shown in Figure 1.3 provides information about when it might be unwise to assume normality when using an Individual Measurement control chart. The notes are labeled using the format Statistics Note Chapter #.#. For example, Figure 1.3 shows Statistics Note 3.4, which implies that it is the fourth statistics note in Chapter 3, Control Charts for Variables, of this book. The Statistics Notes are contained within two vertical lines with a control chart logo, as shown in Figure 1.3.

Figure 1.3 Sample Statistics Note from Chapter 3

Statistics Note 3.4: There are times when the normality assumption should not be ignored, and a more appropriate distribution should be used to calculate control limits. Some things to consider include excessive skew in the data, data that is close to a natural boundary condition (for example, 0 or 100), and normal based limits that are non-sensible, such as negative values for a positive quantity. Appropriate distributions are well-documented for many physical and scientific phenomena.

A sample JMP Note is provided in Figure 1.4. This note shows how to transform a response in the launch window when launching an Individuals control chart in the **Control Chart** platform. The notes are labeled using the format JMP Note Chapter #.#. For example, Figure 1.4 shows JMP Note 3.4, which implies that it is the fourth JMP note in Chapter 3, Control Charts for Variables, of this book. The JMP notes are contained within a box with rounded corners with the JMP person logo, as shown in Figure 1.4.

Figure 1.4 Sample JMP Note from Chapter 3

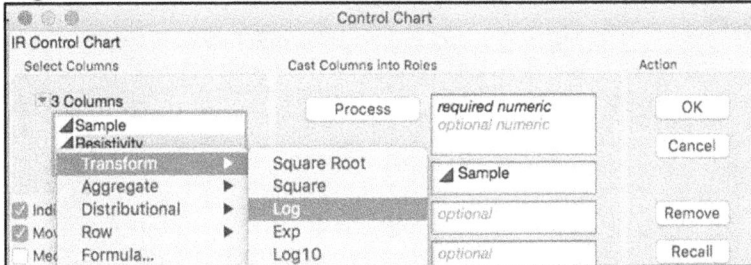

A few other typographical conventions are used in our book. For example, JMP platforms are displayed using bold font and platform paths are shown using an ▶ to identify individual selections for example, **Analyze ▶ Quality and Process ▶ Control Chart Builder**. In the step-by-step directions, JMP table names for example, Chapter 3 – ISQC Table 6.1.jmp, are in a different font type, not bold, from the main text. Column Variable names for example, **Flow Width (µm)**, used to populate fields in dialog windows, are bolded. When populating dialog boxes, we bold the specific dialog field for example, **Process** to be filled in. Within the JMP analysis output, we also bold the banner titles for example, **P of Sample Fraction Nonconforming, pi**, next to the red triangle with the drop-down menu with the corresponding options. We also bold the desired option in the drop-down menu for example, **Tests ▶ All Tests**, needed to complete the example using JMP.

Chapter 2: Overview of Statistical Quality Control Topics and JMP

Overview

This chapter discusses the key concepts and JMP platforms associated with statistical quality control (SQC) topics. More specifically, we review statistical process control (SPC) for continuous and attribute data, measurement system analysis (MSA) for gauge R&R studies, and how to carry out a process health assessment (PHA). We also provide practical advice for how to navigate common challenges when using these techniques, based on our many years of experience. Most of the JMP platforms in the **Analyze ▶ Quality and Process** menu and tips for structuring the JMP tables for the analysis are covered.

Statistical Process Control

SPC Insights

For most people, Shewhart's $\overline{X}$ & Range control chart is the first point of entry into the world of SPC. This foundational chart has taught us several important lessons about using SPC effectively over the years. The first one has to do with the key concept of rational subgroups, also introduced by Dr. Shewhart, in order to minimize the variation within subgroups and maximize the variation between subgroups. The Range chart, shown in Figure 2.1, is used to determine if the variation within a subgroup is consistent from subgroup to subgroup. While the $\overline{X}$ chart is used to determine if, given the within subgroup variation, the subgroup means are consistent over subgroups. In other words, the bottom chart is a measure of *within subgroup* variation and the top chart is a measure of *between subgroup* variation. Prof. Montgomery discusses Rational Subgroups in Section 5.3.4 in ISQC Chapter 5. An excellent illustration of the impact of the improper specification of rational subgroups is the analysis of the Injection Molded Sockets data in Wheeler and Chambers (1992) (see also Ramírez, J. and Ramírez, B. (2009) and Ramírez and Zangi (2014)).

Figure 2.1 X̄ and Range Chart for Flow Width from Chapter 3

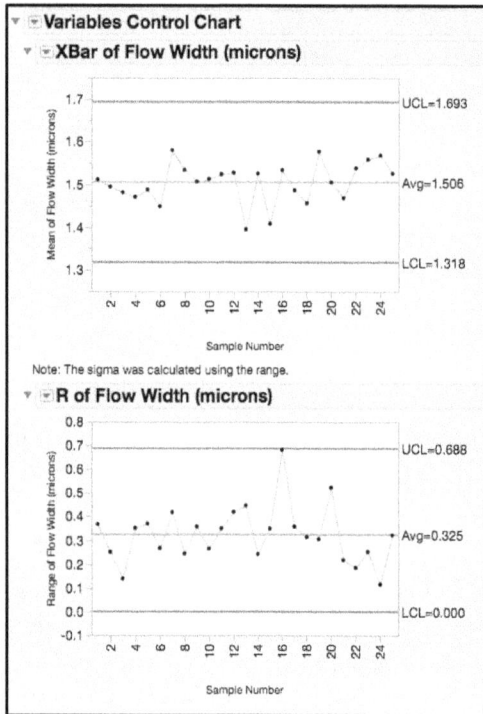

The effective use of the X̄ & Range control chart is also related to the magnitude of the two sources of variation, *between* and *within* a subgroup. In batch processes, for example, rational subgroups are often formed using one production batch, where multiple measurements are taken. Since the between-batch variation can be much larger than the within-batch variation, the control limits on the X̄ chart can be unusually tight, resulting in the majority of the points outside of the limits. It is an image that is difficult to forget. But the three-way (or 3-way) control chart saves the day, by adjusting the limits for the X̄ chart to account for the batch-to-batch variation, using a moving range control chart for the batch means, and preserving the Range control limits in the usual way. ISQC Chapter 6 and Chapter 3 of this book provide additional discussions on this topic.

In some industries, such as, semiconductor, sometimes it is necessary to seek out advanced approaches to extract more value from a monitoring program. One example is the implementation of SPC on equipment to maintain the optimal environment in a ISO Class 1 cleanroom environment. One example is a make-up air handler, which uses Proportional-Integral-Derivative (PID) controllers, with data available at very short time intervals. The use of an Individual Measurement and Moving Range control chart would have about 75% of the points involved in a runs test violations. This is a direct result of using an inappropriate technique to monitor nonstationary autocorrelated output. The use of Time Series control charting techniques to deal with autocorrelation is discussed in Chapter 8 of this book and Chapter 10 in ISQC.

The use of P/NP and C/U charts may also be challenging in practice. For example, when working with wafer yield, which is calculated as the ratio of the number of good die (chips) to the total number of die on a wafer, the binomial distribution is used to derive the control limits for a P chart. However, the observed variation is usually larger than expected under the binomial model and results in control limits that are too tight. This is known as overdispersion, and the P chart has to be adjusted accordingly to reflect this. The beta-binomial distribution can be used to model overdispersion and chart wafer yield (Cantell et. al., 1998). Particle counts can also exhibit the same phenomenon, and the C chart can be modified to account for overdispersion by using a gamma-Poisson distribution. For more details see Chapter 4 of this book and Chapter 7 of ISQC.

Reducing special cause variation by identifying and removing root-causes for signals is by far one of the most difficult activities associated with SPC. This is especially true for batch-based complex processes, where a smoking gun is difficult to locate, and the trail is often cold before we start looking. However, the chances of finding and responding to special cause variation, improves by leaps and bounds when there is "real-time" data and "real-time" monitoring. Another way to increase our ability of achieving stable processes is to combine SPC with the other SQC tools like measurement system analysis (MSA) and process health assessments (PHA).

JMP SPC Platforms

The SPC-related JMP platforms discussed in this book are shown in Figure 2.2, which are found under the **Analyze ▶ Quality and Process** menu. Below is a high-level description and roadmap of these SPC platforms. The first thing to note is that the **Control Chart Builder** and the **Control Chart** platforms both produce variables (XBar, IR) and attributes (P, NP, C, U) control charts. While Chapters 3 and 4 outline the capabilities of both platforms and provide several examples of each one, emphasis is placed on the **Control Chart Builder** since it is a newer platform and may eventually replace the legacy **Control Chart** platform. Also, the **Distribution** platform, not shown here, is used to establish probability based control limits with an appropriate, non-normal distribution.

The small shift detection platforms shown in Figure 2.2 include the **UWMA**, **EWMA**, and **CUSUM**, found in the **Control Chart** platform. The **CUSUM Control Chart** also produces CUSUM control charts and is a newer platform, which may eventually replace the legacy **CUSUM** option, found in the **Control Chart** platform. Finally, the **Pareto Plot** and **Diagram** tools (see Figure 2.2), found in the **Quality and Process** menu, are used to create a Pareto Plot and a Fishbone diagram, respectively and are presented in Chapter 4 of this book.

Figure 2.2 JMP Platforms for Control Charts

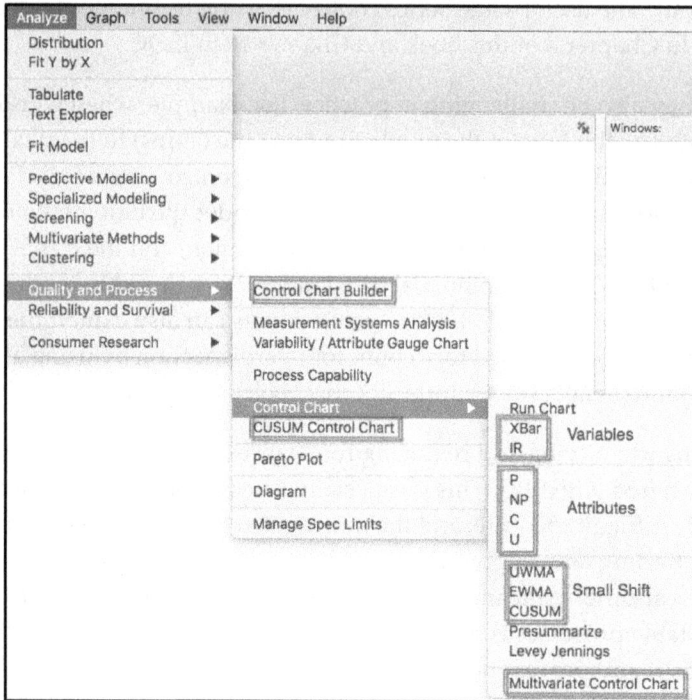

The **Time Series** platform is shown in Figure 2.3. Currently, there is not one JMP tool that will create time series based control charts and it takes several steps to generate them. First, an appropriate Time Series model must be identified and fit using the **Time Series** platform found in the **Specialized Modeling** menu. The residuals are then saved to a JMP table and the chart is made using the **Control Chart Builder**. Alternatively, the forecasts and prediction limits can be saved to a JMP table and the chart is made using the **Graph Builder**. While **Multivariate Control Chart** is created using the **Control Chart** platform (Figure 2.2), two platforms under the **Multivariate Methods** menu, **Multivariate** and **Principal Components** (Figure 2.4) are both illustrated in Chapter 9 of this book, to gain additional insight.

Figure 2.3 Time Series Platform

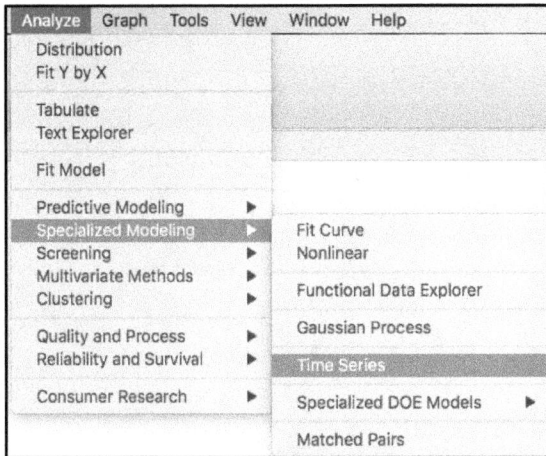

Figure 2.4 Additional Multivariate Platforms

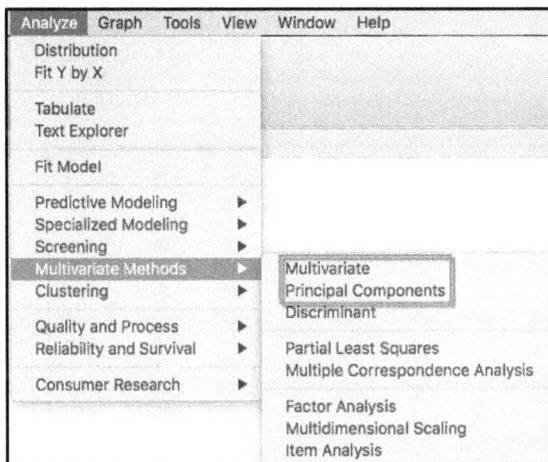

JMP SPC Table Formats

In order to create a control chart, one piece of information is required, that is, the values for the variable being charted. Another piece of information that is desirable but not required, if the sample size is constant, is the subgroup variable. In addition, there are two basic JMP table layouts, horizontal or vertical, that can be used. In a vertical layout, Figure 2.5a, the values for each individual variable are included in one column, Data, and another column is added to identify the Variable Name for every single value. In this layout, one more column can be added to identify the subgroups. In a horizontal layout, Figure 2.5b, each individual variable is placed in its own column, A through J, in this example. If the values have different rational subgroups, then additional columns would be required for each unique subgrouping scheme.

Figure 2.5a Vertical SPC JMP Table Layout

	Variable Name	Subgroup	Data
1	A	1	74.40
2	B	1	90.72
3	C	1	40.62
4	D	1	46.32
5	E	1	1.23
6	F	1	1.30
7	G	1	1.30
8	H	1	204.00
9	I	1	491.00
10	J	1	50.09
11	A	2	81.69
12	B	2	86.29
13	C	2	42.26
14	D	2	44.55
15	E	2	1.16
16	F	2	1.12
17	G	2	1.16
18	H	2	202.00
19	I	2	482.00
20	J	2	51.01

Figure 2.5b Horizontal SPC JMP Table Layout

Chapter 2 - Horizontal Layout.jmp

	Subgroup	A	B	C	D	E	F	G	H	I	J
1	1	74.40	90.72	40.62	46.32	1.23	1.3	1.3	204	491	50.09
2	2	81.69	86.29	42.26	44.55	1.16	1.12	1.16	202	482	51.01
3	3	72.05	82.24	37.06	46.21	1.23	1.22	1.23	201	490	53.31
4	4	77.93	84.95	37.59	44.62	1.26	1.33	1.33	202	495	52.89
5	5	73.24	85.08	41.82	44.33	1.13	1.21	1.21	197	499	52.68
6	6	73.80	87.75	43.82	47.83	1.27	1.2	1.27	201	499	47.53
7	7	80.19	90.73	38.02	42.10	1.11	1.27	1.27	198	507	48.78
8	8	70.75	86.74	39.89	42.45	1.14	1.21	1.21	188	503	51.21
9	9	72.89	88.04	37.32	40.83	1.13	1.27	1.27	195	510	56.02
10	10	74.07	82.19	36.04	46.01	1.11	1.26	1.26	189	509	52.95
11	11	72.31	78.63	40.44	41.06	1.13	1.2	1.2	195	510	53.04
12	12	73.46	86.19	33.54	43.79	1.12	1.27	1.27	192	510	47.93
13	13	80.00	90.40	38.77	50.21	1.08	1.14	1.14	196	515	48.81
14	14	77.56	78.61	41.04	42.40	1.11	1.21	1.21	194	513	52.77
15	15	82.82	95.05	36.58	40.93	1.14	1.22	1.22	196	520	48.46
16	16	78.28	83.25	39.80	40.30	1.12	1.17	1.17	199	518	49.99
17	17	74.76	90.50	41.21	47.83	1.07	1.15	1.15	197	517	44.07
18	18	79.69	86.50	42.77	48.30	1.05	1.15	1.15	197	526	51.08
19	19	70.77	81.94	38.05	45.40	1.09	1.26	1.26	192	525	44.41
20	20	78.20	83.36	40.28	44.52	1.14	1.16	1.16	195	519	51.80

In this section, we discuss the most efficient way to create univariate variables and attributes control charts using JMP. We want to distinguish between creating univariate control charts for one variable at time versus creating univariate control charts for multiple variables at one time. For the former, the horizontal JMP table layout in Figure 2.5b is best, since there is no straightforward way to isolate the results for one variable in a vertical JMP table layout.

It is often desirable to create control charts for multiple variables at one time, using one SPC tool and having the charts contained within the same output window. This is as a matter of efficiency, especially when there are many variables to monitor. For this scenario, the following distinctions are made in subsequent discussions:

- JMP table layout (vertical or horizontal): As was described above, in a vertical layout all data is in one column and there are additional columns to identify the variable name and subgroup number. In the horizontal layout, each variable has a unique column.
- Control chart types (same or mixed): This reflects the control chart types that will be used for all the variables of interest. For example, same implies that the same control chart type will be used for all variables, for example, XmR chart. Mixed means that several control chart types will be used for the variables of interest, for example, XmR and $\bar{X}$ & R control charts.
- Monitoring Phase (I or II): Phase I monitoring implies that control limits will be estimated from the baseline data and applied to the same data; while Phase II monitoring means that new data will be added and monitored using saved control limits.

The ability to generate multiple univariate control charts for variable data at one time, using the **Control Chart Builder** and **Control Chart** platforms, is summarized in Table 2.1. The information is presented for the control chart types (Same or Mixed) and JMP table layouts (vertical or horizontal). There are more options available to generate the same type of control chart for multiple variables. For example, with a vertical JMP table layout, the **Control Chart Builder** can generate the same control chart type for multiple variables in one pass, either by using the Process Name variable in the **Phase** zone, or with the **By** button. With a JMP horizontal table layout, charts can be generated one-at-a-time by selecting a column and clicking the **New Y Chart** button. With the legacy **Control Chart** platform and a vertical JMP table layout is simple, we just use the **By** option and enter the data variable and subgroup variable in the appropriate fields. With a horizontal JMP table layout multiple variables are included in the **Process** field.

It is not easy to generate multiple control charts of mixed type using JMP. One way to accomplish this is to use the legacy **Control Chart** platform with a JMP vertical table layout. One caveat is that for mixed chart types involving XmR and XBar & R or XBar & S charts, all the XBar charts must have a range (R) or a standard deviation (S) chart, but not a mix. From experience, mixed charts are a fairly common scenario and the tip in Table 2.1 is quite useful. With the **Control Chart** platform **XBar** must be selected, and the **Subgroup** field and **By** field must be appropriately populated. Before the output is displayed the following warning message, Figure 2.6, appears.

Figure 2.6 Control Chart Platform Warning Message for XBar for Mixed Chart Types

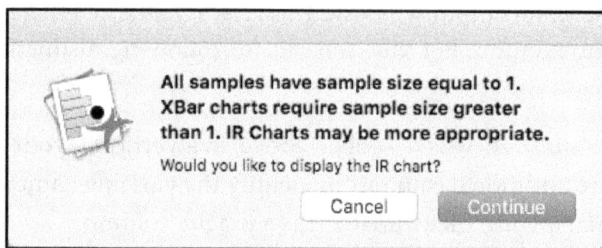

However, just click **Continue** and the mixed type of control charts will be produced.

Table 2.1 JMP Table Layouts for Phase I Monitoring of More Than 1 Variable

Control Chart Types Produced in One Pass	Control Chart Builder		Control Chart	
	Vertical Layout	Horizontal Layout	Vertical Layout	Horizontal Layout
Same, e.g., all XmR control charts	Yes. Use the **By** button, or **Phase** zone.	Yes. Drag subgroup variable to **Subgroup** zone and then select all variables and click the **New Y Chart** button.	Yes. Use the **By** button. Note, for XBar charts with non-constant subgroup sizes, the subgroup variable must be used for **Sample Label**.	Yes. Select all variables that need to be charted and click **Process**. Use the same subgroup variable in the **Sample Label**. By adding missing values to a given variable values, different subgroup sizes can be handled.
Mixed, e.g., XmR and X̄ & R control charts	No.	No.	Yes. Select **XBar**. Note the subgroup variable must be used for **Sample Label** and the variable name must be used in the **By** field.	No.

Phase II monitoring requires hard coded limits and the ability to add new data to the JMP data table. Once again, for a one variable-at-a-time approach, the horizontal layout of the data is the easiest way to carry out Phase II monitoring. Control chart types and limits should be saved to the **Column Properties** for each variable and, when new data are added, these limits will be used when the appropriate control chart tool is launched.

As is shown in Table 2.2, like Phase I, there are more options for producing multiple control charts using saved limits and new data, when all the control charts are of the same type. When working with a vertical JMP table layout in the **Control Chart Builder** platform, the **Get Limits** option is needed. This table can be created using the **Phase Chart** option and selecting **Save Limits ▶ In New Table**. The user will be prompted for the JMP table containing the variable name, chart type, and control limits. For mixed chart types, it is not possible to use a saved limits table.

Table 2.2 JMP Table Layouts for Phase II Monitoring of More Than 1 Variable

Control Chart Types Produced in One Session	Control Chart Builder		Control Chart	
	Vertical Layout	**Horizontal Layout**	**Vertical Layout**	**Horizontal Layout**
Same, e.g., all XmR control charts	Yes. Use the **By** button and **Get Limits**. The limits table is also in a vertical layout with a column identifying the chart variables. Use the **Phase** zone and **Get Limits**.	Yes. Use **Get Limits**. The limits table is also in a horizontal layout. Drag subgroup variable to **Subgroup** zone and then select all variables that have limits saved as a **Control Limits** column property, and click the **New Y Chart** button.	Yes. Use the **By** button and **Get Limits**. The limits table is also in a vertical layout with a column identifying the chart variables.	Yes. Use **Get Limits**. The limits table is also in a horizontal layout. Select all the variables that have limits saved as a **Control Limits** column property, and click **Process**. Select the subgroup variable and click **Sample Label**.
Mixed e.g., XmR and $\bar{X}$ & R control charts	No.	No.	No.	No.

For multivariate control charts for a single process with multiple measurements per subgroup, a horizontal JMP table layout works the best. In this case, each column represents one of the multiple measurements per subgroup. For several variables where the same multiple measurements are taken, a vertical JMP table layout, with a variable denoting the process where the measurements come from, can be used. The process variable can be placed in the **By** or **Group** fields of the dialog window.

Measurement System Analysis

MSA Insights

Successful measurement system studies are predicated on the desire to quantify, understand and reduce the contribution of measurement system variation, and in so doing, reduce the overall observed variation. This is idea is captured in the following equation:

$$\sigma^2_{Observed} = \sigma^2_{Part} + \sigma^2_{Measurement} \tag{1}$$

In this equation, the observed, or total, variation is the variation that is present in the measurements that we take on our products or processes. This total variation consists of the true part or process variation plus the variation due to the measurement system. It is important to understand the impact of the transmission of measurement system variation to the observed variation, to get the most out of our statistical quality control efforts.

For many years, the AIAG Gauge R&R approach (AIAG (2010)) was the golden standard for how to analyze data from a Gauge R&R study. Following a step-by-step form, hand calculations were performed, using a myriad of equations, so one could arrive at a conclusion about the quality of the gauge. In Step 13 (see Wheeler, (2009)), the %GRR was calculated using the combined gauge Repeatability and Reproducibility, which was then multiplied by 6, divided by (USL – LSL) and multiplied by 100. This quantity is referred to as the Precision to Tolerance, or "P/T" ratio, and was used to classify the quality of the gauge. Granted, this was years ago (the first AIAG edition was in 1990), but with this approach, it was easy to get lost in the calculations and potentially miss out on some important findings about the measurement system.

Measurement System Analysis (MSA) and Evaluating the Measurement Process (EMP) offer significant enhancements over the more traditional measurement studies. In the EMP approach, the focus is on the measurement system, which is to be thought of as another manufacturing process step – and not just as a measurement device or gauge, that is used by operators. Figure 2.7 shows an $\bar{X}$ & Range chart applied to the results from a measurement study. Did you know that a Shewhart control chart can also be use to evaluate measurement performance? The Range chart shows the test-retest error by part and operator, indicating that George might have higher and inconsistent test-retest error. In the $\bar{X}$ chart, the control limits are determined using the within subgroup variation or measurement error, and for a measurement system with good discrimination, we expect most of the points to be out of control. This is the only situation we know of where we want the $\overline{X}$ chart to completely out of control! We can also use the $\overline{X}$ chart to see if the operators have a significant offset, bias, and if they have the same part-to-part patterns (George is off). MSA studies are included in Chapter 5 of this book and in ISQC Chapter 8.

Figure 2.7 $\overline{X}$ Chart & Range Chart for EMP Study – Sample JMP Data '2 Factors Crossed'

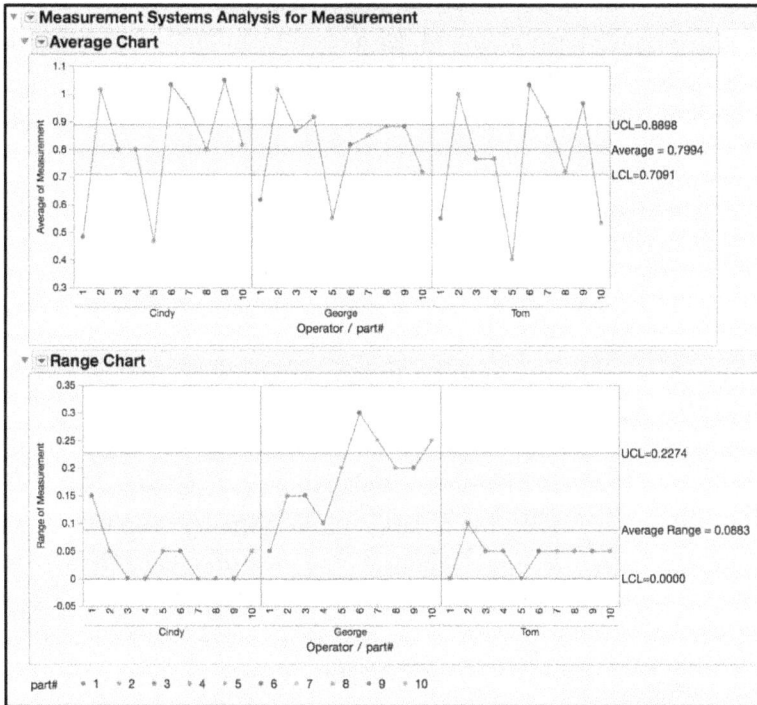

Another important difference in the EMP approach is how the "Operator" effect is treated. In a traditional Gauge R&R analysis, *Operator* is a random component, because, in theory, there are an infinite number of potential operators that can be included in the study. In contrast, in an EMP study, *Operator* is considered as a fixed term, because we do not have an infinite pool of operators. We have a small, and finite, number of operators like Joe, Sue, and Mary that use a particular measurement device. In an EMP we focus on the impact of *Operator* on the average performance and the measurement error, to see if we can improve the measurement system by standardizing on the *Operator* with the best technique. Therefore, an EMP study should be run to inform the development and improvement of the measurement system, and to qualify or validate it.

Destructive samples make it challenging to conduct a measurement system study. One way around this dilemma is to see if a "golden standard" is available, through NIST or the vendor. Another is the use of 'sibling samples'. For example, in measuring the break force of a laminated rolled good, samples used for the repeatability and reproducibility components can strategically be taken in a close proximity to each other on the laminate and treated as if it were the "same" part. When neither of these options is available, the "reproducibility" variance component can be estimated by randomly assigning parts to operators, days, and so on.

There is a perception that in order to get a good estimate of test-retest error, each operator should take as many repeated measurements of the same part, under the same conditions, as is possible. With a carefully designed MSA this is not needed. For example, if 3 operators take 4 measurements on each of 5 parts, then the partitioning of degrees-of-freedom is shown in Table 2.3. In this case, the error, or repeatability component, is estimated with an excessive number of degrees-of -freedom, $3\times5\times(4-1)=45$, while operator and part only have 2 and 4 degrees of freedom, respectively.

Table 2.3 Degrees of Freedom for MSA Study: 3 Operators, 5 Parts, and 5 Repeats

Source	Degrees of Freedom
Operator	2
Part	4
Operator*Part	8
Error	45
Corrected Total	59

By reducing the number of repeats and increasing the number of parts, there are more degrees of freedom for the part-to-part variation. For the same total number of measurements, n = 60, a study with 3 operators, that take 2 measurements on each of 10 parts, results in the partitioning shown in Table 2.4. The repeatability is still estimated well with $3\times10\times(2-1)=30$ degrees of freedom, and the part variability is better estimated with 9 degrees-of-freedom rather than 4.

Table 2.4 Degrees of Freedom for MSA Study: 5 Operators, 6 Parts, and 2 Repeats

Source	Degrees of Freedom
Operator	2
Part	9
Operator*Part	18
Error	30
Corrected Total	59

Determining the overall quality of the measurement system is the most challenging part of an MSA analysis. The comparison of the measurement system error to the tolerance, or specification width, using the P/T ratio is well known. It is a useful, but restrictive, comparison because it only looks at the ability to correctly disposition good product as acceptable, and bad product as unacceptable. Yes, this is important because it deals with customer requirements and business needs. However, we also need to be able to detect signals in the presence of noise. For example, we want to be able identify import factors in a statistically designed experiment, and detect process changes on a statistical process control chart. Both of these activities can be hindered if the measurement system error is too large. In the EMP approach, the overall classification of a measurement system is determined according to its ability to detect process shifts on a control chart. For all approaches mentioned, make sure the typical sources of variation were included in the selection of the *parts*. If they were underrepresented, then the outcome might appear bleaker than it actually is.

JMP MSA Platforms

JMP has two platforms that perform MSA types of analyses, using industry standard nomenclature and calculations. These are highlighted in Figure 2.8 and are accessed from the **Analyze ▶ Quality and Process** menu. The **Measurement System Analysis** platform is used for Gauge R&R and EMP studies using continuous variables. The user must identify the **MSA Method** (EMP or Gauge R&R) in the launch window before any output is produced. The **MSA Method** will inform the default JMP output and available options from within the output window. For the Gauge R&R method, the AIAG analysis approach is used and the default output and options follow the same format as those produced in by the **Variability / Attribute Gauge Chart** platform. In addition, if a standard is available, it can be included in the analysis. The EMP method uses the approach outlined by Wheeler (2006). A control chart like the one shown in Figure 2.7 is the default output and eight additional options are available to produce other EMP elements, like, the EMP Results, Bias Comparison chart and the Test-Retest Error Comparison chart.

The **Variability / Attribute Gauge Chart** carries out a Gauge R&R for continuous or attribute variables. The user must identify the **Chart Type** (Variability or Attribute) in the launch window before any output is produced. The **Chart Type** will inform the default JMP output and available options from within the output window. For the Variability selection, the AIAG analysis approach is used and the default output and options follow the same format as those produced in by the **Measurement System Analysis** platform. Once again, if a standard is available, it can be included in the analysis. In an Attribute Gauge R&R, comparisons between appraisers are made, and the agreement and effectiveness are evaluated. Plots and output is produced following this approach, many of which use some type of agreement statistic, such as, Kappa. If a standard is available, it can also be included in this type of MSA.

Figure 2.8 JMP Platforms for Measurement System Analysis

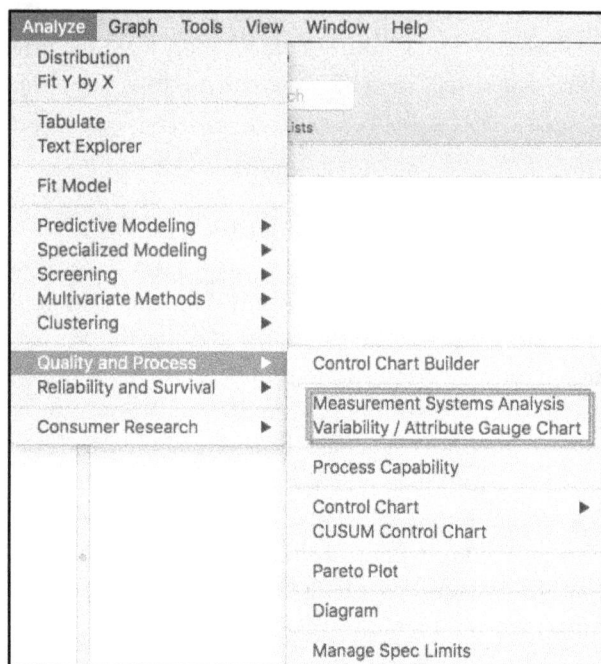

JMP MSA Table Formats

An MSA study is a designed experiment with factors and levels for each factor. Therefore, to assess the impact of each factor on the parameter of interest, the factors and their levels need to be included in the JMP table. For example, if more than one operator was included in the study, there needs to be column in the JMP table for *Operator* and all rows should be filled in with the operator's name associated with each measurement value. There should also be a column for the *Part* identifier and, if this was a non-destructive test method, it is very important that each part has the same identifier for each operator. If a standard was used, then there should be a column the table that contains the true value of the *standard*. Multiple MSA studies can be combined in the same JMP table by using the vertical table layout and including a column for the parameter name. This was discussed extensively in the previous discussion for JMP SPC Table Formats.

An example of a typical JMP table for an MSA for a variable gauge is shown in Figure 2.9. We recommend using obvious naming conventions for the MSA factors, such as *Operator*, *Part*, *Lab Location*, and so on, and the *standard*, since this will facilitate populating the launch windows and aid in interpreting the output. Make sure that the modeling type for the MSA factors are set to a nominal scale because these will be used to estimate variance components or comparisons between the levels. The JMP table in Figure 2.9 has a standard, which provides the true value for each measurement taken. Finally, every measurement in every row must have complete information, as is shown in Figure 2.9. We are sure that all these tips are obvious but thought it might be helpful to reinforce them.

Figure 2.9 Sample JMP Table Format for a Variable MSA

	Operator	part#	Measurement	Standard
1	Cindy	1	0.5	0.5
2	Cindy	1	0.55	0.5
3	Cindy	1	0.4	0.5
4	Cindy	2	1.05	1
5	Cindy	2	1	1
6	Cindy	2	1	1
7	Cindy	3	0.8	0.8
8	Cindy	3	0.8	0.8
9	Cindy	3	0.8	0.8
10	Cindy	4	0.8	0.8
11	Cindy	4	0.8	0.8
12	Cindy	4	0.8	0.8
13	Cindy	5	0.45	0.5
14	Cindy	5	0.5	0.5
15	Cindy	5	0.45	0.5

2 Factors Crossed.jmp — Locked File /Library/Applicatio — Notes Example data to use for Variability Chart — EMP Measur...ems Analysis — JMP Applica... MSA Charts — Columns (4/0): Operator, part#, Measurement, Standard — Rows: All rows 90

The JMP table format is a bit different for an attribute gauge R&R. These JMP tables require a horizontal layout, where the *part* classification for each operator must be included in its own column. An example is shown in Figure 2.10. The data in this table is from a study where three operators, labeled anonymously as A, B and C, inspected the same 50 parts and classified them as 1 = Pass or 0 = Fail. Note that each part was inspected three times by each operator. Once again, obvious nomenclature should be used for the item inspected, and any standard and operator names, if appropriate. In Figure 2.10, *Part* and *Standard* have the obvious meanings. The data in the first row signifies that for the first part, which has a true value of '1', all three operators classified it as a '1'. They successfully did this two more times, as is shown in rows 2 and 3.

Figure 2.10 Sample JMP Table Format for an Attribute MSA

	Part	Standard	A	B	C
1	1	1	1	1	1
2	1	1	1	1	1
3	1	1	1	1	1
4	2	1	1	1	1
5	2	1	1	1	1
6	2	1	1	1	1
7	3	0	0	0	0
8	3	0	0	0	0
9	3	0	0	0	0
10	4	0	0	0	0
11	4	0	0	0	0
12	4	0	0	0	0
13	5	0	0	0	0
14	5	0	0	0	0
15	5	0	0	0	0
16	6	1	1	1	1
17	6	1	1	1	0
18	6	1	0	0	0
19	7	1	1	1	1

Columns (5/0): Part, Standard, A, B, C

Rows: All rows 150; Selected 0; Excluded 0; Hidden 0; Labelled 0

Similar to a JMP table for a variable gauge R&R, the factors are also set to a nominal modeling type. In Figure 2.10, *Standard*, *A*, *B*, and *C* have all been set to a nominal modeling type and have the red histogram symbol next to their names in the Columns area. For binary classifications of 0 or 1, such as the ones shown in Figure 2.10, a continuous modeling type will produce the same output. However, for binary responses, we think it is best to stick with the nominal modeling type for attribute studies.

Process Health Assessment

PHA Insights

Wheeler and Chambers (1992) introduced the concept of *the four possibilities for any process* using Four Process States: Ideal, Threshold, Brink of Chaos and State of Chaos. In this classification, the state of a process is identified using two dimensions of process performance—the process capability, or the ability to meet acceptance limits, and the process stability, the ability to maintain a state of statistical control. The capability of a process is often determined using P_{pk}, while the stability of a process is determined using Shewhart charts and runs tests. However, Shewhart charts do not lend themselves to quantify process stability, especially if runs tests are applied. For example, can a stable process have any runs violations? Is having one violation worse than having multiple violations? Are runs above the centerline worse than having 2 of 3 points fall outside of 2σ? What about the possibility of false alarms?

B. Ramírez and G. Runger (2006) developed the Stability Ratio (SR) test as a way to quantify and test for process stability. It is used to classify a parameter as stable, exhibiting common cause variation, or as unstable, and symptomatic of special cause variation. The Stability Ratio (SR) is the ratio of long-term variation to short-term variation,

$$\text{Stability Ratio} = \frac{\sigma^2_{Long\ Term}}{\sigma^2_{Short-Term}} \qquad (2)$$

The long-term variation is estimated using the sample variance of all the data, and the short-term variation is estimated from the within subgroup variation in a control chart using, for example, $\overline{R}/d_2$ or $\overline{MR}/d_2$. This is analogous to Sir Ronald Fisher's fundamental Analysis Of Variance (ANOVA) principle of comparing the between variation to the within variation. Like ANOVA, a significance test can be performed, assuming the SR approximately follows an F-distribution.

The introduction of the Stability Ratio test is a breakthrough for conducting what we call a *Process Health Assessment* or *PHA*. Not only does it provide a consistent way to classify a parameter as stable or not, but it also has a quantitative measure that could be placed on a stability continuum and used to prioritize improvement efforts. In a process health assessment, each process parameter is classified as

1. Capable or incapable

   Based on comparing the confidence interval for P_{pk} to a given value like 1 or 1.33

2. Stable or Unstable

   Based on the p-value of the F approximation to the Stability Ratio (SR) test

A *process performance dashboard* (see J. Ramírez 2018) is a visualization of the four process states: Ideal State (stable and capable), Yield Issue (stable and incapable), Process Issue (unstable and capable) and Double Trouble (unstable and incapable). The process performance dashboard inspired the **Process Performance Graph** that was introduced in JMP version 13, in the **Process Screening** platform. It combines the Stability Ratio with P_{pk} to determine the overall health of parameter. An example from Chapter 6 of this book is shown in Figure 2.11 (see also J. Ramírez (2016)).

Figure 2.11 Process Performance Dashboard Example from Chapter 6

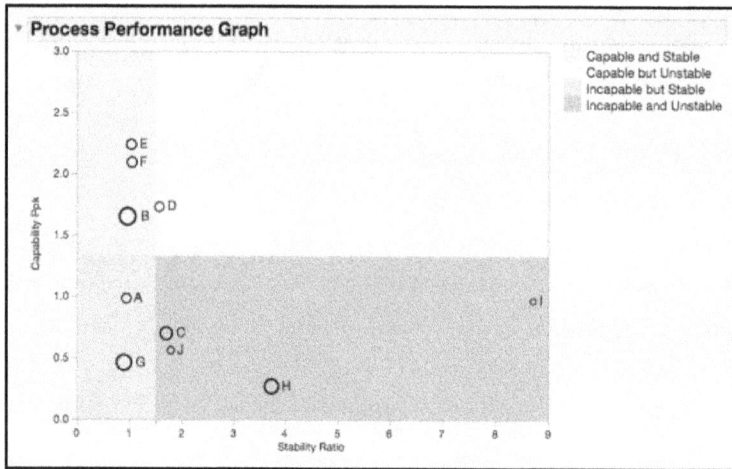

The *Process Performance Graph* can be used to monitor the process health of multiple processes for a given product and manufacturing site, for multiple products at a manufacturing site, or for multiple products and multiple manufacturing sites. The frequency for updating the process health metrics can be determined by the manufacturing volumes at the various sites. While small movements of the points are to be expected with each revision of the Process Performance Graph, it is very revealing when parameters move quadrants and assume a new process health state. The best movement is when parameters congregate in the stable and capable (Ideal) zone.

Quality improvements are more dramatic when there is a comprehensive approach, which includes targeted efforts leveraging all the SQC tools discussed in this chapter. For example, when using SPC, a 'one and done' mentality will not serve you well. After we troubleshoot a signal, we might be tempted to address it, document it and move on to the next one. Variability reduction which is sustained, usually happens through the identification of assignable causes for systematic variation, which can take some time to discover. Make sure you are conducting a periodic review of all the signals and looking for systematic issues. Remember, the best chance of having stable and capable processes, is to design them that way right from the start and do the same for your measurement systems.

JMP PHA Platforms

We uniquely define a PHA platform as a platform that includes both process capability indices and the process stability metric, the Stability Ratio, for single parameter assessments, or the addition of the Process Performance Graph for multiple parameter assessments. The **Control Chart Builder** is used for individual parameter assessments of continuous variables, as is shown in Figure 2.12. This platform is accessed from the **Analyze ▶ Quality and Process** menu. For individual parameter assessments, the variables charts in the **Control Chart Builder** will produce estimates of short- and long-term variation, the Stability Ratio, process capability indices, using short term (C_p, C_{pk}, and so on) and long-term estimates of variability (P_p, P_{pk}, and so on), their confidence intervals, and nonconformance estimates. In order to generate this output, the Spec Limits Column Properties must be populated. The **Distribution** and legacy **Control Chart** platforms are not included here, since they do not include the process stability metric. However, the **Distribution** platform is highlighted in Chapter 5 of this book.

Figure 2.12 JMP Platform for Process Health Assessment for Individual Parameters

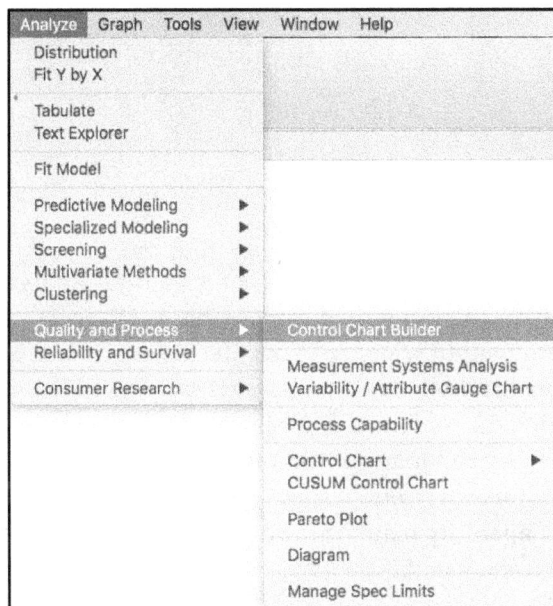

Two platforms are available for use with multiple parameter assessments of continuous variables, as is shown in Figure 2.13. It is a good idea to populate the Spec Limits Column Properties in the JMP table, prior to launching them. The **Process Screening** platform, which was introduced in JMP version 13, is accessed from the **Analyze ▶ Screening** menu. For multiple parameter assessments, the output includes a table with summaries in three main areas: Variability (Stability Ratio and long and short-term estimates of variation), Control Chart Alarms (control chart alarm rates) and Capability (out-of-specification rates, C_{pk} and P_{pk}). The Process Performance Graph can be launched from the default output window. There are many more useful options in the **Process Screening** platform, which are discussed in detail in Chapter 6 of this book.

Figure 2.13 Main JMP Platform for Process Health Assessment for Multiple Parameters

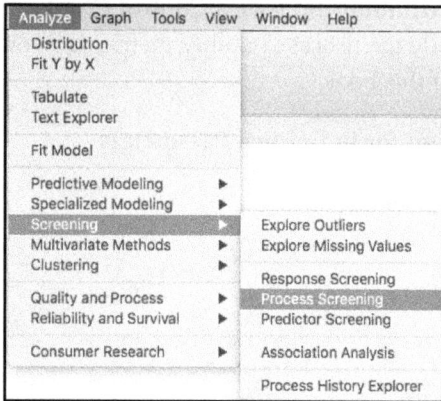

The **Process Capability** platform for multiple parameter assessments is accessed from **Analyze ▶ Quality and Process**, Figure 2.14. Alternatively, it can be launched from within the **Process Screening** platform as an option from the default output window. While the default output is slightly different, depending on where it is launched from, both have access to the same information. There are several plots summarizing the combined process capability assessments, for example, the Goal Plot and Capability Box Plots, summary tables containing specification limits, the Stability Ratio, P_{pk}, and Actual and Estimated % out of spec. The Process Performance Graph can also be produced from the **Process Capability** platform. The options in the **Process Capability** platform are described in Chapters 5 and 6 of this book.

Figure 2.14 JMP Process Capability Platform for Process Health Assessment for Multiple Parameters

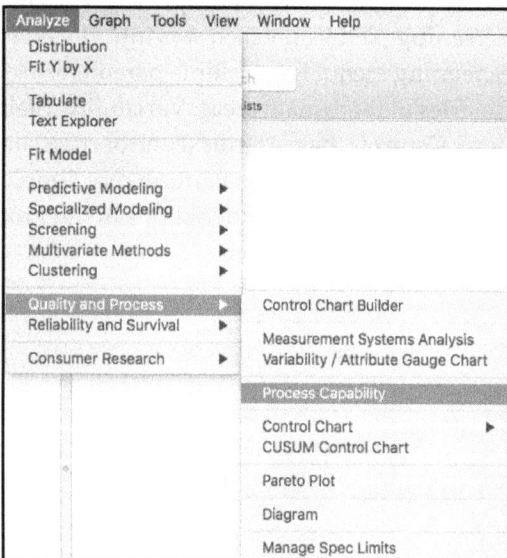

PHA JMP Table Formats

For each parameter assessed, specification limits are required to calculate process capability indices, and a control chart type must be identified to calculate the Stability Ratio. The horizontal layout (see Figure 2.5b) lends itself more readily to this type of analysis. The specification limits for each parameter should be entered in the Spec Limits field in the Column Properties for each parameter. If an XBar control chart is used then a subgroup variable is also needed in the JMP table, for parameters with unequal subgroup sizes. As an alternative, if the subgroup sizes are equal then the subgroup size may be entered in the dialogue window. For multiple parameter assessments, the same control chart type must be entered, and for XBar charts, even the same subgroup size is also required. However, missing values can be used to input different subgroup sizes. We are unaware of how to specify different control chart types, such as, XmR and XBar & Range, for the parameters in the **Process Screening** platform. Similar considerations apply to multiple parameter assessments using the **Process Capability** platform.

Since the specifications limits are required to conduct a capability analysis, the vertical JMP table layout does not lend itself for a process health assessment in JMP. This is because there is no way to incorporate the specification limits for each parameter, in a vertical format. Also, if one uses the **By** button with either the **Process Screening** or **Process Capability** platform with a vertical layout, there is no prompt to enter the limits, parameter by parameter, inside of these platforms.

Chapter 3: Control Charts for Variables

Overview

This chapter illustrates how to generate control charts using examples from Chapter 6, "Control Charts for Variables," of *Introduction to Statistical Quality Control* (ISQC), as well as some of the fundamental ideas behind statistical process control (SPC).

These control chart techniques are presented for data measured on a quantitative scale and are referred to as *variable control charts*. They include the $\bar{X}$ and Range, $\bar{X}$ and Standard Deviation, and Individual Measurement and Moving Range control charts.

Two JMP platforms are highlighted in this chapter: the **Control Chart Builder** and the **Control Chart**.

Variables Control Chart Review

Most books on control charts are partitioned into two buckets: control charts for *variable* data and control charts for *attribute* data. This distinction is important to select the most effective control chart to adequately represent the data of interest. In general, variable data is a measurement that is obtained on a continuous scale, such as temperature, pressure, or thickness. For a thorough discussion of measurement scales, see Chapter 2 in Ramírez and Ramírez (2009).

The most common control charts for *variable* data include the $\bar{X}$ and Range, the $\bar{X}$ and Standard Deviation, and the Individual Measurement and Moving Range (XmR). The first Shewhart control chart, the $\bar{X}$ and Range, is the landmark chart of SPC as we have come to know it today. This chart is appropriate when the natural grouping of the measurements taken in a process is greater than one, also referred to as the subgroup size, n. The $\bar{X}$ chart plots the subgroup averages and is used to understand the homogeneity of a process by determining if the subgroup-to-subgroup averages are consistent, as compared to the within-subgroup variation. The Range chart plots the subgroup ranges (maximum value – minimum value) and looks for consistent within-subgroup variation from subgroup to subgroup.

The $\bar{X}$ and Standard Deviation chart is also used to monitor subgroup averages and within-subgroup variation. However, instead of using the subgroup ranges, the chart displays the sample standard deviation to monitor the variation within each subgroup. It is a more appropriate choice when the number of measurements in a subgroup is larger (for example, $n>=5$). The control limits for the $\bar{X}$ chart are calculated using an estimate of the within-subgroup variation (ranges or standard deviations).

The third chart that is covered in this chapter is the one for individual measurements, referred to as XmR. This chart is appropriate when the natural subgroup size is one, and the data are continuous in nature. For example, if one thickness measurement is taken per hour or per equipment run, then an XmR chart is appropriate. The control limits for this chart are constructed from an estimate of the variation from consecutive moving ranges.

The control charts described here are built on statistical assumptions, including the basic model for the observations $y_i = \mu + \varepsilon_i$, where $\varepsilon_i \sim$ i.i.d. $N(\mu, \sigma)$. Although these charts are robust to moderate departures in these assumptions, we emphasize several examples from ISQC to understand the impact of certain departures on the performance of the chart. For example, a 3-way control chart is used to widen inappropriately tight limits on an $\bar{X}$ chart due to a lack of independence among the subgroup measurements, and probability limits from a lognormal distribution are used to accommodate a skewed distribution.

JMP Variables Control Chart Platforms

Two platforms are used to create variables control charts such as $\bar{X}$ and Range, $\bar{X}$ and Standard Deviation, and XmR charts. One is the legacy **Control Chart** platform and the other one is the **Control Chart Builder**. The **Control Chart Builder** is part of the new generation of JMP quality tools, which makes it easier to design, create, and evaluate control charts. These platforms were introduced in Chapter 2. In this chapter, we focus on the use of these platforms for variables data. Table 3.1 provides a summary of the features we find most useful from both platforms.

Table 3.1 Comparison of Features for JMP Variables Control Chart Platforms

Feature	Control Chart Builder	Control Chart
Control chart types	$\bar{X}$ and Range $\bar{X}$ and Standard Deviation XmR	$\bar{X}$ and Range $\bar{X}$ and Standard Deviation XmR
Save limits	In Column and in new Table	In Column and in new Table
Save summaries	Yes	Yes

Feature	Control Chart Builder	Control Chart
Save sigma	No	Yes
Annotation features	Using the Annotate tool	Using the Annotate tool
Interactivity	Yes	Yes

Note that throughout the remainder of this chapter the term *XBar* is used for $\bar{X}$.

Examples from ISQC Chapter 6

The examples presented here from Chapter 6 of ISQC, and their emphasis, are shown in Table 3.2. The examples are reproduced using JMP as are shown in ISQC. For some examples, additional output not provided in ISQC is shown to illustrate JMP functionality or elaborate on important points considered by the authors.

Table 3.2 Summary of Examples from Chapter 6 of ISQC

ISQC Example Number	JMP Table Name	JMP Platform Control Chart Types	Key Points
6.1 Flow Width	Chapter 3 – ISQC Table 6.1, 6.2	Control Chart XBar and Range and Phase Chart	Apply runs tests and save limits to new data. Add box plots to graphs. Produce an OC curve for an XBar chart and create a Phase Chart.
6.3, 6.4 Piston Ring Diameter	Chapter 3 – ISQC Table 6.3, 6.4	Control Chart Builder XBar and Standard Deviation	Save control limits. Use variable subgroup sizes for control limits.
6.5 Loan Processing Cost	Chapter 3 – ISQC Table 6.6	Control Chart IR (XmR)	Create an XmR chart and apply runs tests.
6.6 Resistivity of Silicon Wafers	Chapter 3 – ISQC Table 6.8	Control Chart IR (XmR) and JMP Script	Fit normal and lognormal distributions to data and generate lognormal probability limits.
6.11 Vane Height of Aerospace Casting	Chapter 3 – ISQC Table 6.11	Control Chart Builder 3-way chart and Variability/ Attribute Gage	Create control limits for hierarchical data and perform variance components analysis.

ISQC Example 6.1 Flow Width

In this example, we show how to construct an XBar and Range chart using the legacy **Control Chart** platform in JMP because this platform has the ability to add box plots to the control chart. The data in Table 6.1 of ISQC consists of Flow Width measurements for a hard-bake process used with photolithography in semiconductor manufacturing. For each of 25 runs, a single measurement is taken on five wafers. Therefore, the natural subgroup is one run of the process equipment, and the subgroup size is five wafers or $n = 5$.

The following steps illustrate how to construct the control chart using the **Control Chart** platform:

1. Open the JMP table Chapter 3 - ISQC Table 6.1.jmp, which has variables called *Sample Number, Wafer Number,* and *Flow Width (microns)*. Sample Number is the subgroup variable and Flow Width (microns) is the measurement.

2. Select **Analyze ▶ Quality and Process ▶ Control Chart ▶ XBar** (Figure 3.1).

   Figure 3.1 JMP Menu Selections for XBar and Range Chart

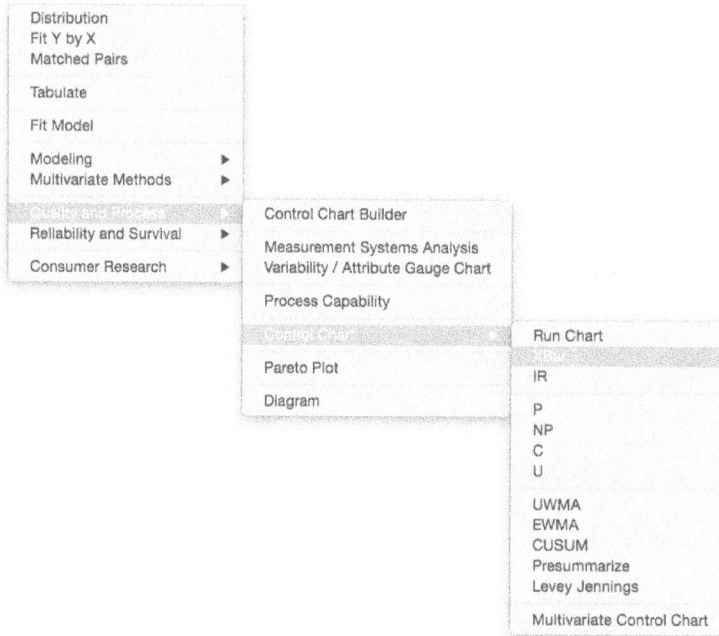

3. When the XBar control chart launch window appears (Figure 3.2), select **Flow Width (microns)** as the **Process** variable. Then select **Sample Number** and click **Sample Label** to identify the subgroup variable.

Figure 3.2 Column Selections for XBar and Range Chart

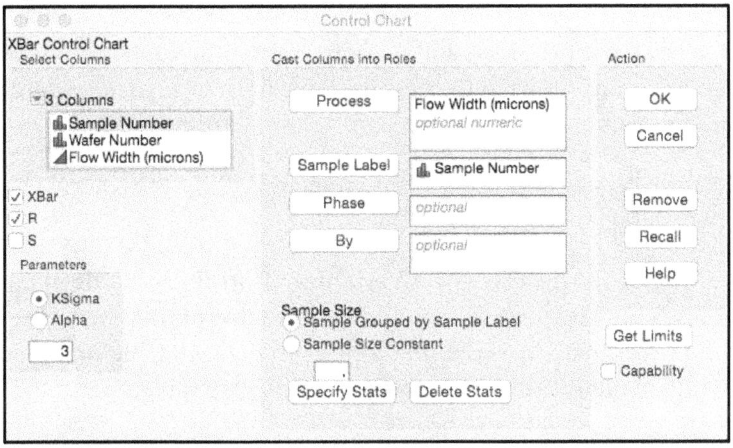

4. Click **OK** to create the control chart.

Figure 3.3 XBar and Range Control Chart for Flow Width

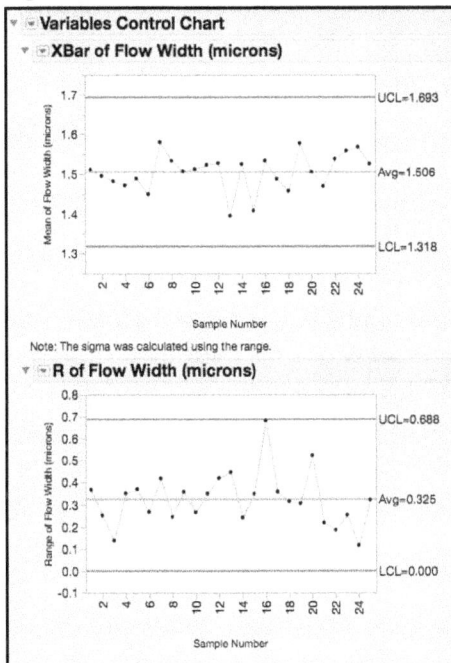

The chart in Figure 3.3 corresponds to Figure 6.2 in ISQC. The Range chart is interpreted first, and because there are no points beyond the control limits, the within-subgroup variation appears to be consistent. Similarly, there are no points outside of the limits for the XBar chart, and we can say that the process is in a state of control. Note that the process capability analysis for this data, which is discussed in ISQC Chapter 6, is presented in Chapter 5 in this book.

New Data Added

ISQC Table 6.2 shows 20 additional runs of the process with five wafers each, for a total of 100 observations. We want to see if the process is still in control using the control limits established from the first 25 runs. To apply the limits to the new data, we have to save the control limits that were established for the previous chart.

The following steps illustrate how to save the control limits in JMP for the control chart in Figure 3.3 and apply them to new data:

1. Click on the red trianno, bogle next to the **Variables Control Chart** title at the top of the window. This brings up a menu (Figure 3.4). There are two options available to save the limits: **in Column** saves them to a column property in the JMP table containing the original data, while **in New Table** saves them to a new JMP table.

   Figure 3.4 Saving Control Limits for Variables Charts

   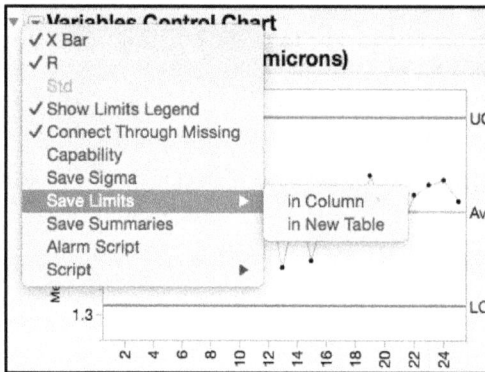

2. To save the limits to a column, select **Save Limits ▶ in Column**. To view the limits, just right-click on the column heading name Flow Width (microns) in the JMP table and select **Control Limits** from the **Column Properties** drop-down menu. In addition, the value of the estimated standard deviation used to compute the limits is also saved as a Sigma column property. The saved column properties are shown in Figure 3.5.

Figure 3.5 Control Limits Saved in Column Properties

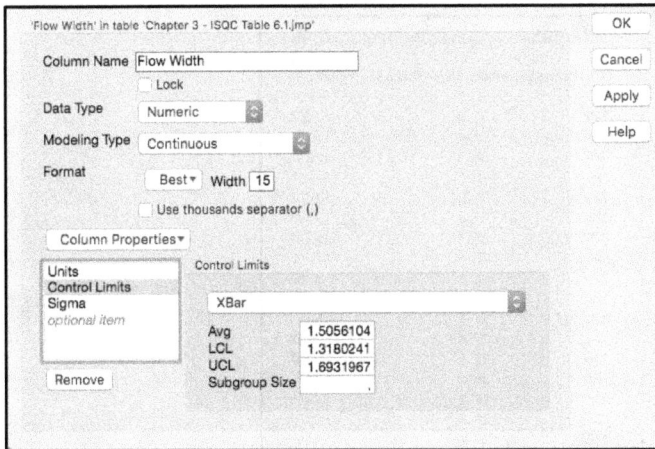

3. To update the control chart with new data, open a new JMP table and paste the new data to it. Save the table as Chapter 3 - ISQC Table 6.2.jmp. The new table has 100 (20 subgroups of size, $n = 5$) rows of data.

4. Make sure Chapter 3 - ISQC Table 6.1.jmp is open and selected. From the main menu bar, select **Tables ▶ Concatenate**. The Concatenate dialog box appears with the Chapter 3 - ISQC Table 6.1.jmp added to the window on the right (Figure 3.6a). To add the new data, select Chapter 3 - ISQC Table 6.2.jmp from the selection list and click **Add**. Enter the output table name as Chapter 3 - ISQC Tables 6.1 and 6.2 and click **OK**. A portion of the resulting table is shown in Figure 3.6b.

Figure 3.6a Concatenating Two JMP Tables

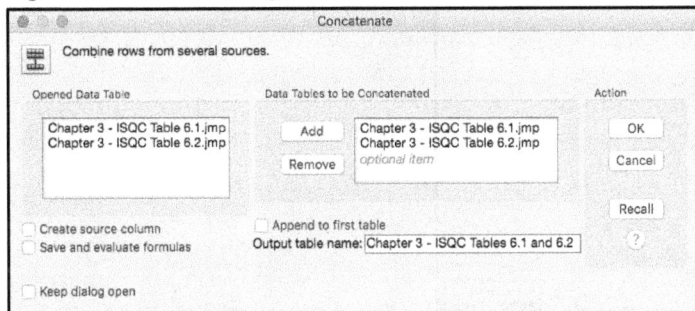

Figure 3.6b Concatenated JMP Tables 6.1 and 6.2

	Sample Number	Wafer Number	Flow Width (microns)
122	25	2	1.3663
123	25	3	1.624
124	25	4	1.3732
125	25	5	1.6887
126	26	1	1.4483
127	26	2	1.5458
128	26	3	1.4538
129	26	4	1.4303
130	26	5	1.6206
131	27	1	1.5435
132	27	2	1.6899
133	27	3	1.583
134	27	4	1.3358
135	27	5	1.4187
136	28	1	1.5175
137	28	2	1.3446
138	28	3	1.4723
139	28	4	1.6657
140	28	5	1.6661
141	29	1	1.5454
142	29	2	1.0931

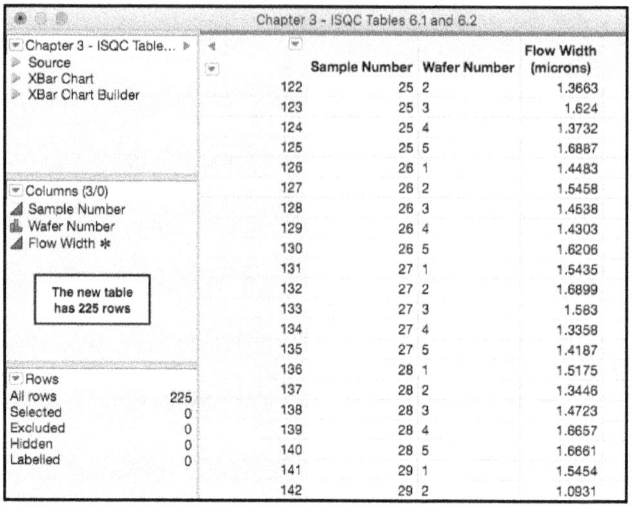

5. *Optional*: Alternatively, to add new data to the chart, click the JMP table Chapter 3 - ISQC Table 6.1 and select **Rows ▶ Add Rows** from the main JMP menu bar. A window appears and you can enter the number of rows we want to add (100) to the table and then click **OK**. Now select all the new rows that were added, rows 126 to 225, and copy and paste new rows of data from a data source, such as Excel, or another JMP table.
6. To view the control chart with the new data, repeat steps 2 through 3 from the previous example. The limits on the control chart should be the same limits shown in Figure 3.3.

Figure 3.7 Updated XBar and Range Control Chart for Flow Width

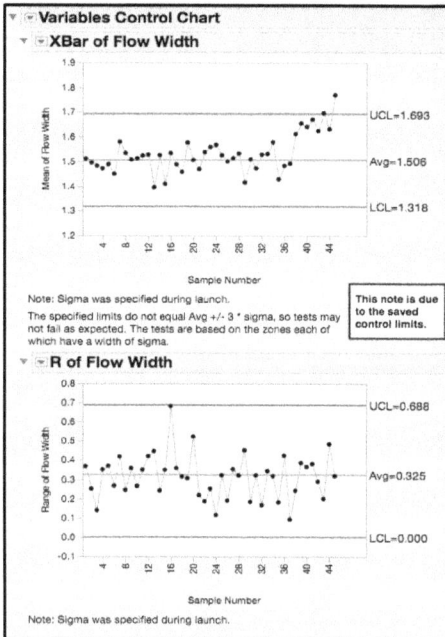

The chart in Figure 3.7 corresponds to Figure 6.4 of ISQC. The Range chart implies that the within-subgroup variation is consistent because no points are above the upper control limit (UCL). However, there are two points above the UCL for the XBar chart and there appears to be a run of points above the centerline. To aid in the visual assessment for patterns among the subgroup averages, a variety of runs tests can be turned on in the chart. The commonly used Western Electric rules are listed as Test 1, Test 2, Test 5, and Test 6. Refer to the online documentation for a complete description of the different runs tests.

JMP Note 3.1: Click on the point above the UCL in the XBar chart to identify it in the JMP table.

---

7. To turn on runs tests, click on the red triangle next to the **XBar of Flow Width (microns)** title bar and select **Show Zones.** Then go back and select **Tests ▶ All Tests** (Figure 3.8). This applies the eight Nelson runs tests, as it is shown in Figure 3.9.

Figure 3.8 Turn on Runs Tests for XBar Chart

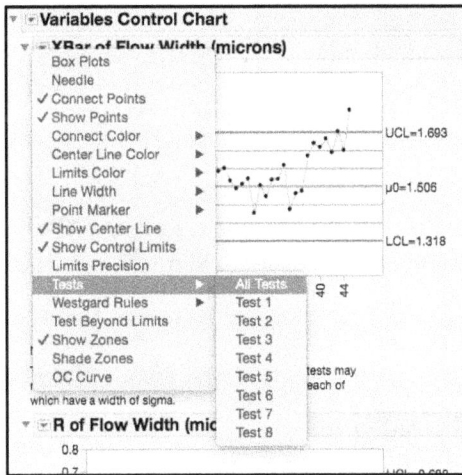

Figure 3.9 XBar Chart for Flow Width with Runs Tests

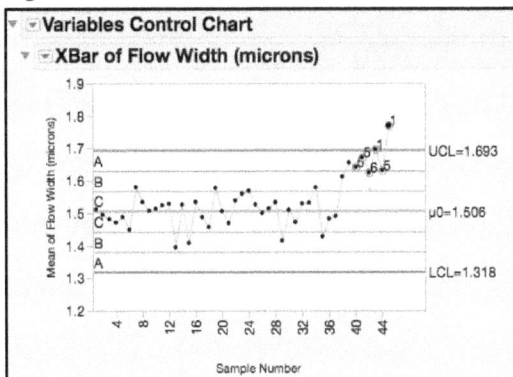

The visual assessment is confirmed, with two points labeled above the UCL and violations for Test 5 and Test 6 (Figure 3.9). It might be helpful to look at the distribution for each subgroup to further evaluate the new data. This is accomplished by adding box plots for each subgroup, which shows the range of the five measurements.

JMP Note 3.2: Select a point in the chart and right-click to see the Chart Options menu. This menu allows you to apply Tests, Test Beyond Limits, Show Zones, and so on.

8. To add box plots to the XBar chart, click on the red triangle next to the **XBar of Flow Width (microns)** title bar and select **Box plots**, as shown in Figure 3.10a. The control chart with box plots is shown in Figure 3.10b.

Figure 3.10a Adding Box Plots to Control Chart

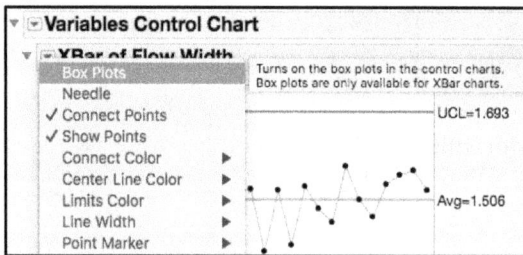

Figure 3.10b XBar Chart for Flow Width with Box Plots

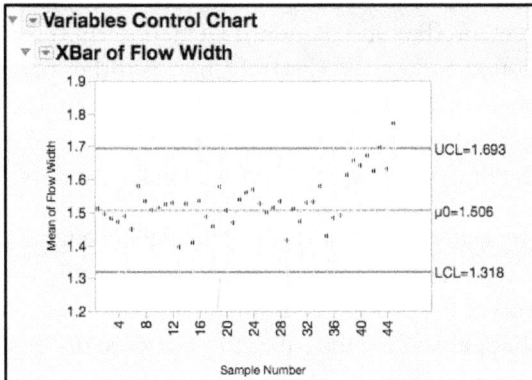

9. The default box plots in Figure 3.10b are light gray. To change the color of the box plots to black, use the script Box Plot Line Color Black.jsl. With the control chart window active, open the script Box Plot Line Color Black.jsl, and run it by pressing CTRL-R (Command-R on a Mac).

10. The script changes the color of the boxes to black as shown in Figure 3.10c, which corresponds to Figure 6.5 of ISQC Chapter 6.

> **JMP Note 3.3**: The default color of the boxes can be changed to black with the JMP script, Box Plot Line Color Black.jsl

Figure 3.10c Updated XBar Chart for Flow Width with Black Box Plots

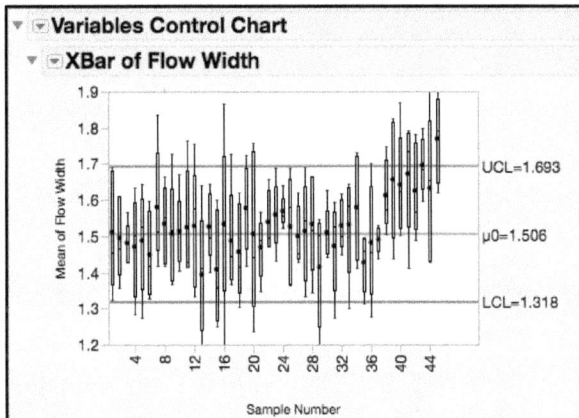

Note that with the exception of the box plots, the XBar and Range chart in this example can be generated using the **Control Chart Builder**. The next example showcases the **Control Chart Builder**.

ISQC Example 6.3 Piston Ring Diameter

In this example, we show how to construct an XBar and Standard Deviation chart using the **Control Chart Builder** in JMP. As mentioned earlier, the **Control Chart Builder**, part of the new generation of JMP quality tools, makes it easier to design and evaluate control charts. The **Control Chart Builder** has a similar drag-and-drop interface to the **Graph Builder**, which allows the user to quickly visualize charts and change the limits calculations, for example, on the fly.

The data set consists of inside diameter measurements for forged automobile engine piston rings (ISQC Table 6.3). For each of 25 runs, a single measurement is taken on five piston rings. Therefore, the natural subgroup is one run of the process equipment, and the subgroup size is $n = 5$. Later in this example, we show what happens when the subgroup sizes are unequal.

The following steps illustrate how to construct the control chart using the **Control Chart Builder** platform:

1. Open Chapter 3 – ISQC Table 6.3.jmp, which has variables called *Sample Number, Piston Ring Number,* and *Inside Diameter (mm)*. Sample Number is the subgroup variable and Inside Diameter (mm) is the measurement.

2. Select **Analyze ▶ Quality and Process ▶ Control Chart Builder** (Figure 3.11). The **Control Chart Builder** launch window appears.

Figure 3.11 Launching Control Chart Builder

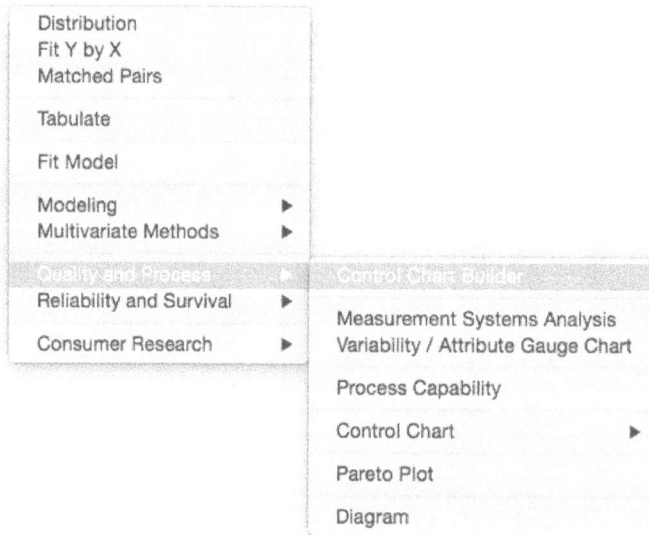

3. Drag **Sample Number** from the left-hand window to the **Subgroup** zone (X axis). Similarly, drag **Inside Diameter (mm)** from the left-hand window to the **Y zone** (Y axis) (Figure 3.12).

Figure 3.12 Launch Window for Control Chart Builder

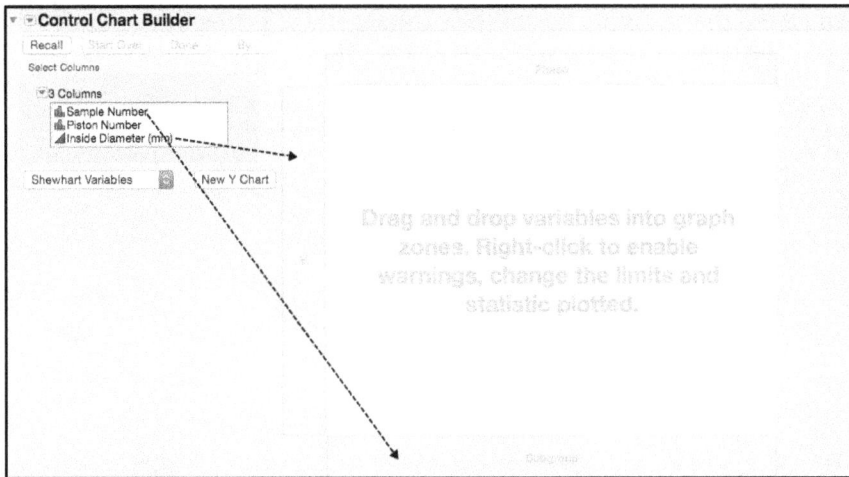

4. The XBar and Range chart appears first in the window. To change it to an XBar and standard deviation chart, select **Standard Deviation** from the drop-down list next to **Sigma** and under **Limits[1]**. Then select **Standard Deviation** from the drop-down list next to **Statistic** and under **Points[2]**. Finally, when the chart has all the required features, click **Done** in the upper left corner of the window (Figure 3.13).

Figure 3.13 XBar and S Chart for Piston Data

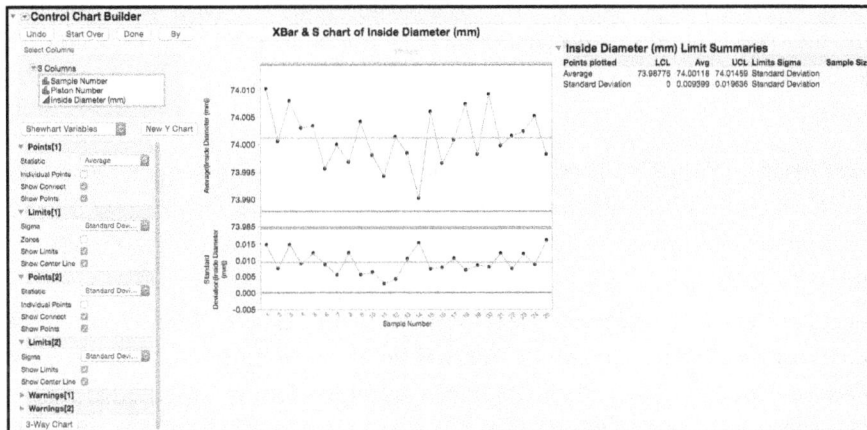

Figure 3.13 corresponds to ISQC Figure 6.17. In this figure, the standard deviation chart is plotting the sample standard deviation for each subgroup and shows that the within-subgroup variation is consistent, with no points outside of the control limits. The XBar chart is plotting the subgroup averages and does not show any points outside of the control limits. If these limits are appropriate to apply to future subgroups, then they can be saved to the column properties in the JMP table using the **Control Chart Builder**.

5. The control limits can be saved to the column properties of the JMP table by clicking on the small red triangle at the top of the window and selecting **Save Limits ▶ in Column** (Figure 3.14).

Figure 3.14 Saving Control Limits for Piston Data XBar and S Chart

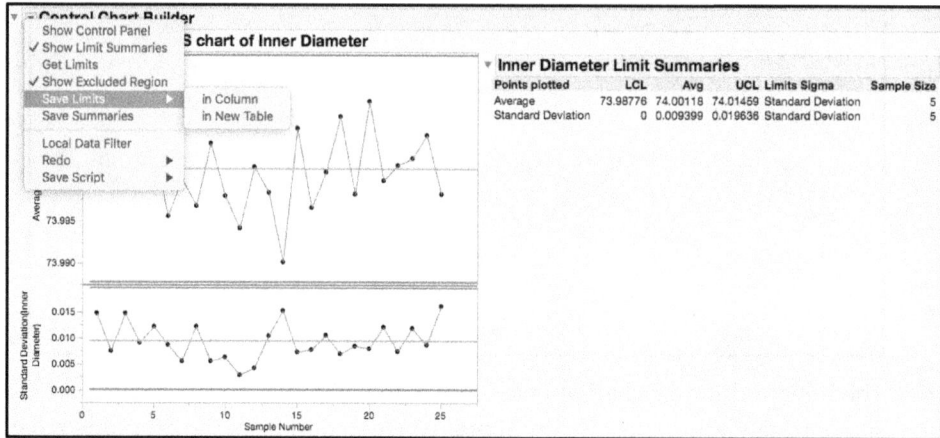

ISQC Example 6.4 Piston Ring Diameter

Sometimes the subgroup size is not constant in the data that we want to plot on a control chart. This could occur for a variety of reasons; for example, perhaps data could not be obtained for all samples in a subgroup or the sampling scheme might depend on a dynamic production schedule. In any event, the control limits for both charts will be impacted since they depend on the subgroup size, n. The **Control Chart** platform will automatically adjust the control limits for variable subgroup size.

1. Open Chapter 3 – ISQC Table 6.4.jmp, which has variables called *Sample Number, Piston Ring Number,* and *Inside Diameter (mm)*. This data is similar to Chapter 3 – ISQC Table 6.3.jmp; however, some of the subgroup results have been removed to create variable subgroup sizes.

2. Select **Analyze ▶ Quality and Process ▶ Control Chart ▶ XBar**. When the XBar Control Chart launch window appears, select **Inside Diameter (mm)** as the **Process** variable. Then select **Sample Number** and click **Sample Label** to identify the subgroup variable.

3. Click **OK** to create the control chart. Figure 3.15 corresponds to ISQC Figure 6.18.

Figure 3.15 XBar and S Chart for Piston Data with Variable Subgroup Sizes

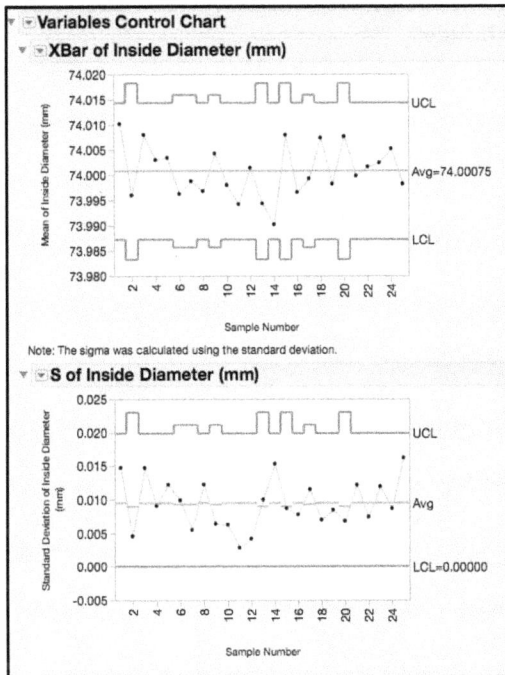

Statistics Note 3.1: The reason the control limits vary is because they are a function of the sample size, and in Figure 3.15 the number of samples in each subgroup varies.

ISQC Example 6.5 Loan Processing Costs

In this example, we show how to construct an Individual Measurement and Moving Range (XmR) chart using the **Control Chart Builder** platform in JMP. The data set consists of the cost of processing loan applications at a bank. The quantity tracked is the average weekly processing costs, which represents the ratio of the total weekly cost and the number of loans processed during the week. Because the bank is interested in the average weekly cost, the natural subgroup is one week of data and the subgroup size is $n = 1$. When the subgroup size $n = 1$, an XmR control chart is an appropriate choice.

The following steps illustrate how to construct the control chart using the **Control Chart Builder**:

1. Open Chapter 3 – ISQC Table 6.6.jmp, which has variables called *Weeks* and *Cost x*. Weeks is the subgroup variable and Cost x is the measurement.

2.   Select **Analyze ▶ Quality and Process ▶ Control Chart Builder** (Figure 3.16).

Figure 3.16 Launching Control Chart Builder

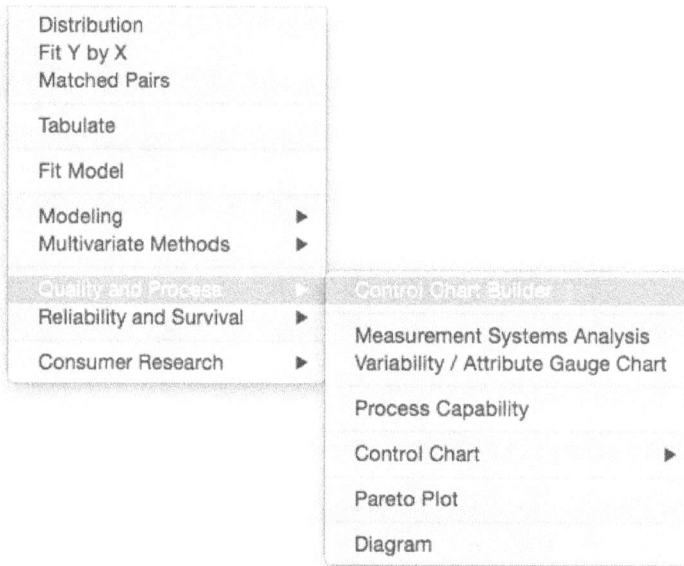

3.   Drag **Cost x** from the left-hand window to the **Y** zone (Y axis). Similarly, drag **Weeks** from the left-hand window to the **Subgroup** zone (X axis) (Figure 3.17).

Figure 3.17 Launch Window for Control Chart Builder

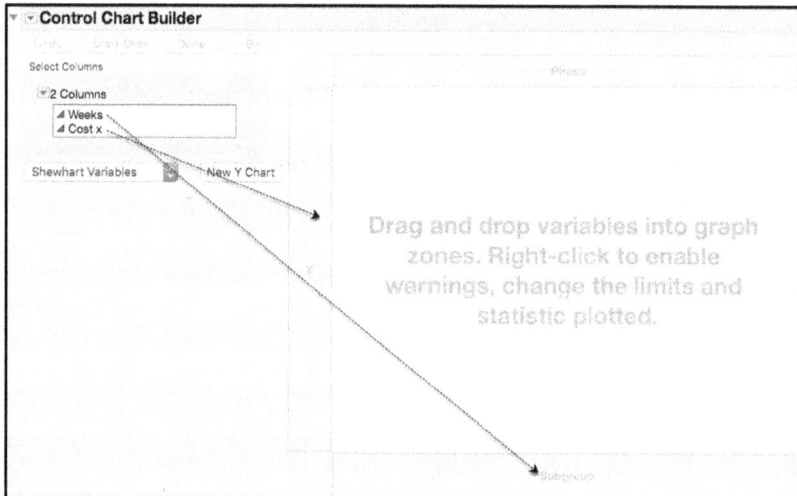

The Individual and Moving Range chart appears in the window, reflecting how quickly charts can be generated with the **Control Chart Builder**. The **Points[1]** menu shows **Individual** under **Statistic**, while the **Limits[1]** menu shows **Moving Range** under **Sigma**. The control limits appear under the **Cost x Limit Summaries** report.

Figure 3.18 Individual & Moving Range Chart for Cost x Data

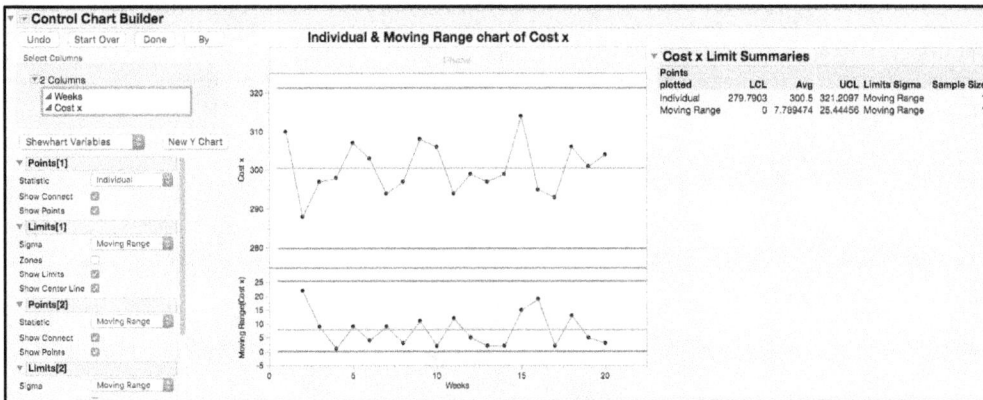

Figure 3.18 corresponds to ISQC Figure 6.19. This chart can also be generated using the **Control Chart** platform.

1.   Select **Analyze ▶ Quality and Process ▶ Control Chart ▶ IR** (Figure 3.19).

Figure 3.19 Launching Control Chart for XmR

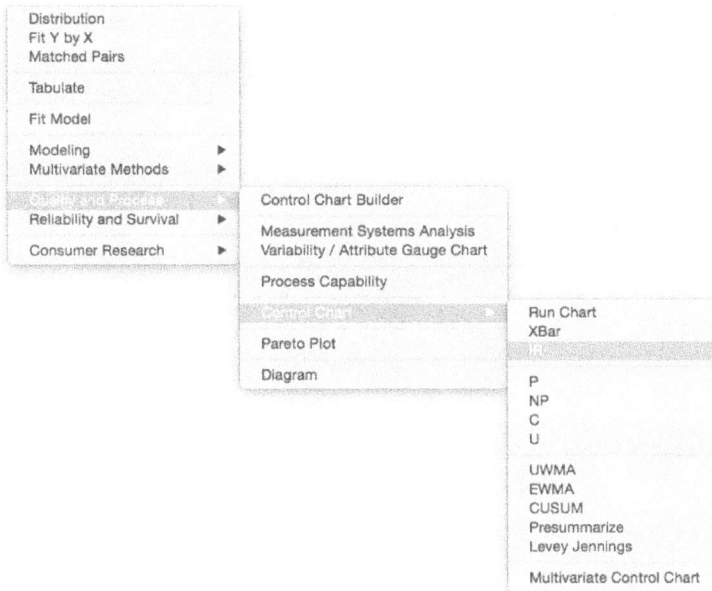

2. When the IR Control Chart launch window appears, select **Cost x** as the **Process** variable. Then select **Weeks** and click **Sample Label** to identify the subgroup variable.

Figure 3.20 Launch Window for XmR for Cost x

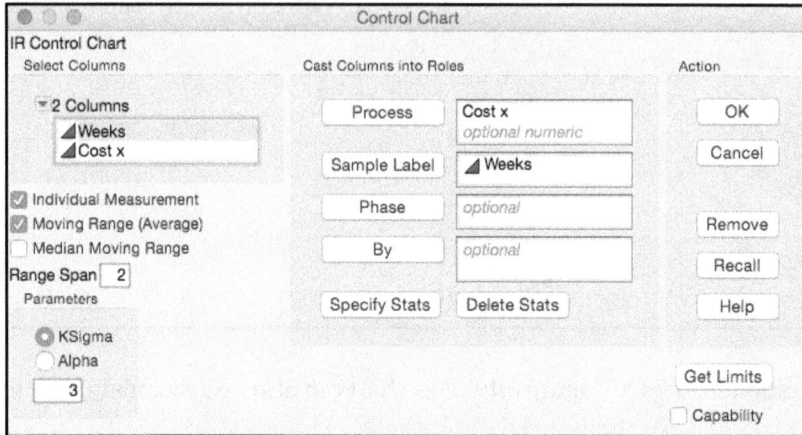

In this launch window, there are several options available for determining how the standard deviation is calculated, which is used to calculate the control limits for the individuals chart. The default settings (that is, **Moving Range (Average)** and a **Range Span** = 2) are selected by default when the launch window appears. The standard deviation can be derived using the average moving range or the median moving range, as shown by the options located in the left-hand side of the window. Although the average moving range is typically used, the median moving range might be more robust to outliers. The span of the moving ranges can also be specified. Once again, while a span of 2 is the most common choice, a larger span might be used to incorporate more variation into the estimate.

3. To produce the control chart, click **OK**.

Figure 3.21 XmR Control Chart for Cost x Data

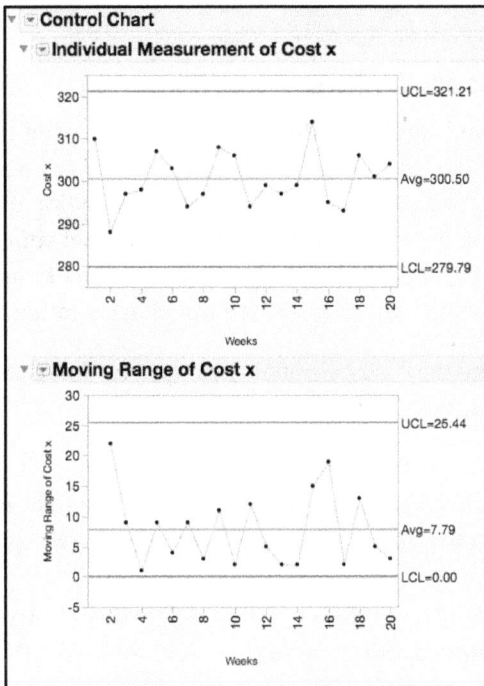

In Figure 3.21, the Moving Range chart monitors the short-term variation by plotting the consecutive differences in two adjacent results. For this parameter, there are no points that exceed the upper control limit. The Individual Measurement control chart monitors the process mean and, because there are no points outside of the control limits, the process output is stable.

Statistics Note 3.2: Some people do not find value in evaluating the Moving Range chart because it is thought to provide redundant information that can be obtained from the Individuals control chart. However, there is a subtle difference in the monitoring objectives for the two charts. It is possible to have a moving range that exceeds the upper control limit, but it does not produce a signal on the Individual Measurement control chart and vice versa. For example, in the chart in Figure 3.21, it is possible that the difference between two consecutive costs can exceed 25.44, with both costs falling within 279.79 and 321.21 (for example, (315 – 285) = 30). Conversely, two consecutive points might exceed the UCL but be in control in the Moving Range chart with a UCL of 25.44 (for example, (324 – 323) = 1).

---

4. More subtle shifts in the mean can be examined by turning on the runs tests. This is accomplished by clicking on the red triangle next to the **Individual Measurement of Cost x** title bar and selecting **Show Zones** and then going back and selecting **Tests ▶ All Tests**.

This applies the eight Nelson runs tests. Because there are no violations, the control chart is not reproduced here.

ISQC Example 6.6 Resistivity of Silicon Wafers

In this example, we show how to construct an Individual Measurement and Moving Range (XmR) chart using the **Control Chart** platform in JMP. The data set consists of resistivity measurements of 25 silicon wafers after an epitaxial layer is deposited in a single-wafer deposition process. Because a resistivity measurement is taken on a single wafer per run, the natural subgroup size is $n = 1$. The assumption of normality is explored, and a lognormal transformation of the data is used for process monitoring using an XmR chart. Note that in JMP the XmR chart is called the IR chart.

The following steps illustrate how to evaluate the distributional assumptions and construct the control chart using the **Control Chart** platform:

1. Open Chapter 3 – ISQC Table 6.8.jmp, which has variables called *Sample Number, Resistivity,* and *log(Resistivity)*. Sample Number is the subgroup variable, Resistivity is the measurement, and log(Resistivity) is the natural log transformation of the Resistivity measurements.

2. Select **Analyze ▶ Distribution**. A launch window appears. Select **Resistivity** and **log(Resistivity)** and select **Y, Columns** to add them to the window. Click **OK** when you are done. The histograms are shown in Figure 3.22.

Figure 3.22 Distributions for Wafer Resistivity Data

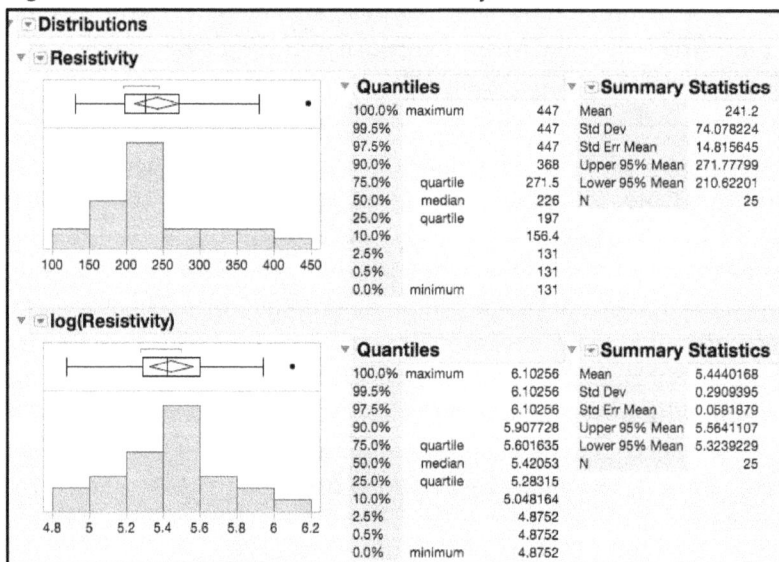

3. To add normal probability plots to Figure 3.22, click on the red triangle at the top of the window next to the **Resistivity** label while holding down the Ctrl key and select **Normal**

Quantile Plot from the drop-down menu. Normal quantile plots are added above each histogram.

4. To rearrange the graphs, click on the red triangle next to **Distributions** at the top of the window, select **Arrange in Rows,** enter **2** in the dialog box that appears, and click **OK**.

Figure 3.23 Normal Probability Plots for Resistivity and log(Resistivity)

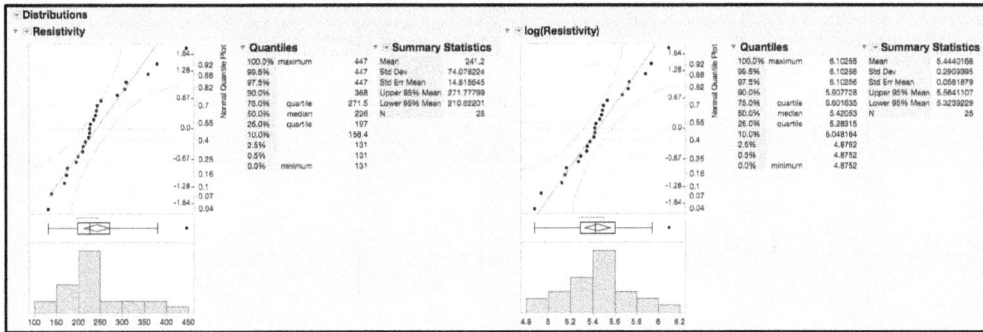

The normal probability plots are displayed above the histograms for each response (Figure 3.23). These displays can be evaluated to determine the best fit for the data at hand. If the data in the probability plot falls along the straight line and within the provided bands, then the distribution is a good approximation to the data set. For wafer resistivity, the normal plot for the log-transformed data appears to provide a slightly better fit than that for the untransformed data. In addition, the histogram and the box plot are more symmetrical around the mean for the log-transformed data, suggesting a departure from the normal distribution. In Figure 3.23 the normal probability plot for Resistivity corresponds to ISQC Figure 6.22, while the one for log(Resistivity) corresponds to ISQC Figure 6.23.

There are two ways to control chart wafer resistivity using the lognormal distribution. The first way involves charting the log, base e, transformed data using an XmR control chart, which is shown here. A second approach, using lognormal probability limits, is discussed in the Statistical Insights section in this chapter.

5. From the main menu, select **Analyze ▶ Quality and Process ▶ Control Chart ▶ IR**.
6. A launch window appears. Select **log(Resistivity)** as the **Process** variable. Then select **Sample** and click **Sample Label** to identify the subgroup variable. Click **OK** when you are done. The chart is shown in Figure 3.24.

Figure 3.24 XmR Chart for log(Resistivity)

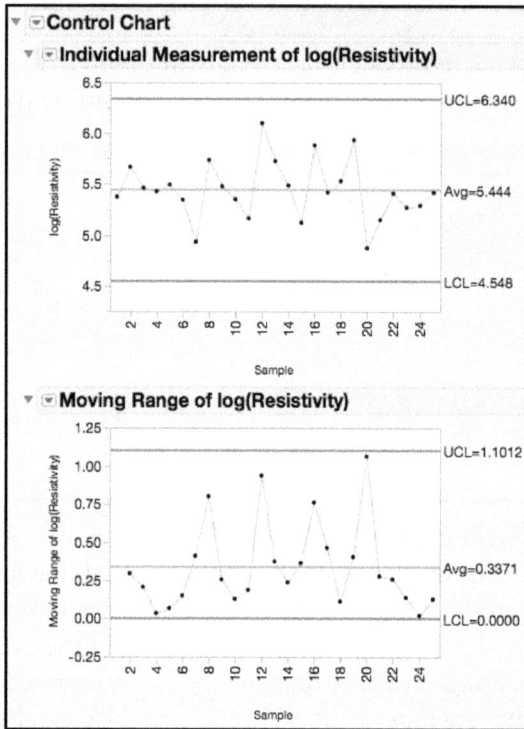

JMP Note 3.4: You can transform the response right from the control chart by right-clicking on the response Resistivity and selecting **Transform ▶ Log**.

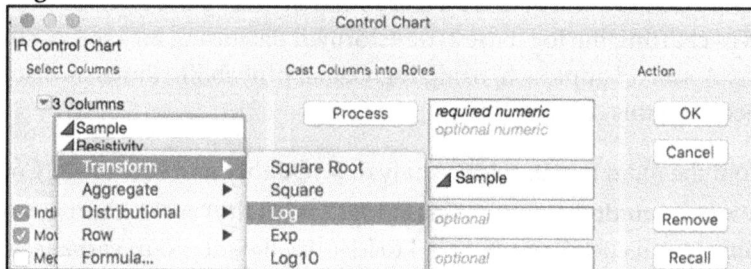

The Moving Range chart of the transformed data, shown in Figure 3.24 (ISQC Figure 6.24), no longer has any points above the upper control limit and the Individual chart shows no violations of runs tests.

We can also use the **Control Chart Builder** to generate the control chart in Figure 3.24, as follows:

1. From the main menu, select **Analyze ▶ Quality and Process ▶ Control Chart Builder**.

2. Drag **log(Resistivity)** from the left-hand window to the **Y** zone (Y axis). Similarly, drag **Sample** from the left-hand window to the **Subgroup** zone (X axis). Figure 3.25 shows the XmR chart.

Figure 3.25 Control Chart Builder XmR Chart for log(Resistivity)

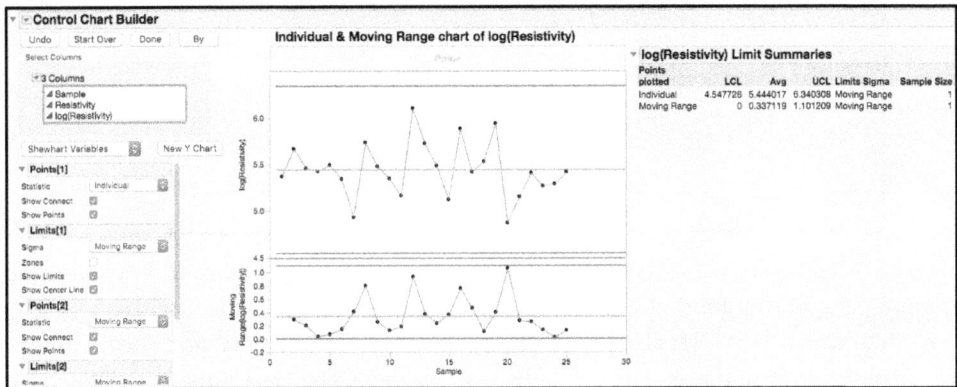

ISQC Example 6.11 Vane Height of an Aerospace Casting

In this example, we show how easy it is to construct a 3-way chart using the **Control Chart Builder** platform in JMP. The data set consists of vane height measurements on 20 aerospace castings, which are used in a gas turbine jet aircraft engine. Data on vane heights are collected by randomly selecting five vanes on each casting produced. Since five measurements are taken per casting, the natural subgroup size is $n = 5$. An XBar and Standard Deviation control chart is adapted for this sampling scheme using a 3-way chart.

The following steps illustrate how to create a 3-way chart using the **Control Chart Builder** platform:

1. Open Chapter 3 – ISQC Table 6.11.jmp, which has variables called *Sample Number, Vane,* and *Vane Height*. Sample Number is the subgroup variable and Vane Height is the measurement.

2. Select **Analyze ▶ Quality and Process ▶ Control Chart Builder**. A launch window appears. Drag **Sample Number** from the left-hand window to the **Subgroup** zone (X axis). Similarly, drag **Vane Height** from the left-hand window to the **Y** zone (Y axis). The XBar and Range charts appear first in the window. To change it to an XBar and standard deviation chart, select **Standard Deviation** from the drop-down list next to **Sigma** and under **Limits[1]**. Then select **Standard Deviation** from the drop-down list next to **Statistic** and under **Points[2]**. Finally, select **Standard Deviation** from the drop-down list next to **Sigma** and under **Limits[2]**. To turn on the runs tests, right-click in the control chart and select **Warnings ▶ Tests ▶ Test 1**.

Figure 3.26 XBar and Standard Deviation Chart for Vane Height Data

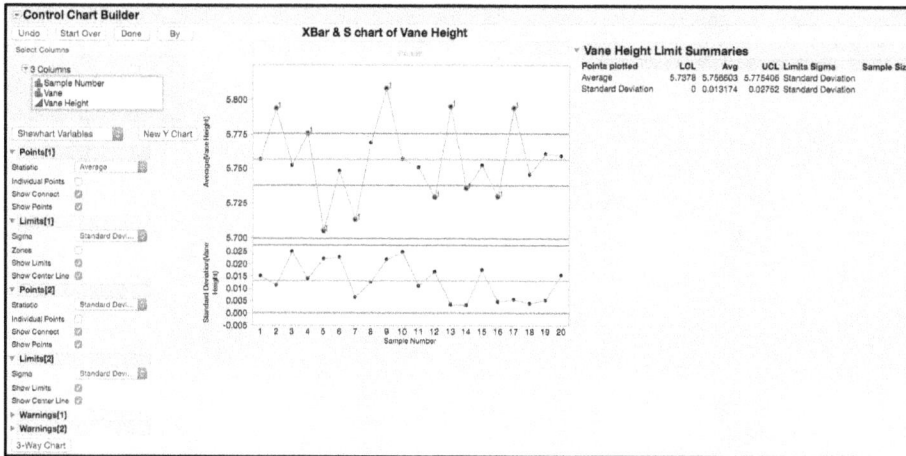

In Figure 3.26, the standard deviation chart displays the sample standard deviation for each subgroup and shows that the within-subgroup variation is consistent, with no points outside of the control limits. The XBar chart is plotting the subgroup averages and shows many points outside of the control limits. The control limits for the XBar chart must be adjusted to include the subgroup-to-subgroup variation, using a 3-way control chart.

3. Click **3-Way Chart** at the lower left-hand side of the window. Select **Moving Range on Means** from the drop-down list next to **Statistic** and under **Points[2]**. Then select **Moving Range** from the drop-down list next to **Sigma** and under **Limits[2]**. Finally, select **Standard Deviation** from the drop-down list next to **Statistic** and under **Points[3]** and click **Done** when you are finished.

Figure 3.27 3-Way Control Chart for Vane Height Data

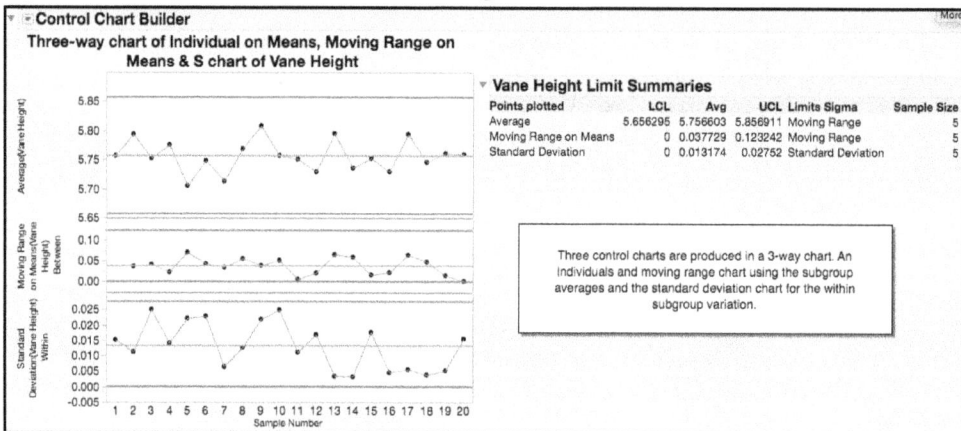

The 3-way chart in Figure 3.27 (ISQC Figure 6.27) gets its name because three control charts are displayed. The first two charts are Individual Measurement and Moving Range charts for the subgroup averages, which display the short-term Casting-to-Casting variation, and the third chart is the standard deviation chart for the within-casting variation. The control limits for the subgroup averages are much wider than the ones previously calculated, and there are no longer any out-of-control points. The control limits for the standard deviation chart remain unchanged from the ones presented previously.

Statistical Insights

In this section, we elaborate on some of the examples provided in ISQC Chapter 6. The examples highlighted in this section include several important concepts we have encountered over our many years of applying SPC successfully to a variety of industries. For most of these examples, additional output not provided in ISQC is included to illustrate JMP functionality or further elaborate on important points.

Operating Characteristic Curve

The operating characteristic (OC) curve (see ISQC Section 6.2.6) shows the probability, β, of not detecting a mean shift (ISQC equation 6.19) with the next subgroup when three sigma limits are used. The curve is usually shown with β on the Y axis and the mean shift, k, on the X axis. For the control chart shown in Figure 3.3 (ISQC Figure 6.2), the OC curve is easily obtained by right-clicking on any chart point and selecting **OC Curve** from the resulting menu or by clicking on the red triangle next to **XBar of Flow Width** and choosing **OC Curve** from the menu (Figure 3.28).

Figure 3.28 Control Chart Options for Flow Width

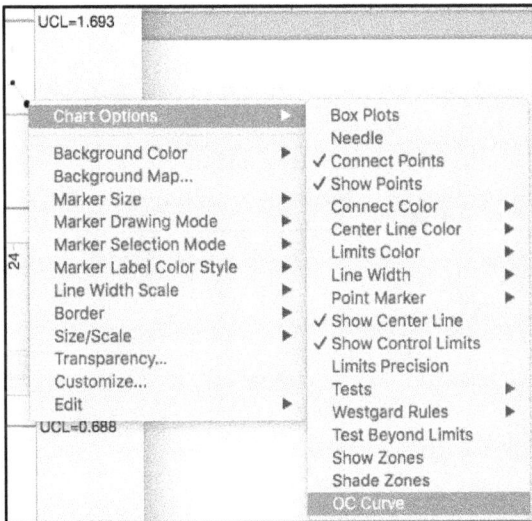

JMP produces a 2-sided OC curve, corresponding to the lower and upper control limits (Figure 3.29). At the top of the OC curve, we see the control chart parameters: target or CL = 1.51, LCL = 1.31, UCL = 1.69, sigma = 0.14, and subgroup size n = 5. At the target value of 1.51, the probability of not detecting a shift is 1, but as we move away from the target in either direction, the probability decreases to 0 at around 1.2 for the lower side and 1.8 for the upper side. The value of 1.8 corresponds to a shift of about 0.3 microns, which is equivalent to 2 standard deviations.

Figure 3.29 OC Curve for the Flow Width XBar Chart.

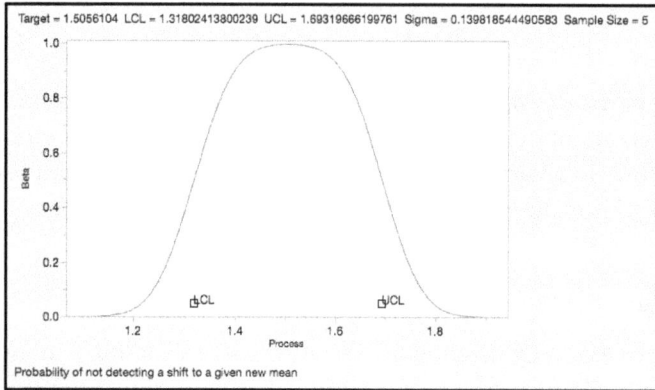

The OC curve can be presented in a more generic fashion, with a target = 0 and standard deviation = 1, by clicking on the **Target** value and entering 0 and on the **Sigma** value and entering 1. The control limits can be calculated as LCL = 0 −k/√n = 0 −3/√5 = -1.34; UCL = 0 + 3/√5= 1.34. We also need to scale the X axis in units of sigma by double-clicking the X axis and setting the **Minimum** =-3.5, the **Maximum** =3.5, and the **Increment** =0.5 in the **X Axis Settings** window, as shown in Figure 3.30a.

Statistics Note 3.3: So why look at the OC curve at all? The OC curve is used to evaluate the sensitivity of the control chart in the units of the data. For example, shifts of 0.3 microns or more will be detected with a high probability ($\beta < 0.044$). However, the probability of not detecting a shift of 0.1 microns is about 93% ($\beta = 0.933$) or the probability of detecting the shift is only 7%.

Figure 3.30a X Axis Setting for OC Curve

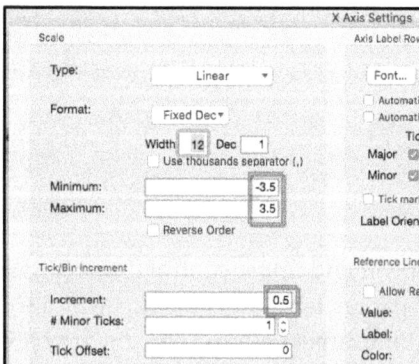

The crosshair tool can be used to read the value of β for a given σ shift. If the shift is 1σ, then for $n = 5$ the probability of not detecting a mean shift is β=0.781 (Figure 3.30b) or the probability of detecting a 1σ shift is only 1-β=0.219. The curve to the right of 0 is the same as the $n = 5$ curve of Figure 6.13 in ISQC Section 6.2.6, from which, for a 1σ shift, as Montgomery points out, "we have β=0.75, approximately."

Figure 3.30b OC Curve for Standardized ±3 Control Limits for $n = 5$

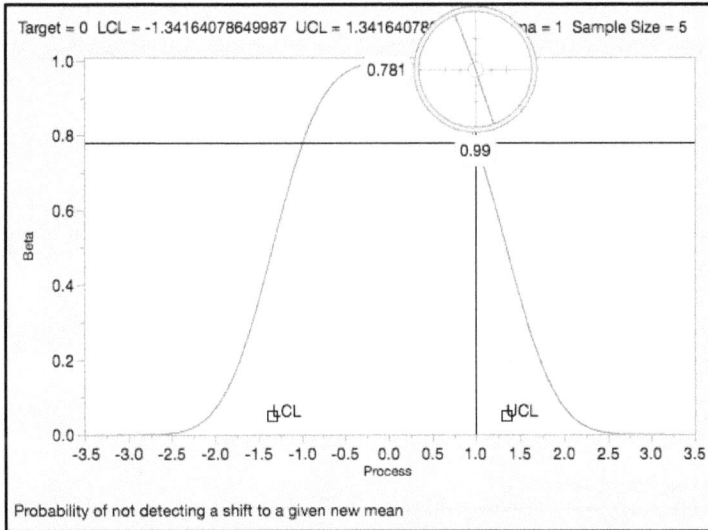

Phase Chart

The additional data in ISQC Example 6.1 showed a shift in the process from an average of 1.506 microns to about 1.56 microns. What if we want to construct a chart that displays two sets of limits, one for the original data and one for the additional data? This can be accomplished using a phase chart, where each phase represents a section of the data. The JMP data set Chapter 3 - ISQC Example 1 Phase Chart.jmp contains the combined data of ISQC Tables 6.1 and 6.2 with an additional column, Source, to denote the source of the data: Original (ISQC Table 6.1) or Additional (ISQC Table 6.2). The following steps illustrate how to create a phase chart using the **Control Chart Builder** platform:

1. Open Chapter 3 - ISQC Example 1 Phase Chart.jmp, which has variables called *Source, Sample Number, Wafer Number,* and *Flow Width*. Sample Number is the subgroup variable and Flow Width is the measurement.

2. Select **Analyze ▶ Quality and Process ▶ Control Chart Builder**. A launch window appears. Drag **Sample Number** from the left-hand window to the **Subgroup** zone (X axis). Similarly, drag **Flow Width** from the left-hand window to the **Y** zone (Y axis). The XBar and Range chart appears first in the window (Figure 3.31).

Figure 3.31 XBar and R Chart for Combined Flow Width Data

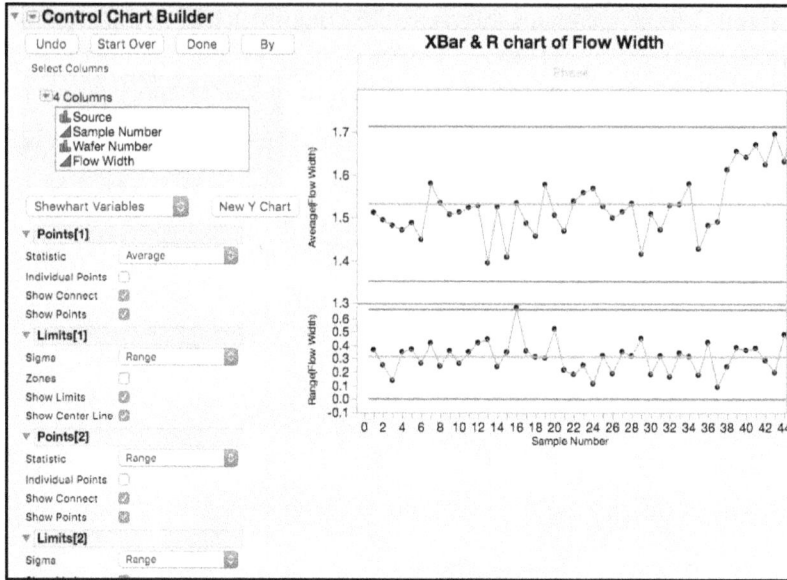

3. To change the chart to a XBar and Range Phase chart, drag **Source** from the left-hand window to the **Phase** zone at the top of the XBar chart, as shown in Figure 3.32.

Figure 3.32 Drag Source to the Phase Area

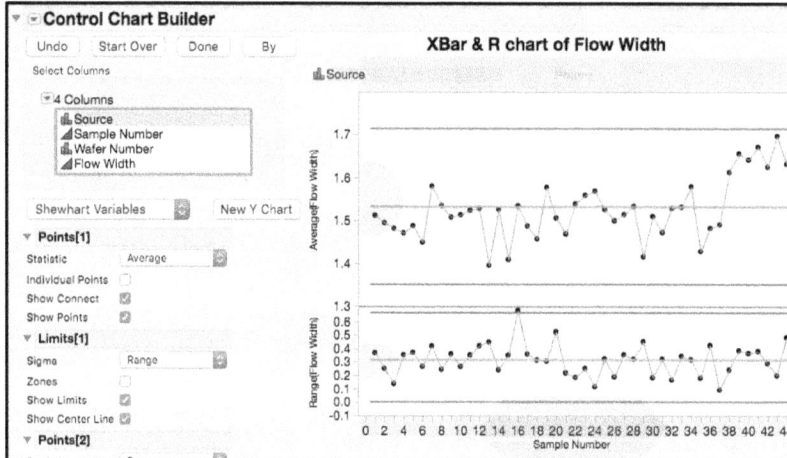

This calculates limits for both the original data and the additional data, as shown in Figure 3.33.

Figure 3.33 Phase Control Chart for Combined Flow Width Data

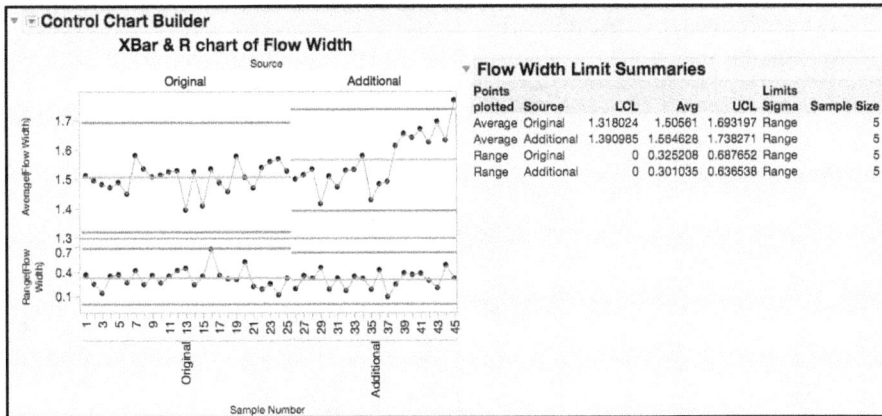

The two Range charts show that the two groups of data have similar average ranges, as shown by the green lines. The XBar charts show that the average of the additional data is larger than the average of the original data, and there is a similar spread between the LCL and UCL. From the Limits Summaries table, we can calculate the average difference between the additional and original data as is 1.565- 1.506 = 0.059 microns.

Lognormal Probability Limits

Even though the process behavior chart is robust to departures from normality (see ISQC Section 6.2.5 or Wheeler and Chambers (1992) Section 4.3), there are situations where the normality assumption should be examined more closely. In ISQC Example 6.6, we demonstrated how to create an XmR control chart using the log-transformed wafer resistivity results (see Figure 3.24). Although the limits might be more appropriate to monitor wafer resistivity, it is more difficult to visually interpret the results in the natural log scale. To alleviate this problem, wafer resistivity might also be charted in the original units using the lognormal distribution. It does require several extra steps to set up the control chart, but interpretation of trends and unusual points is easier.

The following steps show how to evaluate the distribution of the data using the lognormal distribution directly and how to generate a limits table using a script for the Individual Measurement and Moving Range control chart. Note that the limits for the Moving Range chart do not need to be altered to accommodate the skew in the data.

1. Open Chapter 3 – ISQC Table 6.8.jmp.
2. Select **Analyze ▶ Distribution**. A launch window appears. Select **Resistivity** and select **Y, Columns** to add it to the window. Click **OK** when you are done.
3. Click on the red triangle next to **Resistivity** and select **Continuous Fit ▶ Lognormal** from the drop-down menu. The parameters for the fitted distribution appear in the window, with a drop-down menu labeled **Fitted LogNormal**. To add a lognormal probability plot, select **Diagnostic Plot** from the drop-down menu. The parameter estimates, and diagnostic plot are shown in Figure 3.34.

Figure 3.34 Lognormal Probability Plot for Wafer Resistivity

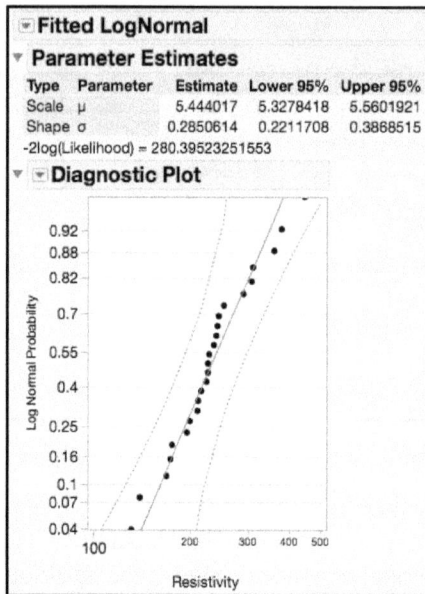

The lognormal probability plot is used to evaluate the appropriateness of the distribution for the data in a similar manner as the normal probability plot. The points should fall along the straight line and be mostly contained within the bands. The control limits are obtained using the parameter estimates for the lognormal distribution and the lognormal quantiles associated with "3σ" probabilities for a normal distribution, 0.135% (LCL) and 99.865% (UCL). These quantiles are easily obtained from JMP.

4. Click on the red triangle next to **Fitted LogNormal** and select **Set Spec Limits for K Sigma**. A dialog box appears (Figure 3.35), which prompts you for the **K value** and the desired limits (one- or two-sided). Tail probabilities corresponding to K standard deviations are computed and the probabilities are converted to quantiles from the lognormal distribution. To calculate 3σ control limits, enter 3 in the **K Value** box, select **Two-Sided for LSL and USL**.

Figure 3.35 Dialog Box for Lognormal Probability Limits

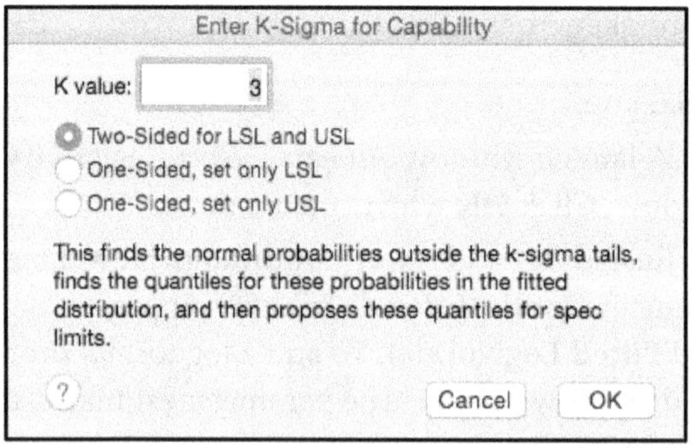

5. Click **OK**. The limits are shown in Figure 3.36.

Figure 3.36 Lognormal Probability Limits for Resistivity Data

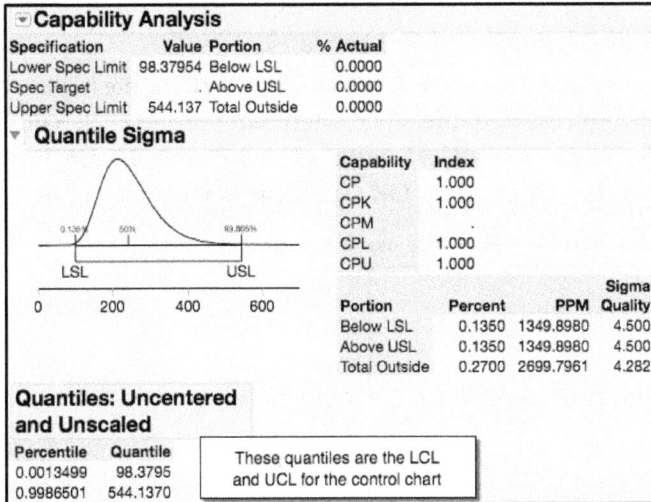

The requested quantiles are displayed and a capability analysis launches using the calculated limits (Figure 3.36). The lognormal quantiles associated with 3σ limits are 98.3795 (LCL) and 543.1370 (UCL). While the K sigma approach for obtaining the limits does not require you to memorize the exact percentiles, it does not automatically produce the centerline (CL) needed for the control chart. To obtain the CL for the control chart, the quantile for the 50% percentile must be calculated from the lognormal distribution.

6. Click on the red triangle next to **Fitted LogNormal** and deselect **Quantiles** and then select **Quantiles**. A dialog box appears allowing up to three probabilities. Enter **0.5** in the first field for the CL and **0.0013499** and **0.9986501** in the other two fields for the LCL and UCL (these two values are the Percentiles in Figure 3.36). Click **OK**.

Figure 3.37 Lognormal Centerline and Limits

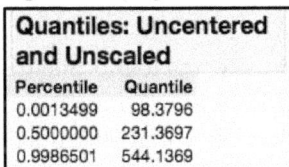

Quantiles: Uncentered and Unscaled

Percentile	Quantile
0.0013499	98.3796
0.5000000	231.3697
0.9986501	544.1369

Now that we have the lognormal limits (Figure 3.37), we must add them as Column Properties for the Resistivity variable in the JMP table and rerun the control chart, as we did in previous examples. Alternatively, we can use a JMP script to automatically calculate the lognormal quantiles and save them to a JMP table, which can be used to create the control chart.

7.   Open Lognormal Quantile Limits.jsl and run the script. The limits are automatically calculated and placed in a JMP table. The table name is Chapter 3 – ISQC Table 6.8 Limits.

8.   Save the table as Chapter 3 – ISQC Table 6.8 Limits.

9.   Select **Analyze ▶ Quality and Process ▶ Control Chart ▶ IR**. In the launch window, select **Resistivity** as the **Process variable**. Then select **Sample** and click **Sample Label** to identify the subgroup variable. Click **Get Limits,** select the control limits JMP table that we saved in Step 8, and click **Open**. Click **OK**. The chart is shown in Figure 3.38.

Figure 3.38 XmR Control Chart for Resistivity using Lognormal Limits

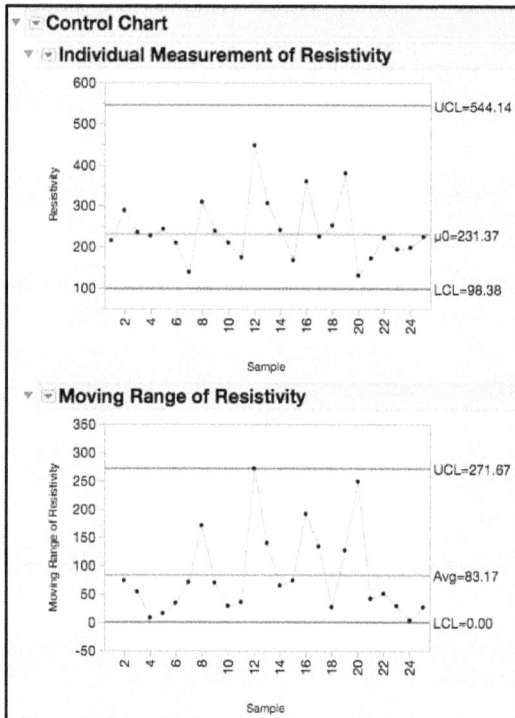

The control limits for the XmR control chart are now adjusted using the lognormal distribution and maintain the original scale of the data. In Figure 3.38, the limits are asymmetric around the centerline, which reflects the skew of the lognormal distribution. Note that only the limits for the Individual Measurement chart have changed. The limits on the Moving Range chart remain the same.

Statistics Note 3.4: There are times when the normality assumption should not be ignored and a more appropriate distribution should be used to calculate control limits. Some things to consider include excessive skew in the data, data that is close to a natural boundary condition (for example, 0 or 100), and normal-based limits that are non-sensible, such as negative values for a positive quantity. Appropriate distributions are well-documented for many physical and scientific phenomena.

The chart in Figure 3.38 can also be generated using the **Control Chart Builder** platform.

1. Select **Analyze ▶ Quality and Process ▶ Control Chart Builder**.
2. Drag **Resistivity** from the left-hand window to the **Y** zone (Y axis). Similarly, drag **Sample** from the left-hand window to the **Subgroup** zone (X axis).
3. Once the chart appears, click on the red triangle next to the **Control Chart Builder** title, and select **Get Limits** as shown in Figure 3.39.

Figure 3.39 XmR Control Chart for Resistivity using Lognormal Limits

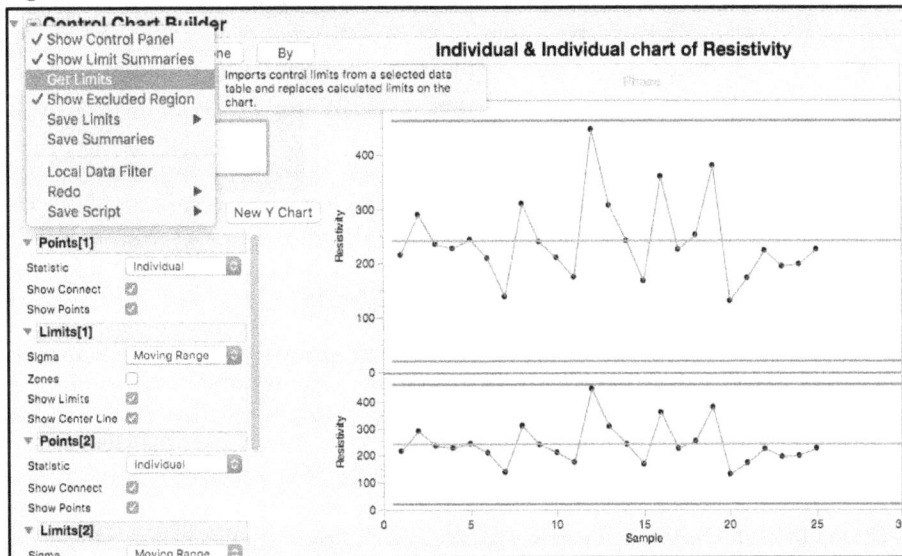

4. Then select the control limits JMP table that we saved in Step 8 in the previous example and click **Open**. Click **OK**. The chart with lognormal limits is shown in Figure 3.40. The Resistivity Limits Summaries report shows a note indicating that limits were imported from a limits file.

Figure 3.40 XmR Control Chart for Resistivity using Lognormal Limits

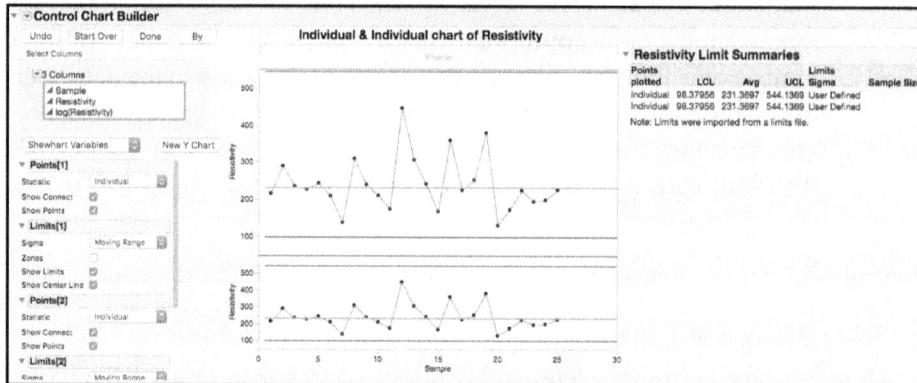

3-Way Control Chart and Variance Components

In ISQC Example 6.11, we observed many subgroup means that were outside of the control limits on the original XBar chart. For this data set, the violations are not due to assignable cause variation but to using the wrong source of within subgroup variation in the calculation of the control limits for the means. For each subgroup, because all five measurements are taken on the same casting, the assumption of independent samples might not hold. A violation of independent samples might deflate the within-subgroup variation and result in limits that are too narrow on the XBar chart.

There are many processes that require the use of 3-way charts to accommodate this issue, and they are usually easy to spot. As is the case with this example, almost all of the subgroup means will be outside of the control limits. However, there might be cases where it isn't as obvious. A variance components analysis (VCA) of the data can add insight into the between- and within-subgroup variation. A VCA can be done in JMP using the **Fit Model** platform within the **Analyze** menu, or the **Variability / Attribute Gauge Chart** platform in the **Quality and Process** menu.

The following steps illustrate how to perform a variance components analysis using the **Variability / Attribute Gauge Chart** platform:

1. Open Chapter 3 – ISQC Table 6.11.jmp, which has variables called *Sample Number, Vane,* and *Vane Height*. Sample Number is the subgroup variable and Vane Height is the measurement.
2. Select **Analyze ▶ Quality and Process ▶ Variability / Attribute Gauge Chart**. A launch window appears.
3. Highlight **Sample Number** in the left-hand window and click **Y, Response**. Similarly, highlight **Vane Height** and click **X, Grouping**. Click **OK**. The default variability chart appears (Figure 3.41).

Figure 3.41 Default Variability Chart for Vane Height

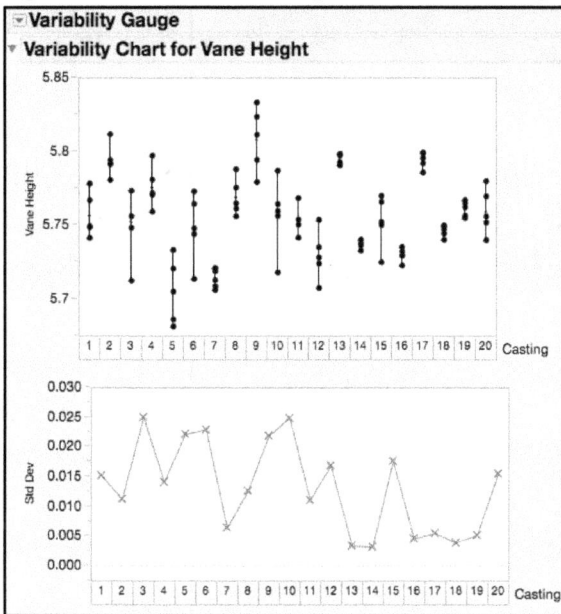

The top graph shows the raw data, organized by subgroup, while the bottom graph shows the standard deviation estimates for each subgroup. There are many options that can be adjusted on the default output.

4. Hold down Alt and click on the red triangle at the top of the window. A menu for all of the available options is displayed (Figure 3.42a). For example, to re-create the XBar and Standard Deviation charts in the Options window, click **Connect Cell Means**, **XBar Control Limits,** and **S Control Limits**. Then click **OK**. The output is shown in Figure 3.42b.

Figure 3.42a Re-creation of the XBar and Standard Deviation Chart

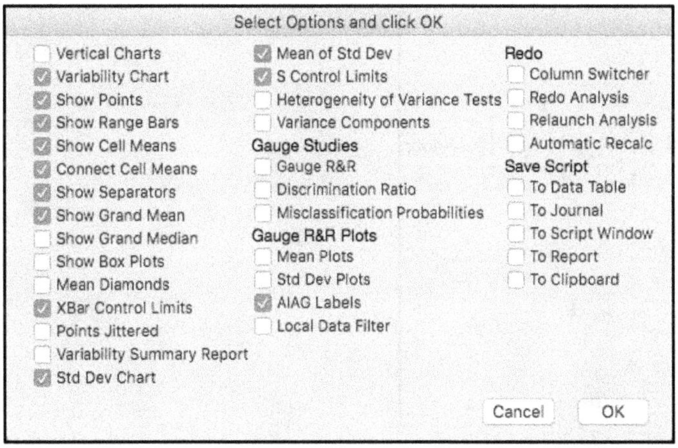

Figure 3.42b Re-creation of the XBar and Standard Deviation Chart

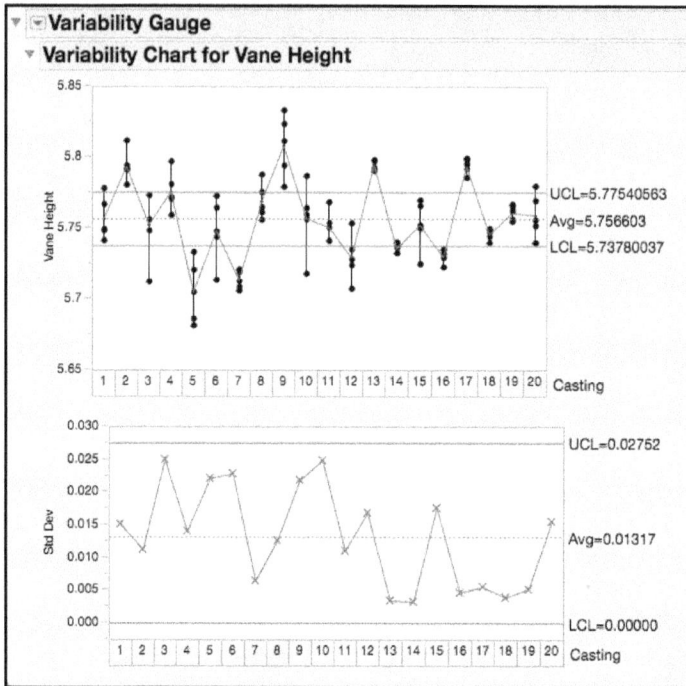

This is a similar chart to the one shown in Figure 3.26. However, because this is a variance components analysis, let's remove the bottom chart and the control limits for the top chart.

5. Hold down Alt and click on the red triangle to launch the Options window. Deselect **XBar Control Limits**, **S Control Limits**, and **Std Dev Chart**. Select **Variance Components** and click **OK**. The results are shown in Figure 3.43.

Figure 3.43 Variance Components Analysis of Vane Heights

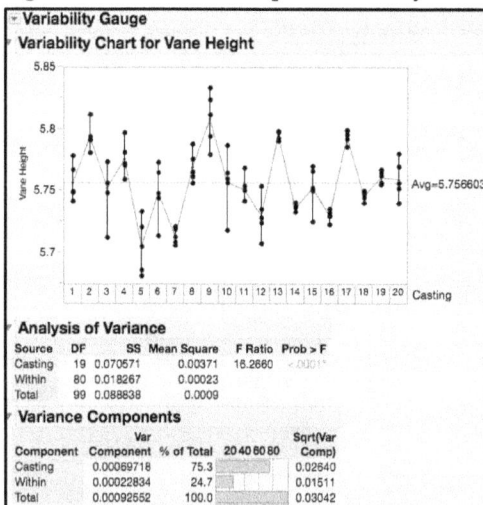

Analysis of Variance

Source	DF	SS	Mean Square	F Ratio	Prob > F
Casting	19	0.070571	0.00371	16.2660	<.0001
Within	80	0.018267	0.00023		
Total	99	0.088838	0.0009		

Variance Components

Component	Var Component	% of Total	20 40 60 80	Sqrt(Var Comp)
Casting	0.00069718	75.3		0.02640
Within	0.00022834	24.7		0.01511
Total	0.00092552	100.0		0.03042

The variance components output includes an Analysis of Variance (ANOVA) table, where subgroups are treated as a fixed effect, and a Variance Components Estimates table, where the subgroups are treated as random effects. The F Ratio in the ANOVA is large and indicates that at least one of the subgroup averages is different from the overall mean, based on the within-subgroup variation. This result supports the many out-of-control points in the XBar chart.

The variance component analysis of this data reveals that the subgroup-to-subgroup (casting-to-casting) variation accounts for 75% of the total variation, while the within-subgroup (vane-to-vane) variation accounts for 25% of the total variation. It is difficult to provide an exact cutoff for what percent of the subgroup-to-subgroup variation will result in the need for a 3-way chart. However, in the authors' experience, when a process is in control and the samples are independent, a VCA will result in a subgroup-to-subgroup variance component that contributes 1% or less to the total variation. If there is assignable cause variation with some out-of-control points, the subgroup variation might account for 25% of the total. Any subgroup contribution of 50% or more to the total variation is highly suspect, and the reader should consider the use of a 3-way control chart.

Statistics Note 3.5: Whenever the subgroup size is > 1, the data hierarchy structure should be investigated to set meaningful control limits for the XBar chart. When the subgroup-to-subgroup variation is substantially larger than the within-subgroup variation, the XBar limits will be too tight and there will be many points outside of the control limits. This occurs in many batch processes, where multiple measurements are taken per each run of a batch (for example, rolled goods, semiconductor wafers, and vats of chemicals).

Rational Subgrouping

Rational subgrouping is a key concept when designing efficient process behavior charts because, depending on how the subgroups are organized, different signals might appear or not appear in the charts. ISQC Section 6.2.2 discusses the important rational subgroups play "in the use of XBar and R charts", and Section 5.6 of Wheeler and Chambers (1992) provides a good description of the issues you might face if the subgroups are not matched to the structure present in the data.

In their article *Designing Insightful Process Behavior Charts*, Ramírez and Zangi (2014) define rational subgrouping as "the process of organizing the data into groups of like things that reflect the context and sources of variation present in the data." They use the injection molding example of Section 5.6 of Wheeler and Chambers to demonstrate how easy it is to use the **Control Chart Builder** to investigate different organizations of the data and how "the careful design of the process behavior chart can reveal patterns that a "default" software chart might mask." They also point out that "even when the charts are designed carefully, it is important to rationally think about the allocation of the different sources of variation to subgroups."

Chapter 4: Control Charts for Attributes

Overview

This chapter illustrates how to generate control charts using examples from Chapter 7, Control Charts for Attributes, of *Introduction to Statistical Quality Control* (ISQC), and includes discussions, tips, and statistical insights on some of the fundamental ideas behind statistical process control (SPC).

These control chart techniques, referred to as *attribute control charts*, are appropriate for data measured on a nominal scale and include P, NP, C, and U control charts.

Two JMP platforms are highlighted in this chapter, the **Control Chart Builder** and the **Control Chart**.

Attributes Control Chart Review

Most books on control charts are partitioned into two sections: control charts for *variable* data and control charts for *attribute* data. This distinction is important because the selection of the most effective control chart depends on the type of the data at hand. In general, attribute data is a measurement that is obtained on a nominal or ordinal scale. Table 4.1 gives brief definitions of the four types of measurement scales. For a thorough discussion of measurement scales, see Section 2.2 in Chapter 2 of Ramírez and Ramírez (2009).

Table 4.1 Summary of Measurement Scales

Measurement Scale	Definition	Examples	JMP Modeling Type
Nominal	Classification into different categories.	Good/Bad Scratch/Wrinkle/Bump Pass/Fail Female/Male	Nominal
Ordinal	Classification into different categories that have an order.	Low/Medium/High Satisfaction Scores (for example, 1, 2, 3, 4, 5) Finishing place in a race (1st, 2nd, 3rd, 4th, and so on)	Ordinal
Interval	Equal distance between intervals. Arbitrary 0.	Temperature (°C/°F)	Continuous
Ratio	Equal distance between intervals. Fixed 0 point.	Temperature (°Kelvin) Age Height (feet) Weight (lbs) Proportions*	Continuous

The most common control charts for attribute data include the P (or NP) and C (or U) charts. The P chart is used to monitor the fraction of nonconforming items. The items might have several characteristics that are inspected or evaluated, but in the end the item is classified as *good* or *bad*. The proportion of the number of bad items to the total items is calculated, p_i, for each subgroup, and this is what is charted on a P chart. Subgroups are formed rationally, as is the case with XBar charts, to maximize the ability of the chart to detect changes in the mean fraction nonconforming. Note that the limits in a P chart depend on the number of items, n_i, per subgroup, and therefore might vary if the n_i vary.

Alternatively, the number of nonconforming units, rather than the fraction of nonconforming, might be charted using an NP chart. The number of nonconforming items in a subgroup, D_i, is totaled and charted on an NP chart. The control limits on this chart also depend on the number of items, n_i, in the subgroup and are constant if n_i remains constant. Both, the P and NP control charts are based on the binomial probability distribution and, as such, are equivalent in their ability to detect shifts in the process. The choice of which to use is a preference in the ease of monitoring a proportion or count.

The C chart is used for monitoring the total nonconformities (or defects) in a subgroup. The charting statistic, c_i, is determined by counting the total number of defects for each unit in a subgroup. This count differs slightly from classifying each item as good or bad. For example, a printed packaging label might be inspected for the total number of misspellings, ink marks, and tears. Once again, subgroups should be formed in a rational way that allows for shifts in the mean to be detected quickly.

The charting statistic for a U chart is derived by dividing the total number of nonconformities, c_i, by the number of items in each subgroup, n_i, to obtain the average number of nonconformities per unit, u_i. The control limits depend on n_i and might vary if n_i varies for each subgroup. The C and U control charts use the Poisson probability distribution to construct the control limits. Once again, because the performance of both the C and U charts is equivalent, the choice of which to use is based on a preference for which charting statistic is easiest to interpret.

The control charts described here are based on statistical assumptions associated with binomial and Poisson distributions, such as independence among the samples and a constant average proportion rate. We discuss several ways to accommodate departures in some of these assumptions, such as using probability limits from more appropriate distributions (for example, the Negative Binomial). A control chart for situations where there is an extremely low failure rate, or rare event, is also discussed.

JMP Attributes Control Chart Platforms

Two platforms are used to create attribute control charts such as, P, NP, C, and U charts. One is the legacy **Control Chart** platform and the other one is the **Control Chart Builder**. The **Control Chart Builder** is part of the new generation of JMP quality tools, one that makes it easier to design, create, and evaluate control charts. These platforms were introduced in Chapter 2. In this chapter, we focus on the use of these platforms for attribute data. Table 4.2 provides a summary of the features we find most useful from both platforms.

Table 4.2 Comparison of Features for JMP Attributes Control Chart Platforms

Feature	Control Chart Builder	Control Chart
Control chart types	P, NP (Binomial) C, U (Poisson)	P, NP (Binomial) C, U (Poisson)
Save limits	In Column and in new Table	In Column and in new Table
Save summaries	Yes	Yes
Launch Window Y or Process variable	Count only	Count or Rate
Annotation features	Using the Annotate tool	Using the Annotate tool
OC Curves	No	Yes

Examples from ISQC Chapter 7

The examples presented in this chapter are from Chapter 7 of ISQC and are shown in Table 4.3. The examples are reproduced using JMP, as are shown in ISQC. For some examples, additional output not provided in ISQC is shown to illustrate JMP functionality or elaborate on important points considered by the authors.

Table 4.3 Summary of Examples from Chapter 7 of ISQC

ISQC Example / Table Number	JMP Table Name	JMP Platform Control Chart Types	Key Points
7.1 Orange Juice Cans	Chapter 4 – ISQC Tables 7.1, 7.2, 7.3	Control Chart P chart Control Chart Builder P/NP Chart Phase Chart	Compute and apply saved limits to new data. Create a phase chart. Generate an OC curve. Adjust for overdispersion.
Table 7.4 Purchase Order Data	Chapter 4 – ISQC Tables 7.4, 7.5	Control Chart P and IR chart Control Chart Builder P Chart	Demonstrate variable subgroup sizes and control limits. Compute a standardized control chart.
7.3 Printed Circuit Boards	Chapter 4 – ISQC Tables 7.7, 7.8, 7.9, Figure 7.14	Control Chart Builder C Chart Pareto Plot Diagram	Create a C chart. Create a Pareto plot and Fishbone Diagram. Generate an OC curve.
7.4 Supply Chain Operations	Chapter 4 – ISQC Table 7.10	Control Chart U Chart	Create a U chart and discuss overdispersion.
7.5 Textile Dyed Cloth	Chapter 4 – ISQC Table 7.11	Control Chart Builder U Chart	Calculate control limits for variable subgroup sizes.
7.6 Valve Failures	Chapter 4 – ISQC Table 7.14	Control Chart Builder Rare Events	Create a control chart for rare events using time between failures.

ISQC Example 7.1 Orange Juice Cans

In this example, we show how to construct a P chart and NP chart in JMP. The data in Table 7.1 of ISQC consists of the number of nonconforming frozen orange juice concentrate cans (D_i), the sample fraction nonconforming (p_i), and the sample number. A subgroup consists of $n = 50$ cans, selected at half-hour intervals over a three-shift period, while the machine is in continuous operation. Each can is inspected to determine if it might leak, either on the side seam or around the bottom joint. For each of 30 subgroups, the number of cans that might leak, out of 50 total cans, is tallied.

The following steps illustrate how to construct a P control chart using the **Control Chart Builder**:

1. Open JMP table Chapter 4 - ISQC Table 7.1.jmp, which has variables called *Sample Number, Number of Nonconforming Cans, Di,* and *Sample Fraction Nonconforming, pi*. In this table, Sample Number is the subgroup variable and, Sample Fraction Nonconforming, pi is the charting statistic.

2. Select **Analyze ▶ Quality and Process ▶ Control Chart Builder** (Figure 4.1).

Figure 4.1 Launching Control Chart Builder

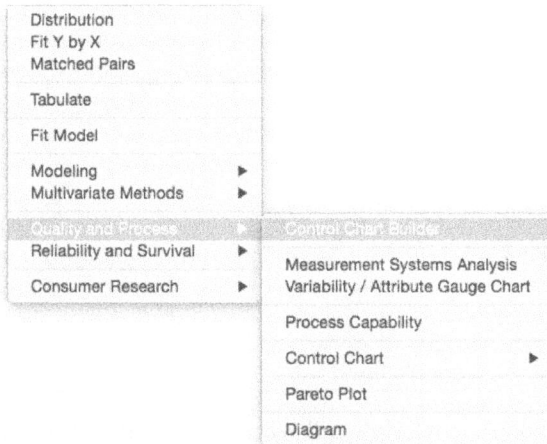

3. Drag **Sample Number** from the left-hand window to the **Subgroup** zone (X axis).
    Similarly, drag **Number of Nonconforming Cans, Di** from the left-hand window to the **Y**
    zone (Y axis) (Figure 4.2). We are selecting the counts, Di, rather than the fraction
    nonconforming, pi, to show how the **Control Chart Builder** calculates the proportions
    from the counts when a sample size is given.

Figure 4.2 Launch Window for Control Chart Builder

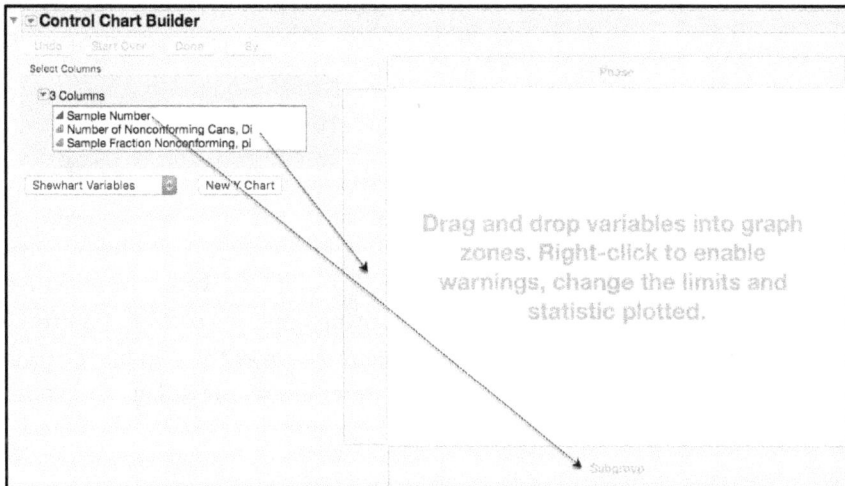

4. A default C chart is created. A few more choices are needed to generate the P chart for the
    fraction nonconforming pi. The different options are located in the options panel to the left
    of the chart. In the **Points** options section, select **Proportion** from the **Statistic** drop-down
    menu; in the **Limits** options section, select **Binomial (P,NP)** from the **Sigma** drop-down
    menu. Figure 4.3a shows the chart before the sample size is entered. The note in the Limits
    Summaries indicates that limits cannot be computed.

Figure 4.3a Intermediate P Chart for Number of Nonconforming Orange Juice Cans

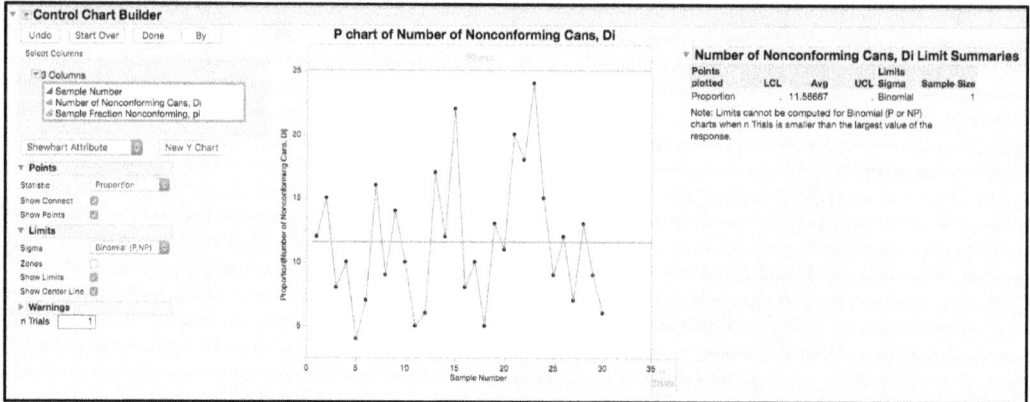

5. To compute the limits, enter 50 in the **n Trials** cell within the **Warnings** options section (Figure 4.3b). The **Control Chart Builder** calculates Sample Fraction Nonconforming, *pi*, from the counts, *Di*, and the sample size of 50.

Figure 4.3b P Control Chart for Fraction Nonconforming Orange Juice Cans

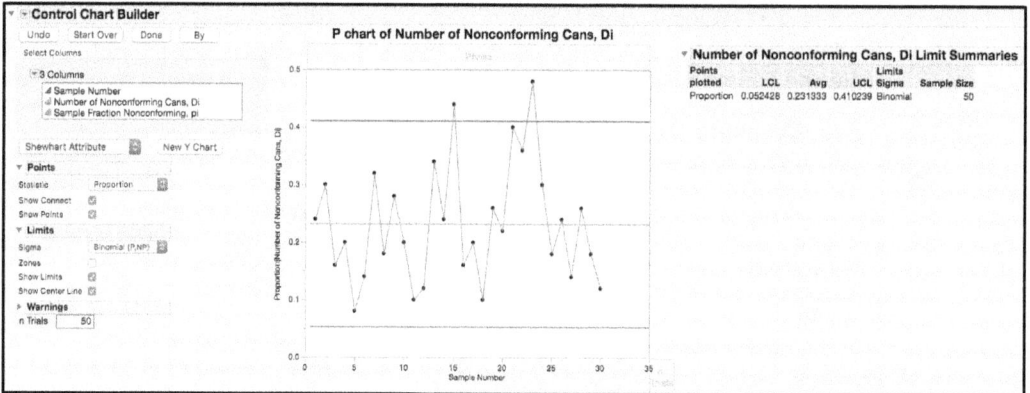

The control chart in Figure 4.3b corresponds to Figure 7.1 in ISQC. Runs tests can be applied to P charts by selecting the appropriate tests from the **Warnings** options section. Figure 4.3c shows the P chart with Test 1 applied to the points. If you mouse hover over the two out-of-control points, the type of test (1) and row number (15 & 23) for the point are displayed, as shown in Figure 4.3c.

Figure 4.3c Test 1 for Fraction Nonconforming Orange Juice Cans

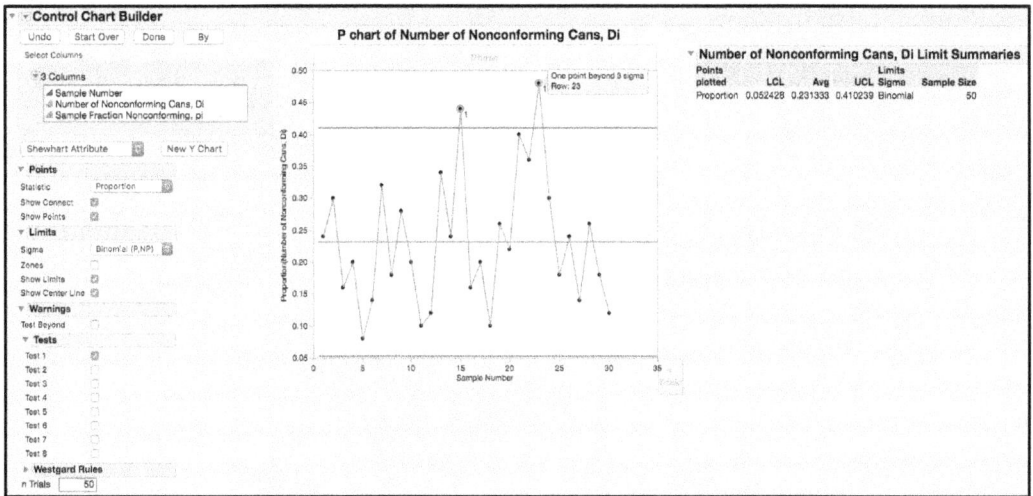

Root causes for these two out-of-control subgroups (see ISQC Figure 7.2) were discovered and samples 15 and 23 were then removed from the data and the control limits were recalculated, as follows:

6. Click on the first point that is above the UCL in Figure 4.3c, sample 15, and then hold down Shift and click on sample 23. From the JMP data table **Chapter 4 - ISQC Table 7.1.jmp**, select **Rows ▶ Exclude/Unexclude**. Note that **Exclude** removes these observations from future calculations, but it does not hide them in the plots. The control chart is automatically updated (see Figure 4.3d).

Figure 4.3d P Chart with Revised Limits for Fraction Nonconforming Juice Cans

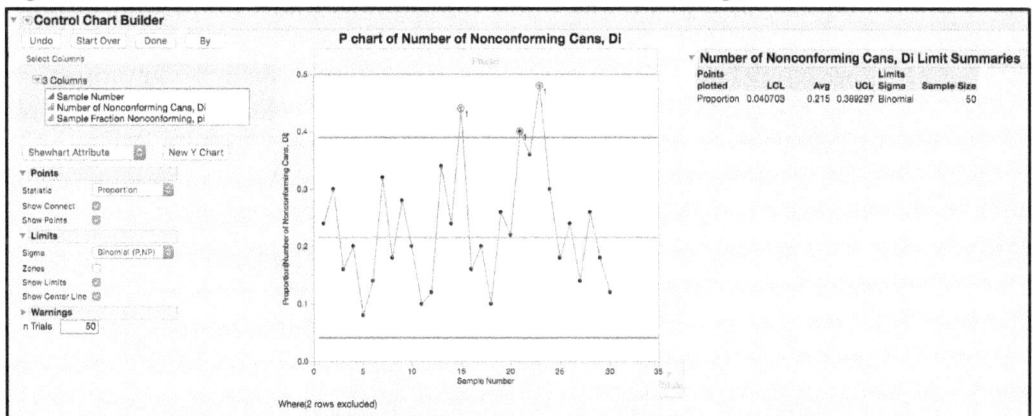

The control limits in Figure 4.3d are updated to reflect the exclusion of the two subgroups and reflects the limits shown in Figure 7.2 of ISQC. The two points are shown in the plot but are grayed out, since we did not hide them when we excluded them. Note in JMP, you can also hide observations, which removes them from any graph.

We now illustrate how to construct the P control charts in Figures 4.3b, 4.3c, and 4.3d using the **Control Chart** platform.

1. Open JMP table Chapter 4 - ISQC Table 7.1.jmp, which has variables called *Sample Number, Number of Nonconforming Cans, Di,* and *Sample Fraction Nonconforming, pi.* In this table, Sample Number is the subgroup variable and Sample Fraction Nonconforming, pi is the charting statistic.

2. Select **Analyze ▶ Quality and Process ▶ Control Chart ▶ P** (Figure 4.4).

   Figure 4.4 JMP Menu Selections for a P Chart

   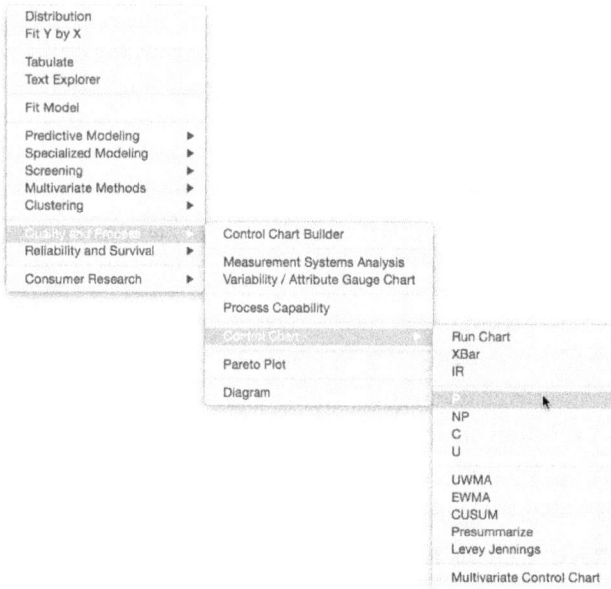

3. When the P chart dialog box appears, select **Sample Fraction Nonconforming, pi** as the **Process** (response) variable. Note, alternatively, that Number of Nonconforming Cans, Di, can be selected for the process variable. Then, select **Sample Number** and click **Sample Label** to identify the subgroup variable. Finally, enter 50 in the field for **Constant Size** (Figure 4.5).

Figure 4.5 JMP Control Chart Menu Selections for a P Chart

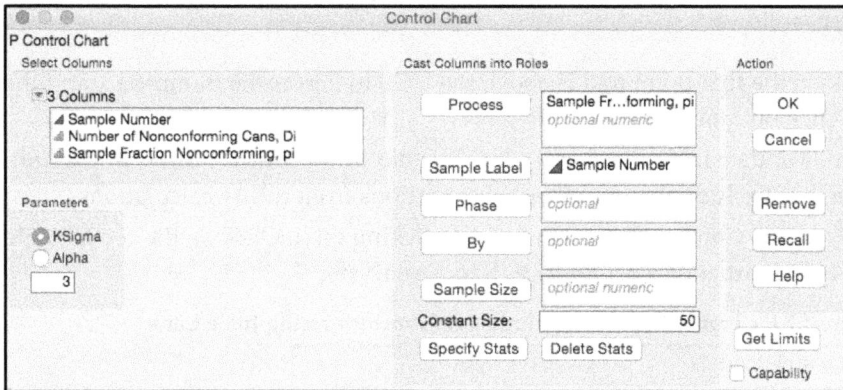

4. Click **OK** to create the control chart shown in Figure 4.6.

Figure 4.6 P Control Chart for Fraction Nonconforming Orange Juice Cans

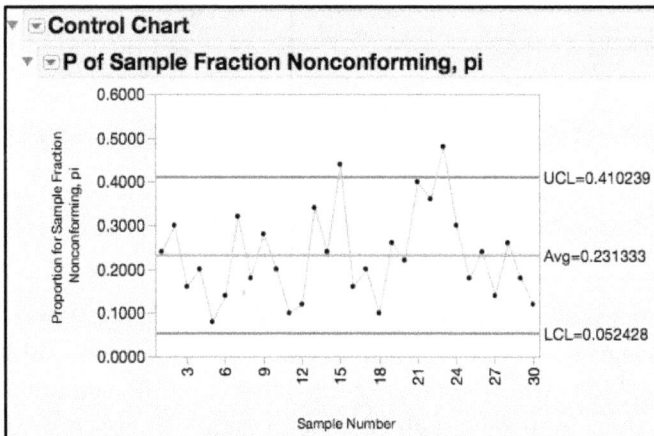

Figure 4.6 corresponds to Figure 7.1 in ISQC. Similar to an XBar and R chart, the first four runs tests (Test 1, Test 2, Test 3, and Test 4) can be applied to P charts by clicking on the red triangle next to the **P of Sample Fraction Nonconforming, pi** title bar and selecting **Tests.** Note Test 5 through Test 8 are grayed out. In ISQC, Montgomery notes that that two points "plot above the upper control limit" UCL = 0.4102. These occur for sample number 15 (p_i = 0.44) and sample number 23 (p_i = 0.48), indicating the process is not in a state of statistical control. Montgomery also points out "these two points must be investigated to see whether an assignable cause can be determined." Upon investigation, it was determined that a new batch of cardboard stock (point 15) and a new, inexperienced operator (point 23) were root causes for these two out-of-control subgroups.

Subsequently, samples 15 and 23 were removed from the data and the control limits were recalculated, as follows:

5. Click on the first point that is above the UCL in Figure 4.6, sample 15, and then hold Shift and click on sample 23.

6. From JMP data table Chapter 4 - ISQC Table 7.1.jmp, select **Rows ▶ Exclude/Unexclude**. Note that **Exclude** removes these observations from future calculations.

7. Activate the Control Chart window by clicking on it. Click on the red triangle next to **Control Chart** and select **Redo ▶ Redo Analysis**.

Figure 4.7 P Chart with Revised Limits for Nonconforming Juice Cans

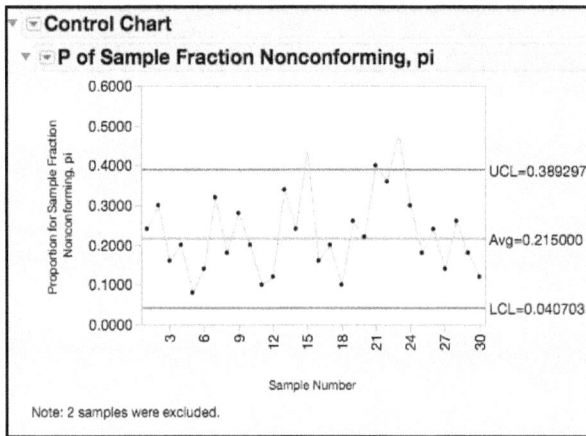

The control limits shown in Figure 4.7 are updated to reflect the exclusion of the two subgroups, matching the limits shown in Figure 7.2 of ISQC. The two points are shown in the plot but are grayed out because we did not hide them when we excluded them (in JMP, you can also hide observations, which removes them from any graph). Although sample 21 now exceeds the UCL, because no root cause was found, it was retained in the data used to calculate new limits.

Saving Control Limits and Applying Them to New Data

ISQC Table 7.2 shows the number of nonconforming cans for 24 additional samples of 50 cans each, for a total of 54 subgroups. We want to see if the process is still in control, using the control limits established from the first 30 runs, excluding subgroups 15 and 23. To apply the limits to the new data, we must save the control limits that were established for the previous chart.

The following steps illustrate how to save the control limits for the control chart in Figure 4.7 and apply them to new data:

1. Click on the red triangle next to the **Control Chart** title at the top of the window (see Figure 4.7) and select **Save Limits**. There are two options available to save the limits; **in Column** saves them to a column property in the JMP table containing the original data, while **in New Table** saves them to a new JMP table (Figure 4.8).

Figure 4.8 Saving Control Limits for Attributes Charts

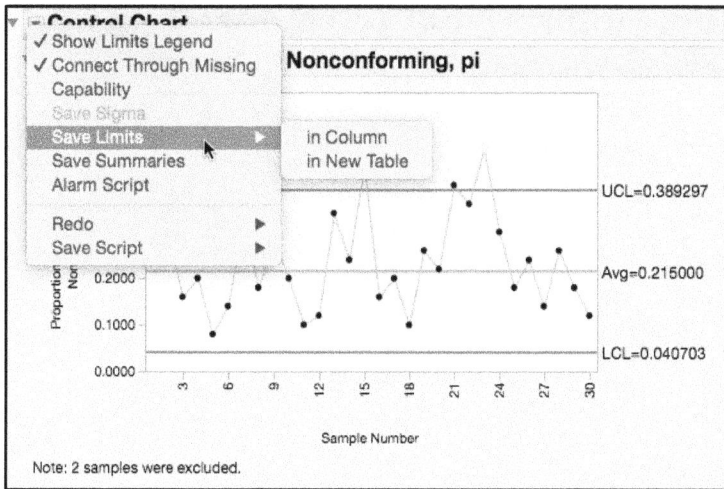

2. To save the limits to a column, select **Save Limits ▶ in Column**. To view the limits, just right-click on the column heading name Sample Fraction Nonconforming, pi in the JMP table and select **Control Limits** from the **Column Properties** drop-down menu (Figure 4.9).

Figure 4.9 Control Limits Saved in Column Properties

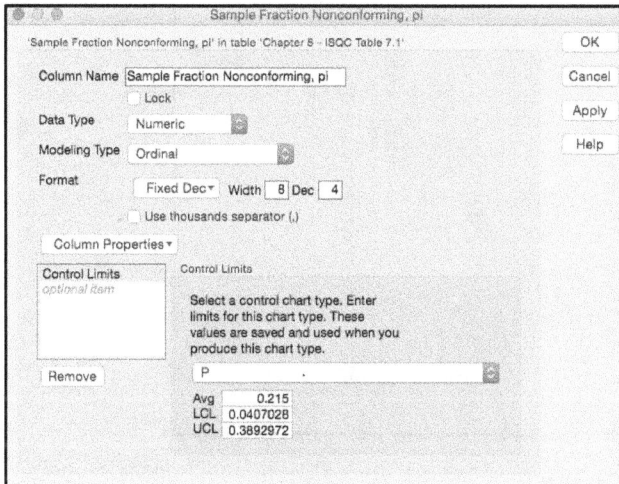

3. To update the control chart with new data, open up Chapter 4 - ISQC Table 7.2.jmp. The new table has 24 rows of data.

4. Make sure Chapter 4 - ISQC Table 7.1.jmp is open and selected. From the main menu bar, select **Tables ▶ Concatenate**. The Concatenate launch window appears with Chapter 4 - ISQC Table 7.1.jmp added to the window on the right. To add the new data, highlight

Chapter 4 - ISQC Table 7.2.jmp from the selection list and click **Add** (Figure 4.10a). Enter the output table name as **Chapter 4 - ISQC Table 7.1 & 7.2** and click **OK**.

Figure 4.10a Concatenating Two JMP Tables

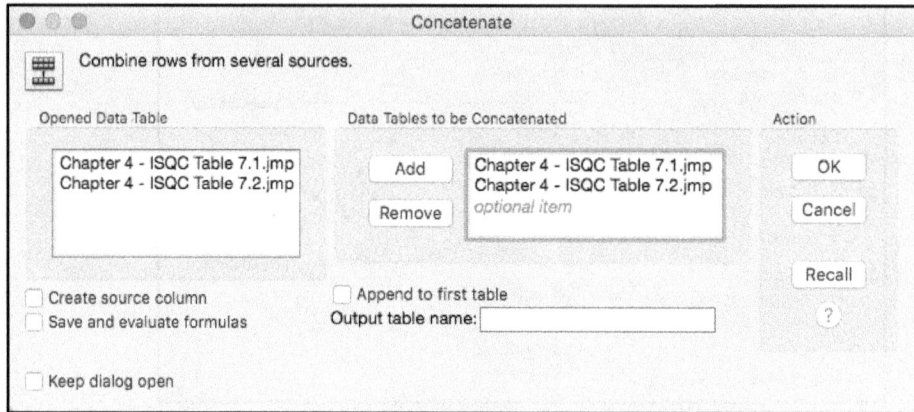

Figure 4.10b Concatenated JMP Tables 7.1 & 7.2

Sample Number	Number of Nonconforming...	Sample Fraction Nonconforming...
1	12	0.24
2	15	0.3
3	8	0.16
4	10	0.2
5	4	0.08
6	7	0.14
7	16	0.32
8	9	0.18
9	14	0.28
10	10	0.2
11	5	0.1
12	6	0.12
13	17	0.34
14	12	0.24
15	22	0.44
16	8	0.16
17	10	0.2
18	5	0.1
19	13	0.26
20	11	0.22
21	20	0.4
22	18	0.36
23	24	0.48
24	15	0.3
25	9	0.18

5. Alternatively, to add new data to the chart, click JMP table Chapter 4 - ISQC Table 7.1 and select **Rows ▶ Add Rows** from the main JMP menu bar. A window appears and you can enter the number of rows you want to add (24) to the table and then click **OK**. Now select all the new rows that were added, rows 31 to 54, and copy and paste new rows of data from a data source, such as Excel or another JMP table.

6. To view the control chart with the new data, repeat Steps 2 through 4 in the previous example. The limits on the control chart, Figure 4.11, should be the same limits shown in Figure 4.7.

Figure 4.11 Updated P Chart for Nonconforming Juice Cans using 54 Subgroups

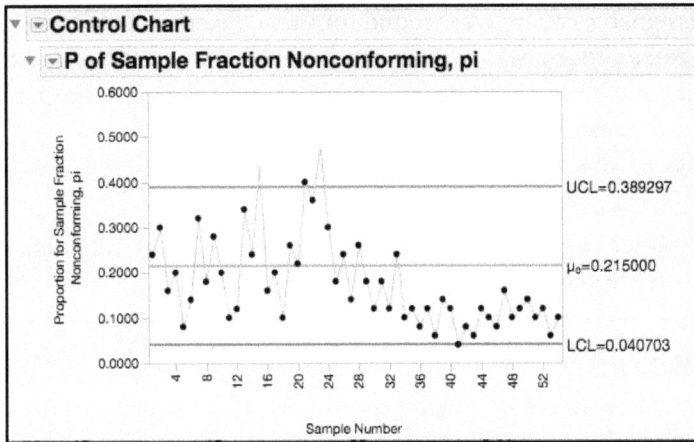

The chart in Figure 4.11 corresponds to Figure 7.3 of ISQC. Because no new points are above the upper control limit, we conclude that the process is in control and operating at an average nonconformity rate of 0.215. However, there is one point below the LCL, and there appears to be a run of points below the centerline. To aid in the visual assessment for patterns among the subgroup averages, the four runs tests can be turned on in the chart: Test 1, Test 2, Test 3, and Test 4. Refer to the online documentation for a complete description of the different runs tests.

> **JMP Note 4.1:** Click on the points above the UCL in the P chart to identify them in the JMP table to exclude them from control limit calculations.

7. To turn on runs tests, click on the red triangle next to the **P of Sample Fraction Nonconforming, pi** title bar and select **Tests ▶ All Tests**. This applies the four Nelson runs tests shown in Figure 4.12.

Figure 4.12 Turn on Runs Tests for P Chart

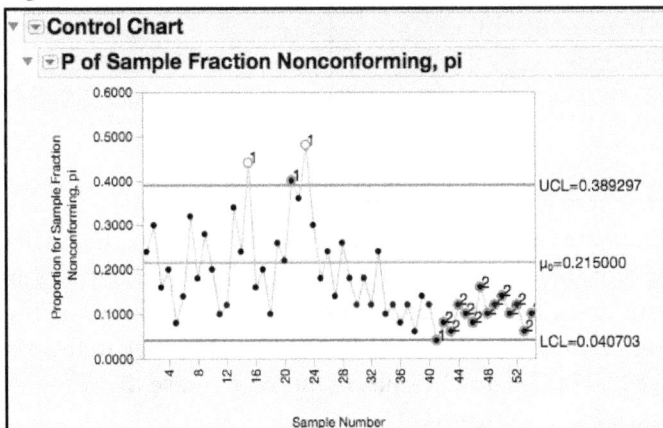

The visual assessment is confirmed, with one additional point labeled below the LCL and several violations for Test 2. No assignable causes were found for these lower nonconforming fractions; however, it is assumed that the process is operating at a mean lower than 0.215, and the limits should be recalculated using the more recent data. A Phase chart is created to more easily visualize the improved performance and compare the two means.

> **JMP Note 4.2:** In the **Control Chart** platform, a P or NP chart can be constructed using either the fraction nonconforming or the total number of defects. Sample sizes are needed for both scenarios.

Creating a Phase Control Chart

The following steps illustrate how to create a Phase control chart for the orange juice data:

1. First, add the Phase variable to the JMP table by clicking Chapter 4 – ISQC Table 7.1 & 7.2.jmp to make it active. Right-click near the top of the table and select **New Columns...** A dialog box appears (Figure 4.13). Enter **Phase** as the **Column Name**, select **Character** as the **Data Type,** and select **Formula** from the **Column Properties** menu. This launches the Formula Editor window.

Figure 4.13 New Column Window to Add Phase Variable

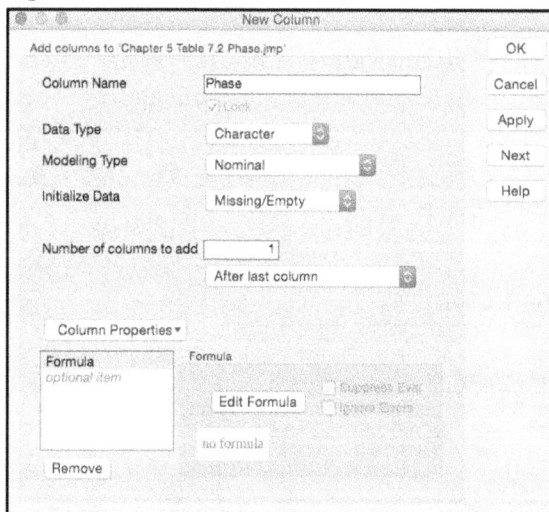

2. In the Formula Editor, in the left-hand panel, click **Conditional** and then select **If**. To build the first condition in the far right-hand panel, double-click **Sample Number** in the **Columns** panel. Return to the left-hand panel and select **Comparison** and then **a <= b**. Double-click in the highlighted box and enter **30**. Then double-click in the **'then clause'** field and enter **"Phase I"**. Finally, double-click in the **'else clause'** field, enter **"Machine Adjustments"**, and click **OK** (Figure 4.14).

> **JMP Note 4.3:** In the **Control Chart Builder,** it is easy to toggle between the P chart and the NP chart by changing the charting Statistic: Proportion or Count. However, the charting statistic is the total number of defects rather than the proportion defective.

Figure 4.14 Formula Editor Window to Add Phase Variable

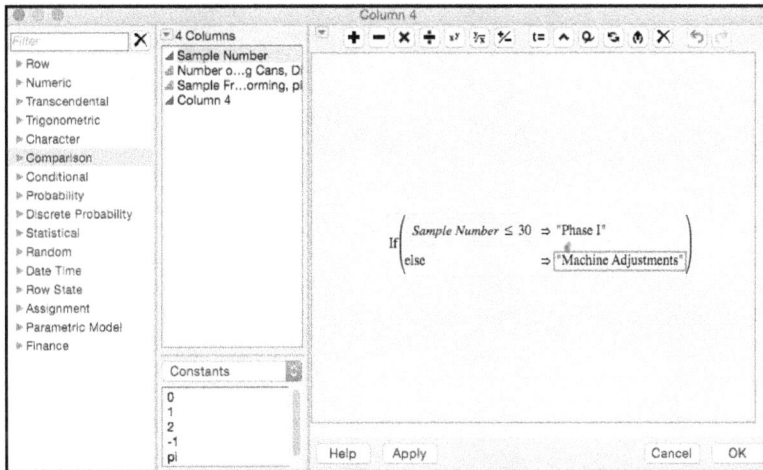

3. To add a new JMP column for the sample size, click Chapter 4 – ISQC Table 7.1 & 7.2.jmp to make it active. Right-click near the top of the table and select **New Columns…** A dialog box appears. Enter **Sample Size** as the **Column Name** and select **Numeric** as the **Data Type.** In **Initial Data Values**, select **Constant** and enter the number 50 (Figure 4.15). This creates a column with the value of 50.

Figure 4.15 Adding Sample Size Column

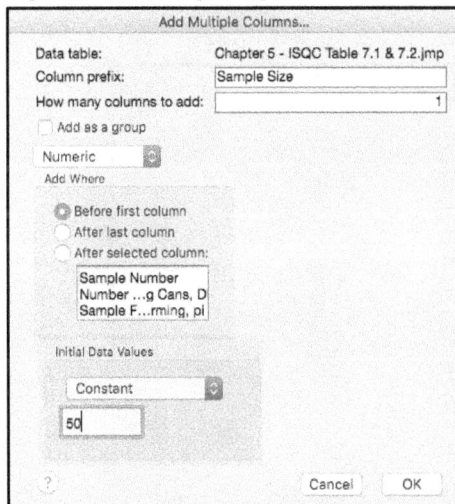

The JMP table has two new columns added to the columns panel that are needed to create the Phase P chart (Figure 4.16). Note the + symbol next to these new variables, which indicates formulas are associated with them.

Figure 4.16 JMP Table with Phase and Sample Size Variables

Sample Number	Number of Nonconforming ...	Sample Fraction Nonconforming...	Phase	Sample Size
1	12	0.24	Phase I	50
2	15	0.3	Phase I	50
3	8	0.16	Phase I	50
4	10	0.2	Phase I	50
5	4	0.08	Phase I	50
6	7	0.14	Phase I	50
7	16	0.32	Phase I	50
8	9	0.18	Phase I	50
9	14	0.28	Phase I	50
10	10	0.2	Phase I	50
11	5	0.1	Phase I	50
12	6	0.12	Phase I	50
13	17	0.34	Phase I	50
14	12	0.24	Phase I	50
15	22	0.44	Phase I	50
16	8	0.16	Phase I	50
17	10	0.2	Phase I	50
18	5	0.1	Phase I	50
19	13	0.26	Phase I	50
20	11	0.22	Phase I	50
21	20	0.4	Phase I	50
22	18	0.36	Phase I	50
23	24	0.48	Phase I	50
24	15	0.3	Phase I	50
25	9	0.18	Phase I	50
26	12	0.24	Phase I	50
27	7	0.14	Phase I	50
28	13	0.26	Phase I	50
29	9	0.18	Phase I	50
30	6	0.12	Phase I	50
31	9	0.18	Machine Adjustments	50
32	6	0.12	Machine Adjustments	50
33	12	0.24	Machine Adjustments	50
34	5	0.1	Machine Adjustments	50
35	6	0.12	Machine Adjustments	50

Columns panel:
- Columns (5/0)
- Sample Number
- Number of Nonconforming Cans, Di
- Sample Fraction Nonconforming, pi
- Phase **+** *
- Sample Size **+**

Rows: All rows 54, Selected 0

4. Select **Analyze ▶ Quality and Process ▶ Control Chart Builder**. The **Control Chart Builder** launch window opens (Figure 4.17).

Figure 4.17 Control Chart Builder Window

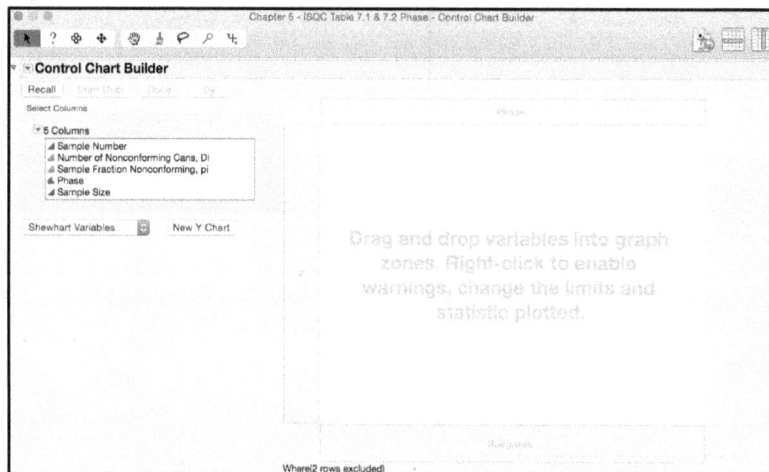

5. Drag **Number of Nonconforming Cans, Di** from the left-hand window to the **Y** zone (Y axis). Similarly, drag **Sample Number** from the left-hand window to the **Subgroup** zone (X axis).

6. The chart type defaults to a Shewhart Attributes C chart. To change it to a P chart, select **Proportion** from the drop-down list next to **Statistics** and under **Points**. Then select **Binomial (P,NP)** from the drop-down list next to **Sigma** and under **Limits**. Next, enter 50 in the box next to **n Trials** at the bottom left-hand side of the window. Alternatively, you can drag and drop the **Sample Size** column into the **n Trials** box. Finally, drag **Phase** from the left-hand window to **Phase** zone at the top of the graph.

Figure 4.18 Control Chart Builder Window Selections for a P Chart with Phases

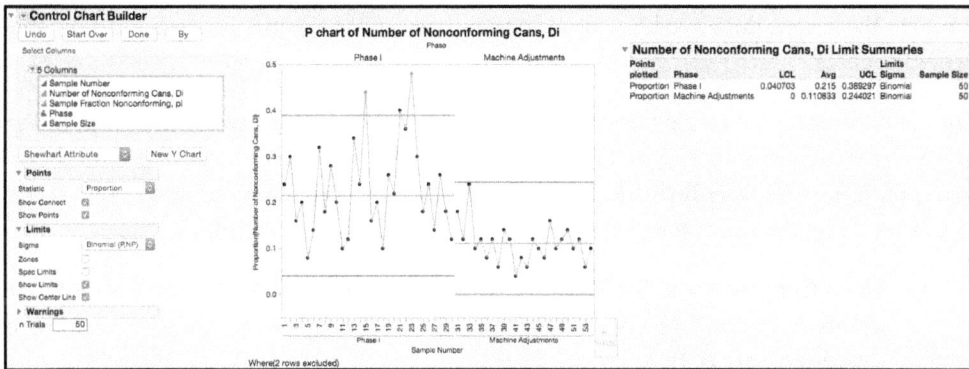

7. Click **Done**. The graph is shown in Figure 4.19.

Figure 4.19 Phase P Chart for Nonconforming Orange Juice Cans

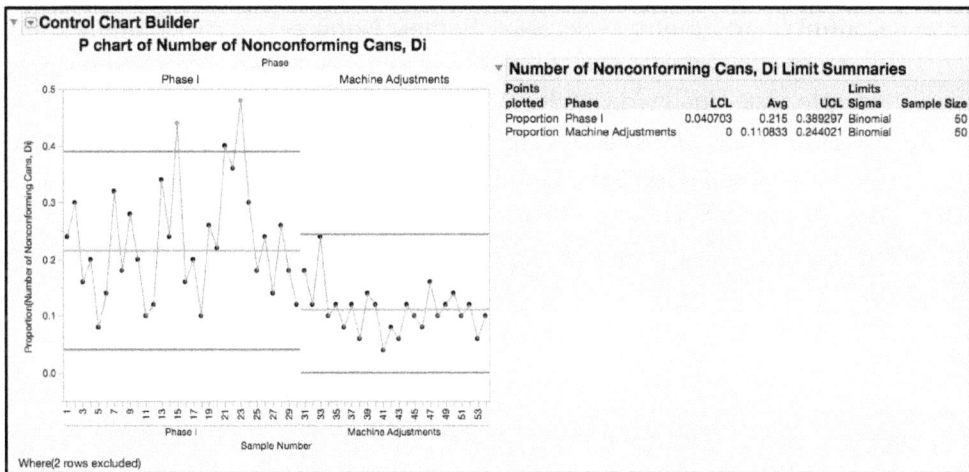

The chart shown in Figure 4.19 is similar to Figure 7.4 in ISQC. In this chart, there are two sets of control limits: one calculated from the first 30 subgroups (minus sample numbers 15 and 23) and the other calculated from subgroups 31 through 54. We can easily see that the overall mean nonconformity rate has decreased from 0.215 to 0.111. This is a substantial reduction and because the process is assumed to be operating in a state of control, the more recent control limits should be applied to future data.

Saving Phase Control Limits and Applying Them to New Data

ISQC Table 7.3 shows the number of nonconforming cans for 40 additional samples of 50 cans each, for a total of 94 subgroups. We want to see if the process is still in control using the recent control limits established from the second 24 runs since the machine adjustments. To apply the limits to the recent data, we have to save the phase control limits that were established for the previous chart.

The following steps illustrate how to save the control limits for the control chart in Figure 4.19 and apply them to new data. Before JMP version 14, the limits for a Phase chart could not be saved using the **Control Chart Builder**, so the **Control Chart** platform is used to save the control limits to a table. Then the final chart is created in the **Control Chart Builder**.

1. Open Chapter 4 – ISQC Table 7.1 & 7.2 Limits.jmp. Note that in this table sample numbers 15 and 23 have been excluded. Select **Analyze ▶ Quality and Process ▶ Control Chart ▶ P**.

2. When the P chart launch window appears, select **Number of Nonconforming Cans, Di** as the **Process** (response) variable. Note for this exercise, Sample Fraction Nonconforming, pi, cannot be used for the process variable because the final chart will be created in **Control Chart Builder**. Next, select **Sample Number** and click **Sample Label** to identify the subgroup variable. Then select **Phase** and click **Phase**. Select **Sample Size** and click **Sample Size**. Finally, click **OK**.

3. To save the new limits, click on the red triangle next to the **Control Chart** title at the top of the window and select **Save Limits ▶ In New Table**. Click on the table, select **File ▶ Save As...** and enter **Chapter 4 – ISQC Figure 7.4 Limits**.

Figure 4.20 Saved Control Limits for Phase P Chart for Nonconforming Cans

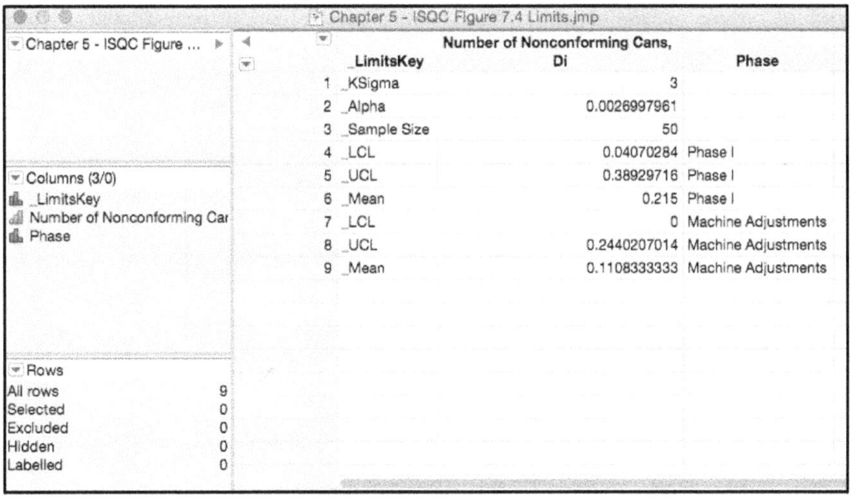
Chapter 5 - ISQC Figure 7.4 Limits.jmp

Chapter 5 - ISQC Figure ...

Columns (3/0)
_LimitsKey
Number of Nonconforming Car
Phase

Rows
All rows 9
Selected 0
Excluded 0
Hidden 0
Labelled 0

	_LimitsKey	Number of Nonconforming Cans, Di	Phase
1	_KSigma	3	
2	_Alpha	0.0026997961	
3	_Sample Size	50	
4	_LCL	0.04070284	Phase I
5	_UCL	0.38929716	Phase I
6	_Mean	0.215	Phase I
7	_LCL	0	Machine Adjustments
8	_UCL	0.2440207014	Machine Adjustments
9	_Mean	0.1108333333	Machine Adjustments

4. Open Chapter 4 – ISQC Table 7.3. The new table has 40 rows of data.
5. Make sure Chapter 4 - ISQC Table 7.1 & 7.2 Phase.jmp is open and selected. From the main menu bar, select **Tables ▶ Concatenate.** The Concatenate dialog window appears with Chapter 4 - ISQC Table 7.1 & 7.2 Phase.jmp added to the window on the right. To add the new data, highlight Chapter 4 - ISQC Table 7.3.jmp from the selection list and click **Add.** In the bottom left corner, check **Save and evaluate formulas.** This automatically fills in the values for the phase and sample size variables in the new table. Enter the output table name as **Chapter 4 – ISQC Table 7.1, 7.2 & 7.3** and click **OK.**
6. Launch the **Control Chart Builder.** Drag **Number of Nonconforming Cans, Di** from the left-hand window to the **Y** zone (Y axis). Similarly, drag **Sample Number** from the left-hand window to the **Subgroup** zone (X axis). Select **Proportion** from the drop-down list next to **Statistics** and under **Points.** Then select **Binomial (P,NP)** from the drop-down list next to **Sigma** and under **Limits.** Next, enter 50 in the box next to **n Trials** at the bottom left-hand side of the window. Finally, drag **Phase** from the left-hand window to **Phase** zone at the top of the graph.

7. To use the previous limits, click on the red triangle next to **Control Chart** and select **Get Limits** and open the JMP limits table created in Step 3, **Chapter 4 - ISQC Figure 7.4 Limits.** Click **Done.**

Figure 4.21 Updated Phase P Chart for Nonconforming Juice Cans

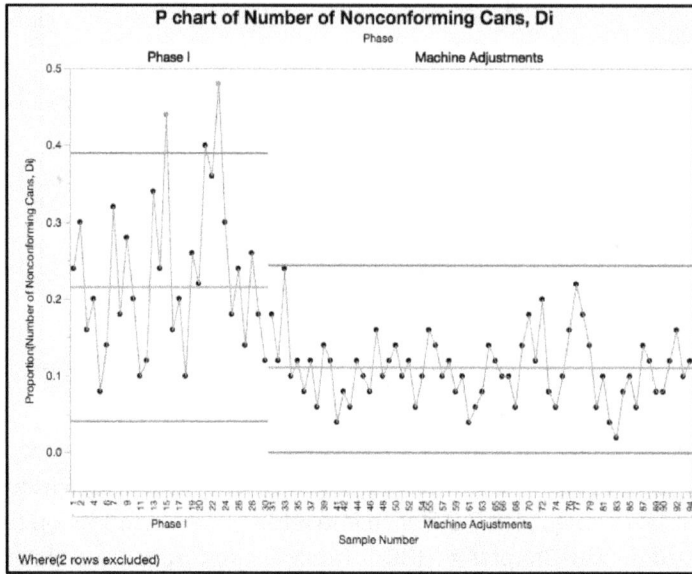

The revised Phase P chart in Figure 4.21 corresponds to Figure 7.5 in ISQC. The sample fraction nonconforming results since the machine adjustments are all within the control limits, indicating that the process is in a state of control.

NP Charts for Nonconforming Orange Juice Cans

In ISQC Example 7.2, a solution is provided to set up an NP control chart for the orange juice concentrate can process in Example 7.1. All sample numbers provided in Table 7.1 are used to construct the limits. In JMP, this can be accomplished in the **Control Chart** platform by selecting **NP** from the drop-down menu and then configuring the launch window in the same way as shown in Figure 4.5, except **Number of Nonconforming Cans, Di** is selected for the **Process** variable.

In the **Control Chart Builder**, it is easy to create an NP chart. However, instead of selecting **Proportion** as the **Statistic** under the **Points** options, leave the default selection of **Count**. The value of 50 is entered in the **n Trials** of the **Warnings** options. The resulting chart is shown in Figure 4.22, and, except for rounding, the LCL and UCL values are close to the ones in the ISQC Example 7.2 solution.

Figure 4.22 NP Chart for Nonconforming Orange Juice Cans

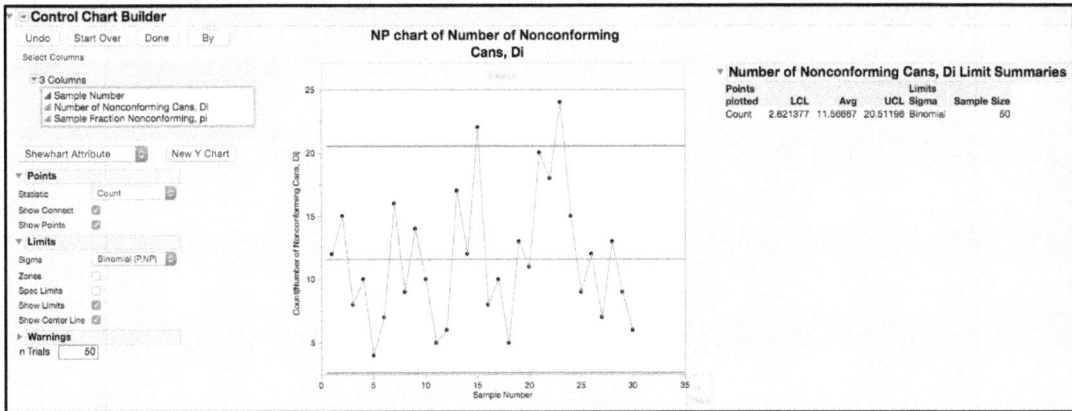

ISQC Table 7.4 Purchase Order Data

In this example, a P chart is constructed using JMP. The data in Table 7.4 of ISQC consists of sample number (*i*), sample size (*ni*), number of nonconforming units (*Di*), and sample fraction nonconforming (*pi*). The data comes from the purchasing group of a large aerospace company, which issues purchase orders to the company's suppliers. A subgroup is defined based on how many purchase orders are issued in a given week, which varies from week to week. A nonconforming unit is a purchase order with an error, such as incorrect part numbers, wrong delivery dates, or inaccurate supplier information. Each week the number of purchase orders with one or more errors is tallied.

The following steps illustrate how to construct a P control chart using the **Control Chart Builder** platform:

1. Open **Chapter 4 – ISQC Table 7.4.jmp**, which has variables called *Sample Number, Number of Nonconforming Units, Di,* and *Sample Fraction Nonconforming, pi*. In this table, Sample Number is the subgroup variable and Sample Fraction Nonconforming, pi is the charting statistic.

2. Select **Analyze ▶ Quality and Process ▶ Control Chart Builder**. The **Control Chart Builder** launch window appears.

3. Drag **Sample Number** from the column select list to the **Subgroup** zone (X axis). Similarly, drag **Number Nonconforming Units, Di** from the left-hand window to the **Y** zone (Y axis). The chart type defaults to a Shewhart Attributes C chart. To change it to a P chart, select **Proportion** from the drop-down list next to **Statistics** and under **Points**. Then select **Binomial (P,NP)** from the drop-down list next to **Sigma** and under **Limits**. Next, drag Sample Size from the left-hand window to the box next to **n Trials**, at the bottom right-hand side of the chart. Finally, click **Tests** and select **Test 1** (Figure 4.23).

Figure 4.23 P Chart Selections for Purchase Order Fraction Nonconforming

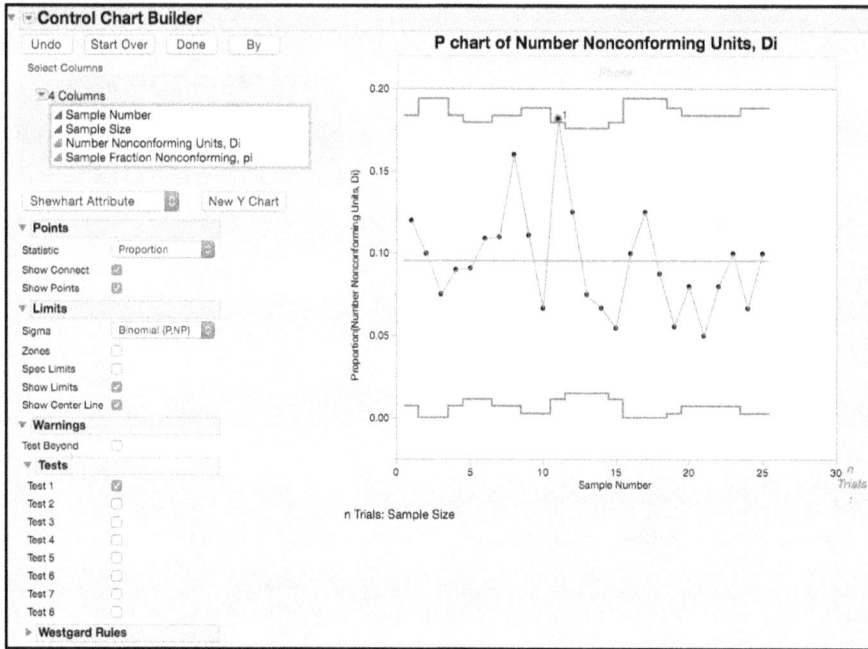

4. Click **Done** to generate the final chart.

Figure 4.24 P Chart for Purchase Order Fraction Nonconforming

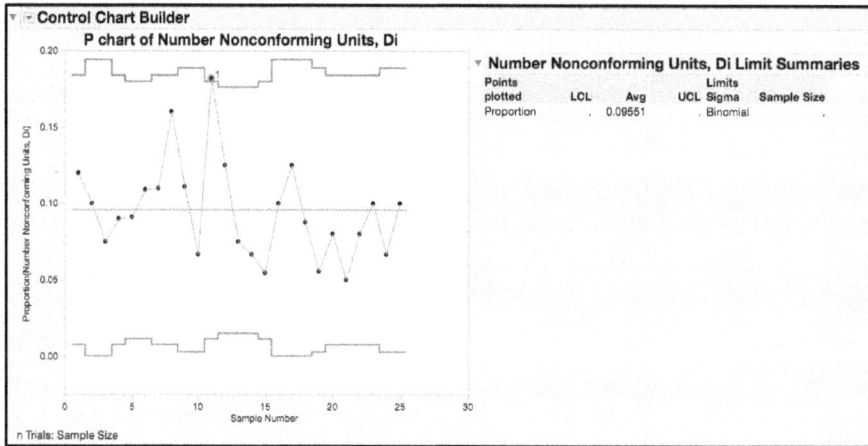

The P chart in Figure 4.24 is similar to the one shown in ISQC Figure 7.6 or Figure 7.7. While the center line is fixed at the overall average, the control limits vary because the number of purchase orders in each subgroup varies between 80 and 120. The tighter limits are associated with the larger sample sizes (for example, $n = 120$), while the wider limits are for the smaller sample sizes (for example, $n = 80$). This platform permits the use of Nelson rules for P charts with variable limits, and there is one point that is just above the UCL for Sample Number 11.

JMP Note 4.4: With varying control limits, the **Control Chart** platform does not permit the use of the Nelson rules, but the **Control Chart Builder** does.

The varying limits might make it more difficult to interpret the charts for trends and out-of-control points. To ease interpretation and manual calculations of charts, the average sample size might be used to construct the control limits for a P chart.

JMP Note 4.5: The **Control Chart Builder** platform does support an average or constant subgroup size with the n **Trials** box, under the **Warnings** options, which appears if the user has not previously added a n Trials variable.

We illustrate the use of the average sample size using the **Control Chart** platform, but this can also be done using the **Control Chart Builder**.

1. Activate Sample Chapter 4 – ISQC Table 7.4.jmp by clicking on it.
2. From the main menu, select **Analyze ▶ Quality and Process ▶ Control Chart ▶ P**.
3. When the P chart launch window appears, select **Sample Fraction Nonconforming, pi** as the **Process** (response) variable. Note that for this exercise, Number of Nonconforming Cans, Di, cannot be used for the process variable. Then select **Sample Number** and click **Sample Label** to identify the subgroup variable. Enter 98, the average sample size, in the field for **Constant Size**.
4. Finally, to reproduce the limits exactly, enter the control process mean, $p = 0.096$. Click **Specify Stats** and enter **0.096** in the **P** field (Figure 4.25).

Figure 4.25 Dialog Window for P Chart Fixed Limits

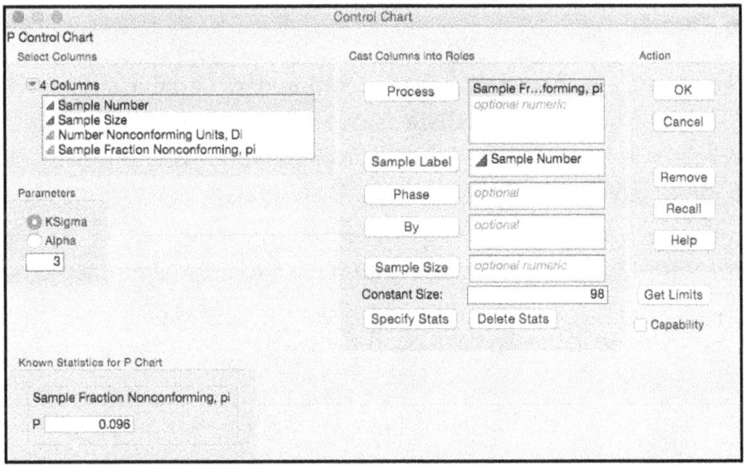

5. Click **OK**. The chart is shown in Figure 4.26.

Figure 4.26 P Chart for Purchase Order Fraction Nonconforming, *n* = 98

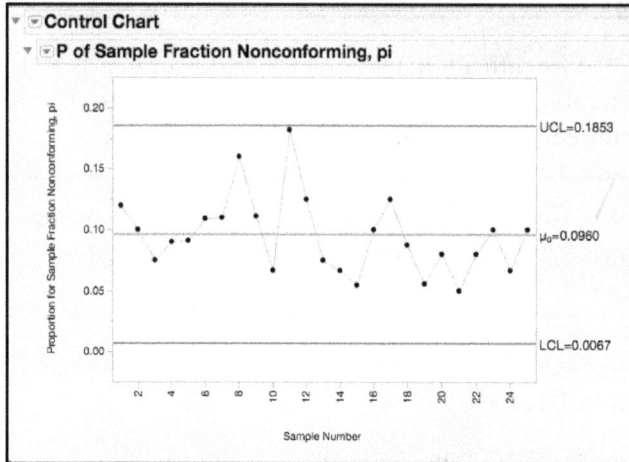

The P chart with fixed control limits in Figure 4.26 is the chart shown in ISQC Figure 7.8. To reproduce the limits exactly, we entered the control process mean, $p = 0.096$, as shown in ISQC. Although no points exceed the control limits, there is one point (Sample Number 11) that is very close to the UCL. Because these limits are approximate, Montgomery recommends calculating the exact control limit when there are points close (below or above) to a control limit.

Statistics Note 4.1: Another approach to accommodate variable subgroup sizes, which also have fixed control limits, is the standardized control chart.

Standardized Control Charts

Standardized control charts are another approach to accommodate variable subgroup sizes, which also have fixed control limits. The charting statistic is created by normalizing each pi by subtracting the overall mean and dividing by its standard deviation. The following steps describe how to construct a standardized chart for the Purchase Order data.

Statistics Note 4.2: The standard deviation used to normalize each p_i is given by $\sqrt{\dfrac{\overline{p}(1-\overline{p})}{n_i}}$, where $\overline{p}$ is the average of all the p_i.

---

1. Activate ISQC Chapter 4 – ISQC Table 7.4.jmp by clicking on it.

2. Right-click near the top of the table and select **New Columns...** A dialog box appears. Enter **Standard Deviation** as the column name and select **Formula** from the **Column Properties** menu. This launches the **Formula Editor** window.

3. In the **Formula Editor**, click on the square root symbol at the top of the window. In the first box, enter 0.096 and then click on the multiplication symbol at the top of the window and enter 0.904 (=1-0.096). Click on the box surrounding the entire multiplication on the inside of the square root sign and then click on the division symbol at the top of the window. Select **Sample Size** from the Columns window to fill in the denominator (Figure 4.27). Click **OK** to close the **Formula Editor** and then click **OK** to close the column's Property window.

Figure 4.27 Formula Editor for Standard Deviation for Purchase Orders Data

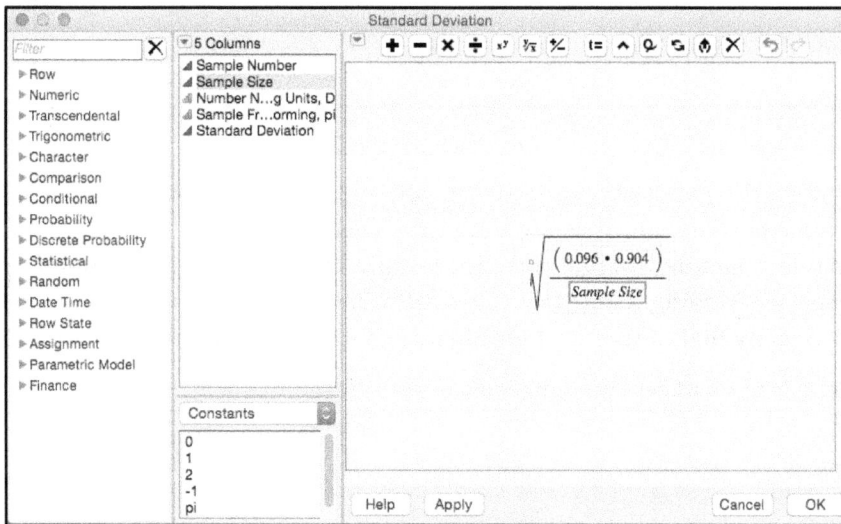

4. Right-click near the top of the table and select **New Columns...** A dialog box appears. Enter zi as the column name and select **Formula** from the **Column Properties** menu. This launches the Formula Editor window.

5. In the Formula Editor, select **Sample Fraction Nonconforming, pi** from the Columns window and then select the minus symbol at the top of the window. Enter 0.096 in the box. Click on the box surrounding the entire subtraction statement and then click on the division symbol at the top of the window. Select **Standard Deviation** from the Columns window to fill in the denominator (Figure 4.28). Click **OK** to close the Formula Editor and then click OK to the close the column's Property window.

Figure 4.28 Formula Editor for Zi for Purchase Order Data

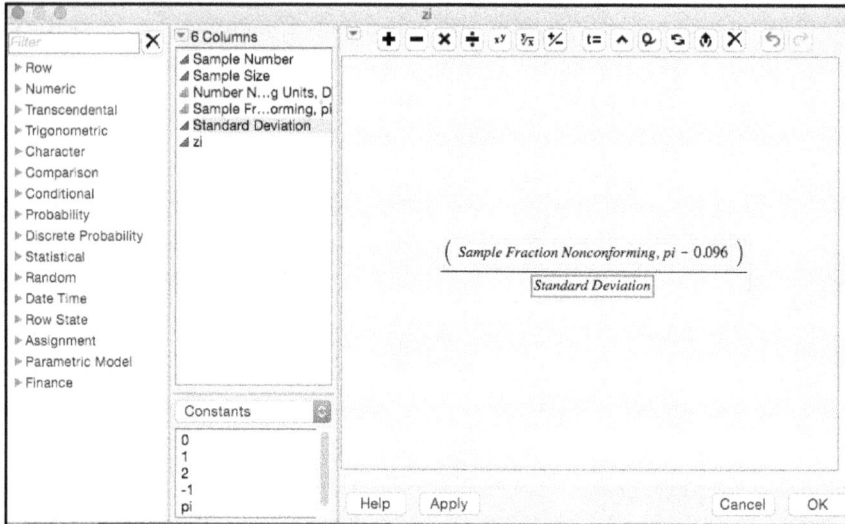

6. Save the table as ISQC Chapter 4 – ISQC Table 7.5.jmp.

7. From the main menu, select **Analyze ▶ Quality and Process ▶ Control Chart ▶ IR**.

8. When the IR chart launch window appears, select **zi** as the **Process** (response) variable. Then select **Sample Number** and click **Sample Label** to identify the subgroup variable. Deselect the **Moving Range (Average)** check box on the left-hand side of the window. Click **Specify Stats** and enter **1** for **Sigma** and **0** for **Mean(measure)** (Figure 4.29).

Figure 4.29 IR Chart Selections for Standardized Purchase Orders

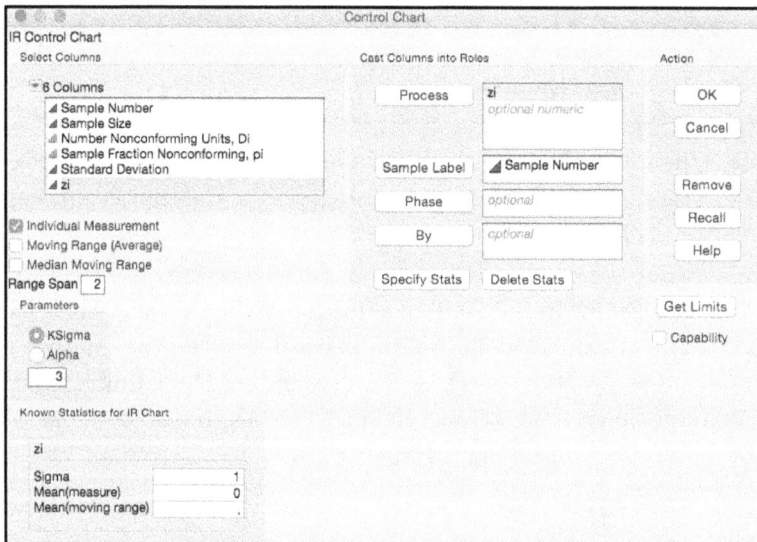

9. Click **OK** when finished. The chart is shown in Figure 4.30.

Figure 4.30 IR Chart for Standardized Purchase Orders Using Zi

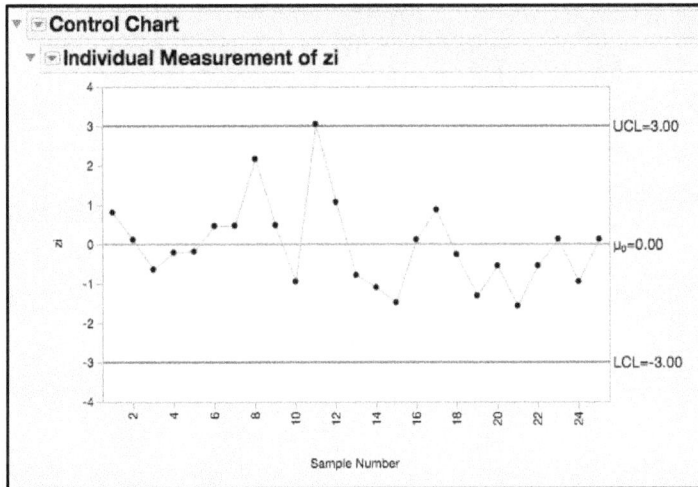

Statistics Note 4.3: P or NP charts are used to monitor good/bad data using the binomial distribution. U or C charts are used to monitor total counts using the Poisson distribution.

The control chart in ISQC Figure 7.10 has been reproduced, Figure 4.30, using an XmR control chart. Because the normalized ps follow a normal distribution with mean = 0 and standard deviation = 1, the control limits are set to -3 and + 3, instead of using the moving ranges. Once again, Sample Number 11 is just outside of the UCL and should be investigated.

ISQC Example 7.3 Printed Circuit Boards

In this example, we show how to construct a C chart using the **Control Chart Builder** platform. The data set consists of the number of nonconformities observed in 26 successive samples of 100 printed circuit boards. The inspection unit (or subgroup) is defined as 100 boards and for each inspection unit, the total number of nonconformities is tallied. In addition, we show how to create a Pareto chart and Fishbone (cause-and-effect) diagram in JMP.

The following steps illustrate how to construct a C control chart using the **Control Chart Builder** platform:

1. Open Chapter 4 – ISQC Table 7.7.jmp, which has variables called *Sample Number* and *Number of Nonconformities*. In this table, Sample Number is the subgroup variable and Number of Nonconformities is the charting statistic.
2. Select **Analyze ▶ Quality and Process ▶ Control Chart Builder**.

3. Drag **Number of Nonconformities** from the left-hand window to the **Y** zone (Y axis). Similarly, drag **Sample Number** from the left-hand window to the **Subgroup** zone (X axis). The chart type defaults to a Shewhart Attributes C chart. Enter **100** in the box next to **n Trials**, at the bottom left-hand side of the window. Click **Done** to produce the final chart, shown in Figure 4.31.

Figure 4.31 C Chart for Printed Circuit Board Nonconformities

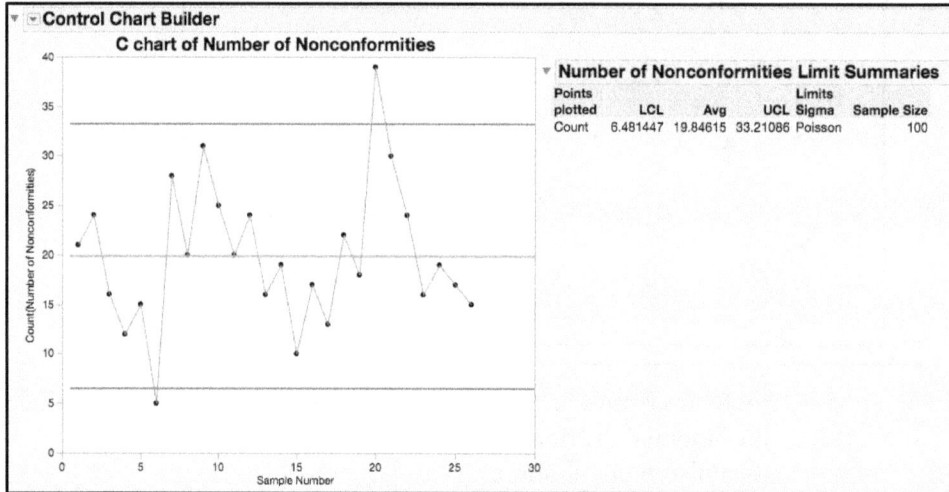

The control chart for the number of nonconformities per sample number for printed circuit boards is shown in Figure 4.31 and can be found in ISQC Figure 7.12. An examination of the chart shows points that exceed the control limits. Sample Number 6 is below the LCL while Sample Number 20 is above the UCL.

Saving Control Limits and Applying to New Data

An investigation revealed potential root causes for both points; therefore, the two points were removed, and the control limits were recalculated for future samples. Twenty more sample numbers, representing the total number of nonconformities for 100 circuit boards, were manufactured and need to be charted. The following steps show how to save the control limits and then apply them to new data:

1. In the C chart shown in Figure 4.31 click the point for Sample Number 6, which is below the LCL and then holding the shift key, click the point for Sample Number 20, which is above the UCL.

2. Click JMP table, Chapter 4 – ISQC Table 7.7.jmp and from the main menu, select **Rows ▶ Exclude/Unexclude**.

3. Go back to the C chart in the Control Chart Builder (shown in Figure 4.31) and from the red triangle, select **Redo ▶ Redo Analysis**. This produces a C chart in another window with the recalculated limits (Figure 4.32).

Figure 4.32 Revised C Chart for Printed Circuit Board Nonconformities

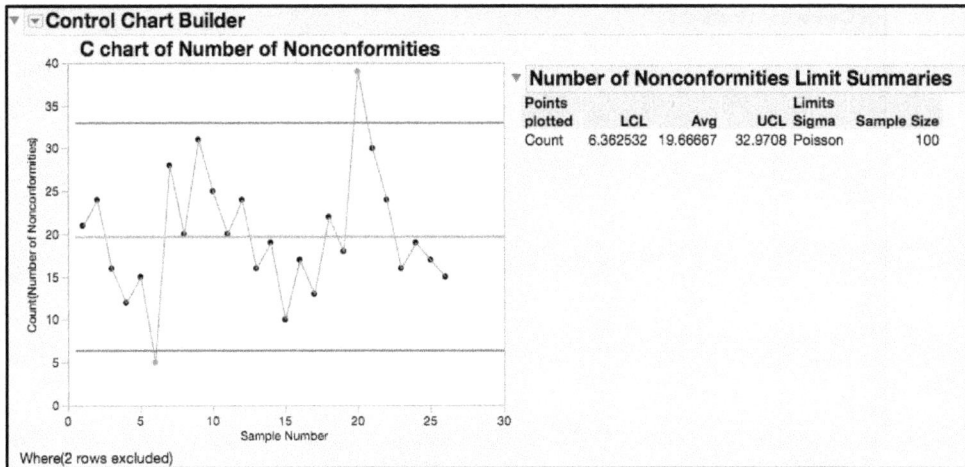

4. To save the control limits to the table, from the red triangle, select **Save Limits ▶ In Column**. The revised control limits are saved as a Column Property for Number of Nonconformities in Chapter 4 - ISQC Table 7.7.jmp.

5. Open Chapter 4 – ISQC Table 7.8.jmp and activate Chapter 4 – ISQC Table 7.7.jmp. From the main menu, select **Tables ▶ Concatenate**. In the new dialog box, click on the second entry in the tables listed in the left-hand side of the window and click **Add**. In the **Output table name** field, enter **Chapter 4 – ISQC 7.7 & 7.8** and click **OK** when finished. The new data set has 46 rows.

6. Select **Analyze ▶ Quality and Process ▶ Control Chart Builder**.

7. Drag **Number of Nonconformities** from the left-hand window to the **Y** zone (Y axis). Similarly, drag **Sample Number** from the left-hand window to the **Subgroup** zone (X axis). The chart type defaults to a Shewhart Attributes C chart. Enter **100** in the box next to **n Trials**, at the bottom left-hand side of the window. Click **Done** to produce the final chart in Figure 4.33.

Figure 4.33 C Chart for Printed Circuit Board Nonconformities Using Saved Limits

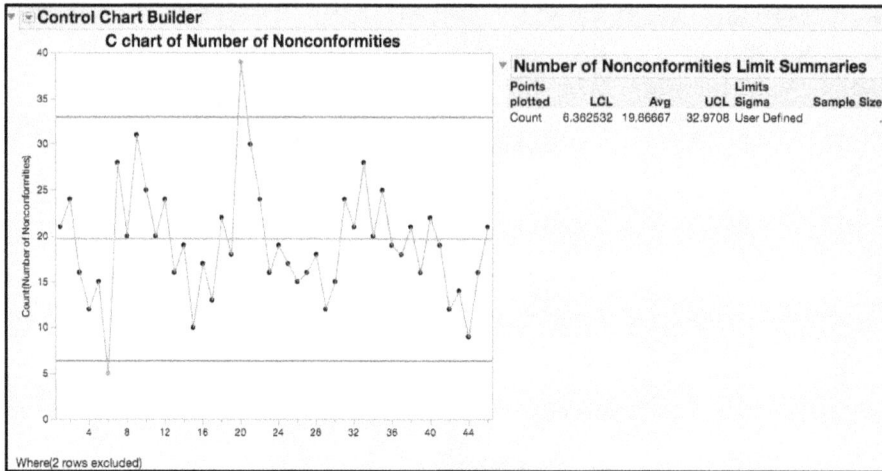

The C chart for the printed circuit board nonconformities in Figure 4.33 evaluates the 20 new subgroups using the limits calculated from the first 26 Sample Numbers, excluding 6 and 20, as shown in ISQC Figure 7.13. The new data indicates the process is in a state of control.

Further Analysis of Nonconformities Using a Pareto Chart and Fishbone Diagram

Further analysis of this type of data includes the use of Pareto charts and Fishbone diagrams. Pareto charts are used to rank order the defect codes to determine the ones that occur with the most frequency. The "80/20 rule" suggests that 80% of the defects are due to 20% of the defect categories, so reducing a few defect categories will go a long way in reducing the overall nonconformity rate. The following steps illustrate how to produce a Pareto chart using JMP:

Statistics Note 4.4: The mean for a binomial variable with parameters n and p is np, and the variance is np(1–p). For a Poisson variable, the mean = variance = λ.

---

1. Open Chapter 4 - ISQC Figure 7.14.jmp. The table includes a column containing the defect categories (*Defect Code*) and a column containing the total number of each defect code (*Frequency*).
2. Select **Analyze ▶ Quality and Process ▶ Pareto Plot**. (Figure 4.34).

Figure 4.34 Launching the Pareto Plot Platform

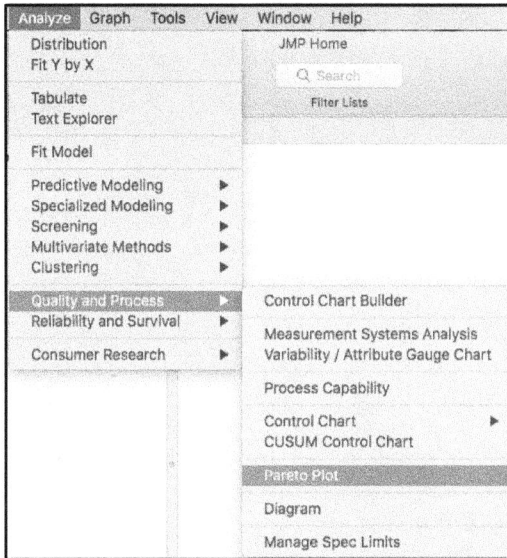

3. In the launch window, select **Defect Code** from the Columns window and click **Y, Cause**. Next, select **Frequency** from the Columns window and click **Freq** (Figure 4.35).

Figure 4.35 Pareto Plot Dialog Selections

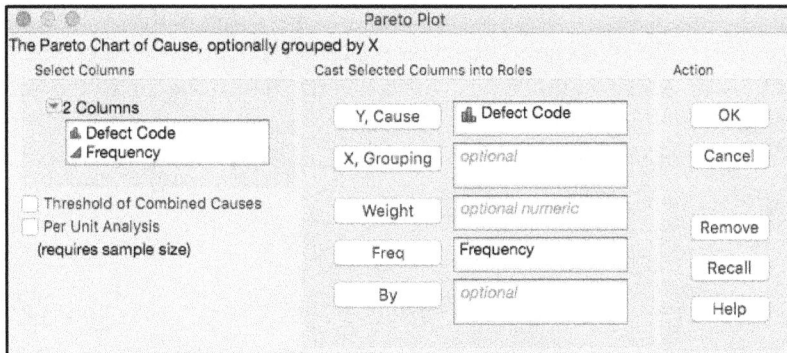

4. Click **OK** when done. The default plot shows the frequency for each defect code (Figure 4.36).

Figure 4.36 Pareto Plot of Printed Circuit Board Defect Codes

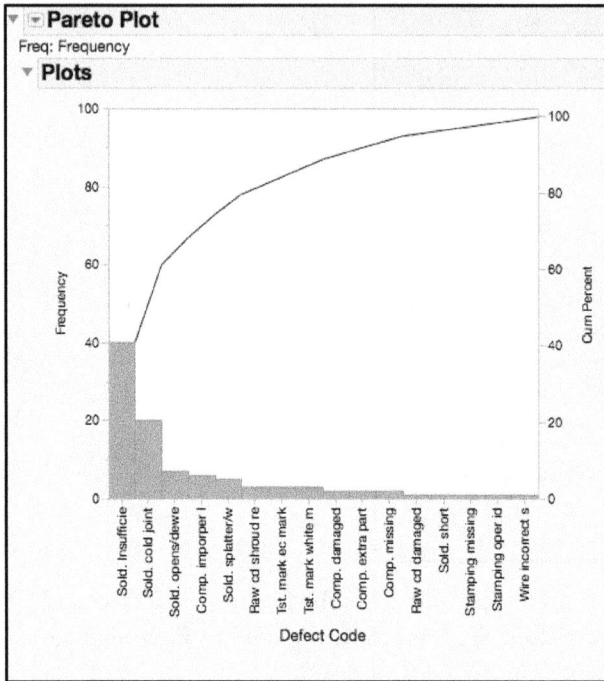

5. From the red triangle next to the **Pareto Plot** label, select **Percent Scale** (to change the Y axis to a percent scale), **N Legend** (to show the total number of defects), and **Category Legend** (to use numbers on the X axis with a legend). From the top of the window, select the "+" icon and holding the cursor, scroll over the graph until the crosshair bar is aligned with 80% on the right-hand Y axis and the cumulative curve (Figure 4.37).

Figure 4.37 Pareto Plot Additional Features of Printed Circuit Board Defect Codes

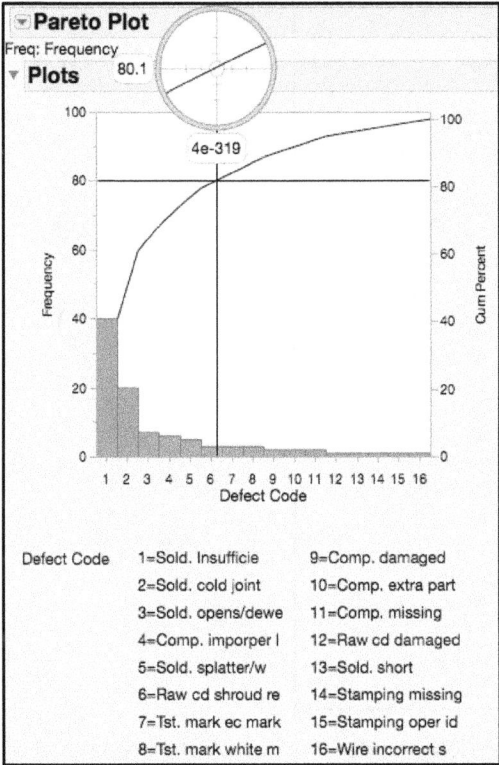

6. From the red triangle, select **Causes ▶ Combine Causes**. In the dialog box, click on the radio button next to **First causes** and then enter **5** in the field immediately to the right. Click **OK** when finished. The plot is shown in Figure 4.38.

Figure 4.38 Pareto Plot 80/20 Rule for Printed Circuit Board Defect

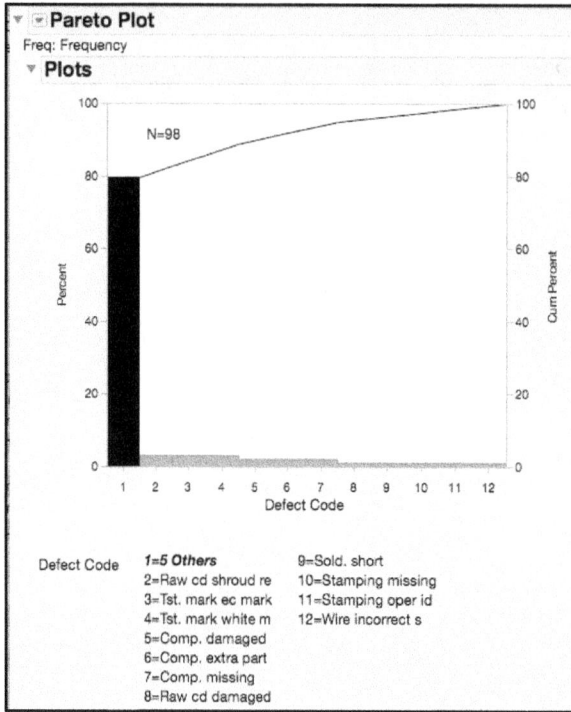

The Pareto plot in Figure 4.36 arranges the defect codes in a bar chart from largest to smallest, similar to ISQC Figure 7.14. Note that the JMP table does not need to be sorted in this order first. The crosshair tool can be used to locate where the cumulative percent is equal to 80% and identify the top defect categories. As shown in Figure 4.37, this occurs after the 5th defect code (31% = 5/16). The **Combine Causes** feature in JMP (see Figure 4.38) illustrates the Pareto 80/20 principle that states that 80% of the effects come from 20% of the causes. Here it was closer to 80/30!

A Cause-and-Effect (fishbone) diagram appears in ISQC Figure 7.15 to help you understand the failure modes of this process. JMP supports the creation of this diagram in the **Quality and Process** menu, using the **Diagram** selection. This ISQC example is currently described in JMP help under **Quality and Process Methods ▶ Cause-and-Effect Diagrams**. The following steps show how to create the diagram:

1. Open the data (Figure 4.39a) by selecting **Help ▶ Sample Data Library ▶ Isikawa.jmp**. Note this data is also in ISQC Table 7.9 or Chapter 4 – Ishikawa.jmp.

Figure 4.39a Data for Printed Circuit Board Defects

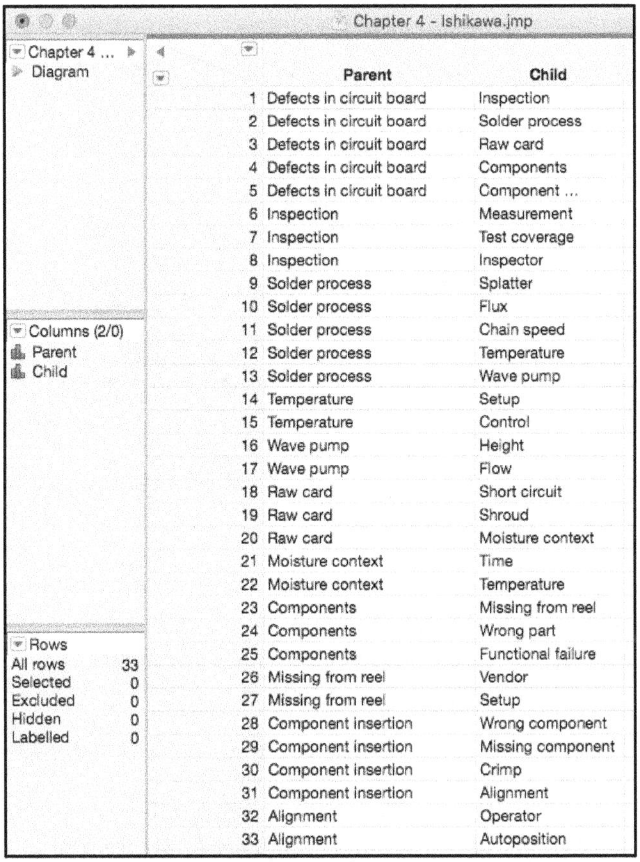

	Parent	Child
1	Defects in circuit board	Inspection
2	Defects in circuit board	Solder process
3	Defects in circuit board	Raw card
4	Defects in circuit board	Components
5	Defects in circuit board	Component ...
6	Inspection	Measurement
7	Inspection	Test coverage
8	Inspection	Inspector
9	Solder process	Splatter
10	Solder process	Flux
11	Solder process	Chain speed
12	Solder process	Temperature
13	Solder process	Wave pump
14	Temperature	Setup
15	Temperature	Control
16	Wave pump	Height
17	Wave pump	Flow
18	Raw card	Short circuit
19	Raw card	Shroud
20	Raw card	Moisture context
21	Moisture context	Time
22	Moisture context	Temperature
23	Components	Missing from reel
24	Components	Wrong part
25	Components	Functional failure
26	Missing from reel	Vendor
27	Missing from reel	Setup
28	Component insertion	Wrong component
29	Component insertion	Missing component
30	Component insertion	Crimp
31	Component insertion	Alignment
32	Alignment	Operator
33	Alignment	Autoposition

2. Select **Analyze ▶ Quality and Process ▶ Diagram**. A launch window, Figure 4.39b, appears and one only has to specify the parent categories, **X, Parent**, and the child categories, **Y, Child**.

Figure 4.39b Diagram Dialog for Printed Circuit Board Defects

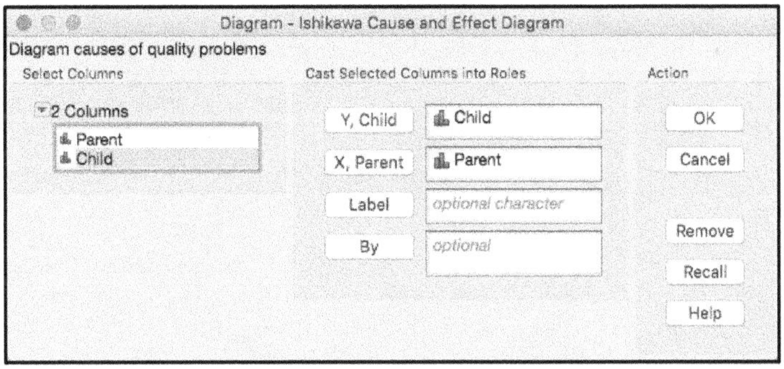

3. After clicking **OK**, the cause-and-effect diagram is generated. The diagram in Figure 4.39c is similar to ISQC Figure 7.15.

Figure 4.39c Cause-and-Effect Diagram for Printed Circuit Board Defects

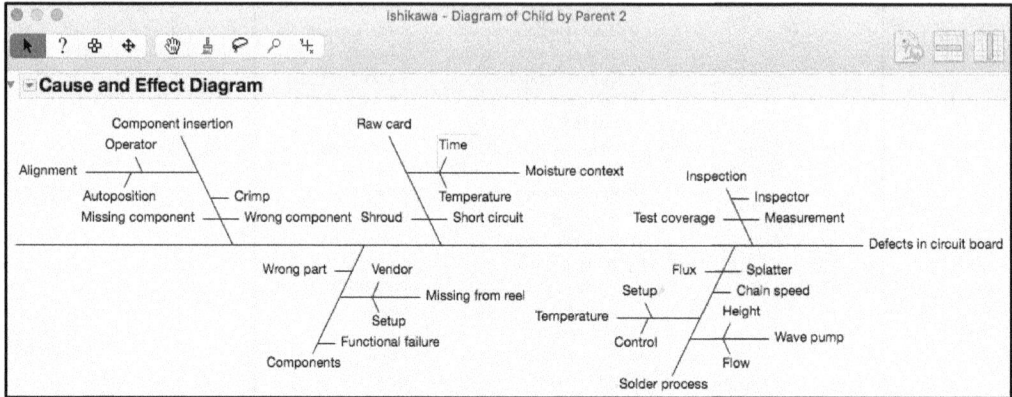

ISQC Example 7.4 Supply Chain Operations

In this example, we show how to construct a U chart using the **Control Chart** platform. The data set consists of the total number of errors observed in 50 randomly selected shipments for 20 consecutive weeks. The inspection unit (or subgroup) is defined as 50 shipments and for each inspection unit, the average number of errors per week is tallied.

The following steps illustrate how to construct a U control chart using the **Control Chart** platform:

1. Open Chapter 4 – ISQC Table 7.10.jmp, which has variables called *Sample Number (week)*, *Sample Size, Total Number of Errors, xi and Average Number of Errors per unit, ui.* Sample Number is the subgroup variable and Average Number of Errors per unit, ui is the charting statistic.
2. Select **Analyze ▶ Quality and Process ▶ Control Chart ▶ U** (Figure 4.40).

Figure 4.40 Launching the U Chart Platform

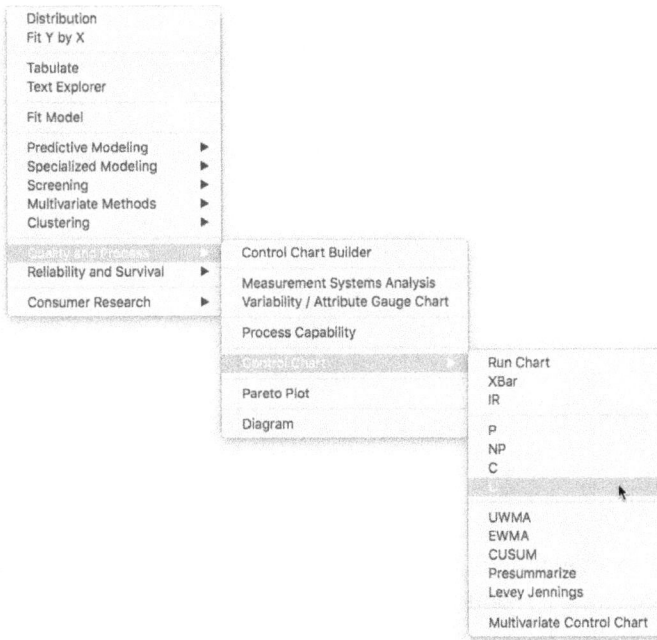

3. A launch window appears (Figure 4.41). Select **Total Number Errors, xi** and then click
    Process. Select **Sample Number (week)** and click **Sample Label**. Then select **Sample Size**
    and click **Unit Size**. Note that by Selecting both the Total Number Errors and Sample Size,
    JMP calculates the Average Number of Errors per unit and plots it on the U chart.

Figure 4.41 U Chart Selections for Shipping Errors

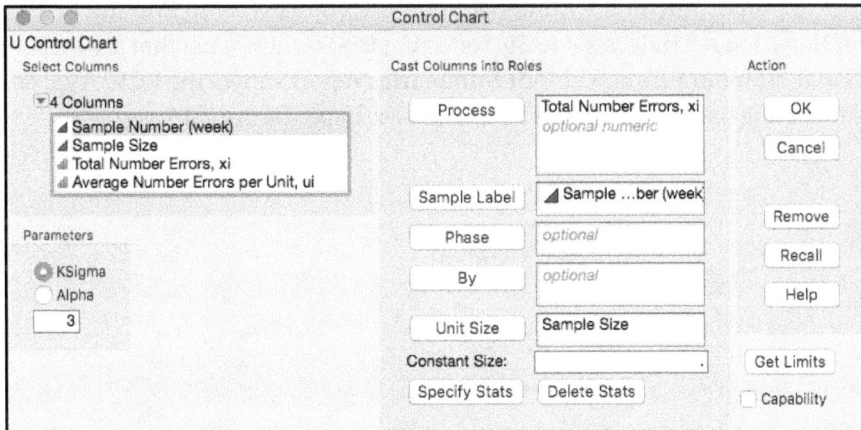

4. Click **OK** when done. The chart is shown in Figure 4.42.

Figure 4.42 U Chart for Shipping Errors

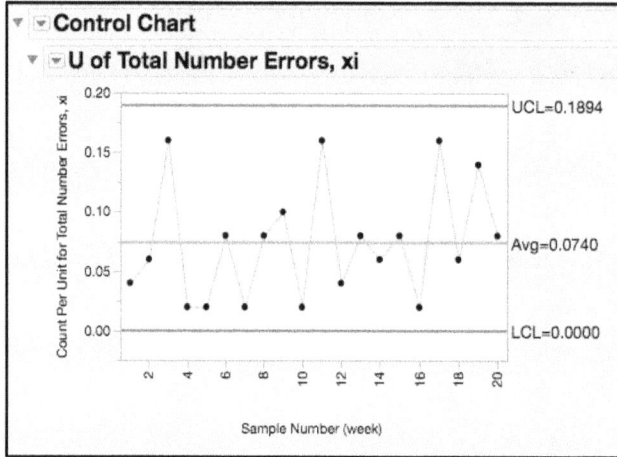

The U chart in Figure 4.42, also shown in ISQC Figure 7.16, does not show any signs of instability and the limits can be used for Phase 2 monitoring. Note that in the **Control Chart** platform, the U chart must be constructed using the total number of errors and the sample size per unit. JMP calculates the averages for each subgroup behind the scenes. If the averages are passed directly as the Process variable, the charting statistics and control limits will be incorrect.

The chart in Figure 4.42 can also be generated by selecting **Analyze ▶ Quality and Process ▶ Control Chart Builder,** using the total number of errors and the sample size per unit. The **Control Chart Builder** platform choices, and chart, are shown in Figure 4.43. In the **Points** options, the **Statistic** is set to **Proportion**; in the **Limits** options, the **Sigma** is set to **Poisson (C,U)**; and in **Warnings** options, the **n Trials** is set to **50**, reflecting the 50 shipments that are inspected each week. The **Total Number Errors, xi Limit Summaries** report shows the LCL, Avg, and UCL, which agree with the limits and centerline line shown ISQC Figure 7.16.

Figure 4.43 Control Chart Builder U Chart for Shipping Errors

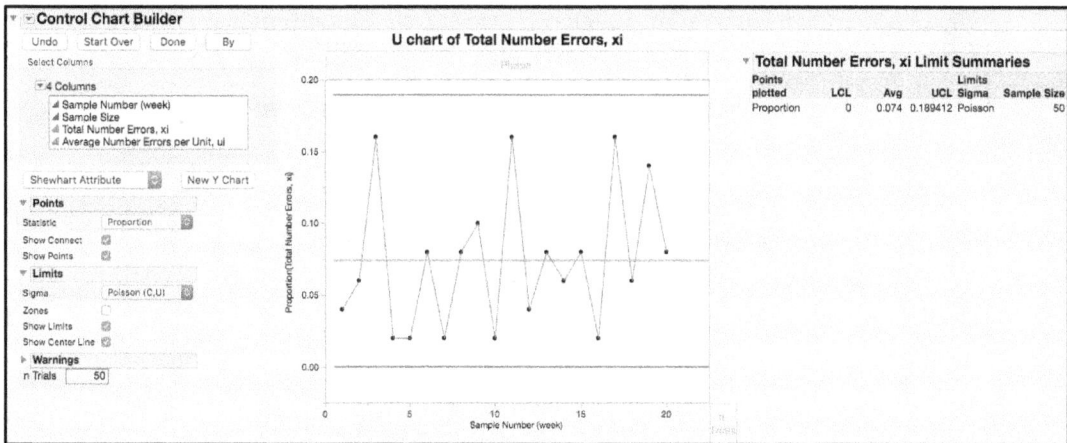

ISQC Example 7.5 Textile Dyed Cloth

In this example, we show how to construct a U chart with varying limits using the **Control Chart Builder** platform. The data set consists of the total number of defects observed per 50 square meters of dyed cloth in a textile finishing plant. The inspection unit (or subgroup) is defined as a roll of fabric, and for each of ten rolls, the total number of defects found in all of the 50-square-meter sections along the roll is tallied.

The following steps illustrate how to construct a U control chart using the **Control Chart Builder** platform:

1. Open Chapter 4 – ISQC Table 7.11.jmp, which has variables called *Roll Number, Number of Square Meters, Total Number Nonconformities, Number of Inspection Units in Roll (n),* and *Number Nonconformities per Inspection Unit.* Roll Number is the subgroup variable and Number Nonconformities per Inspection Unit is the charting statistic.

2. Drag **Roll Number** from the column select list to the **Subgroup** zone (X axis). Similarly, drag **Total Number Nonconformities** from the column select list to the **Y** zone (Y axis). Finally, drag Number of Inspection Units in Roll (n) to the **n Trials** box in the lower right-hand side of the chart.

3. Select **Proportion** in the drop-down menu next to **Statistic** (under the **Points** banner).

4. Click **Done** when finished. The chart is shown in Figure 4.44.

Figure 4.44 U Chart with Varying Limits for Dyed Cloth Nonconformities

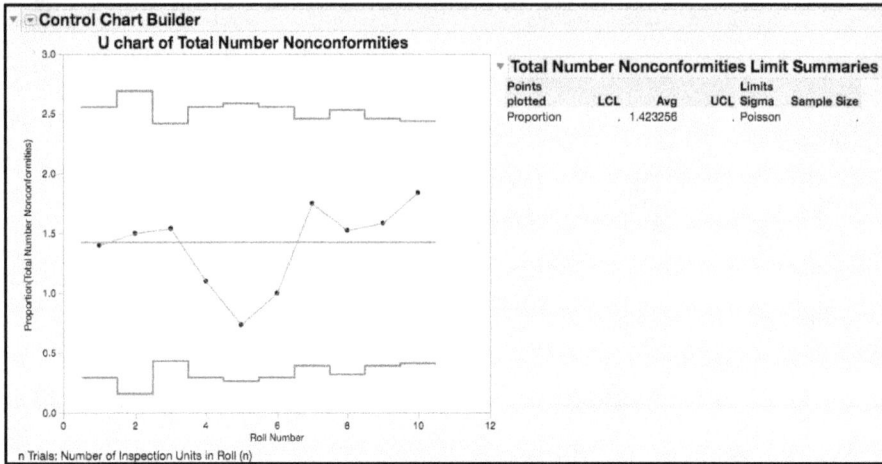

The chart in Figure 4.44 is similar to ISQC Figure 7.17. The control limits vary for each subgroup due to the different unit sizes for each roll. The number of squared meters per roll ranges from 400 m^2 to 650 m^2, resulting in different areas of opportunities for creating defects (the ratio of the total squared meters and 50 m^2). The centerline is fixed at the overall average number of defects calculated from all of the rolls; however, the control limits are unique for each roll. The rolls with the smaller number of inspection units per roll (n) have wider limits, while the rolls with larger n have tighter limits.

ISQC Example 7.6 Valve Failures

In this example, we show how to construct a control chart to monitor the "time between failures" or "time between occurrence" using the **Control Chart Builder** platform. The data set consists of the number of hours between failures (y) of an important valve in a chemical plant for the most recent 20 failures of this valve. The data also includes a transformed variable using a power of the time between failures, $x = y^{0.2777}$. The transformation of the original data was taken because the transformed variable is better approximated by a normal distribution.

The following steps illustrate how to evaluate the distribution for the response using the **Distribution** platform and then construct a control chart using the **Control Chart Builder** platform:

1. Open Chapter 4 – ISQC Table 7.14.jmp, which has variables called *Failure, Time Between Failures (hr)*, and *Transformed Time Between Failures*. Failure is the subgroup variable and Transformed Time Between Failures is the charting statistic.

2. Select **Analyze ▶ Distribution**. In the dialog box, select **Time Between Failures (hr)** and click **Y, Columns**. Then select **Transformed Time Between Failures** and click **Y, Columns**. Click **OK** when finished.

3. From the red triangle next to **Time Between Failures (hr)** banner, select **Continuous Fit ▶ Normal Quantile Plot**. Click the following to deselect them: **Display Options ▶ Quantiles, Display Options ▶ Summary Statistics, Histogram Options ▶ Histogram and Outlier Box Plot**. Repeat for Transformed Time Between Failures.

4. From the red triangle next to the **Distributions** banner at the top of the window, select **Arrange in Rows** and then enter **2** in the field entry. Click **OK**.

Figure 4.45 Normal Probability Plots for ISQC Example 7.6

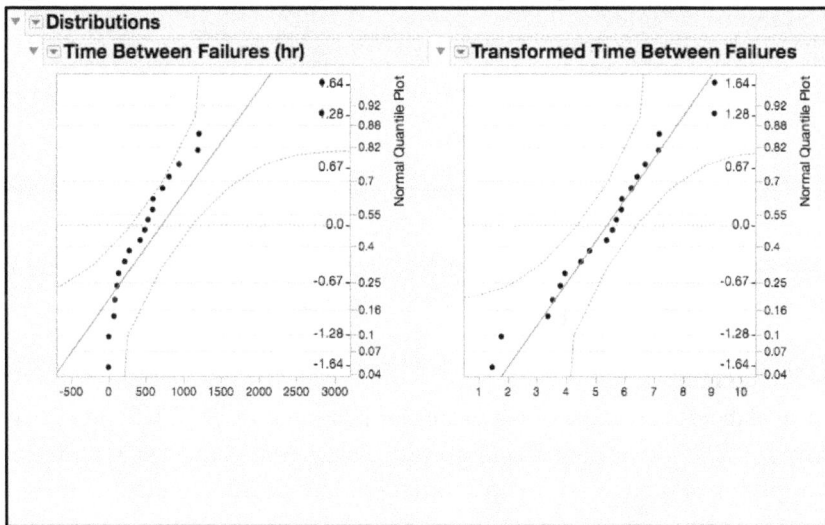

Similar plots in Figure 4.45 are shown in ISQC Figure 7.20 and Figure 7.21. The normal probability plot for time between failures is skewed to the right, indicating that the normal distribution might not be an appropriate approximation. However, the normal probability plot for the transformed variable shows that the points closely align to the straight line, suggesting that the normal distribution is appropriate. To simplify the monitoring strategy, the transformed variable is charted using an XmR control chart.

Statistics Note 4.5: It is not surprising that the normal probability plot for time between failures in Figure 4.45 shows a skewed pattern. Time is a positive quantity, bounded below by zero, and as such might not be approximated by a normal distribution. Apart from a transformation, the time between failures can be usually approximated by an exponential distribution.

---

5. From the main menu, select **Analyze ▶ Quality and Process ▶ Control Chart Builder**.

6. Drag **Transformed Time Between Failures** from the left-hand window to the **Y** zone (Y axis). Click **Done** to produce the XmR chart.

Figure 4.46 XmR Chart for Transformed Time Between Valve Failures

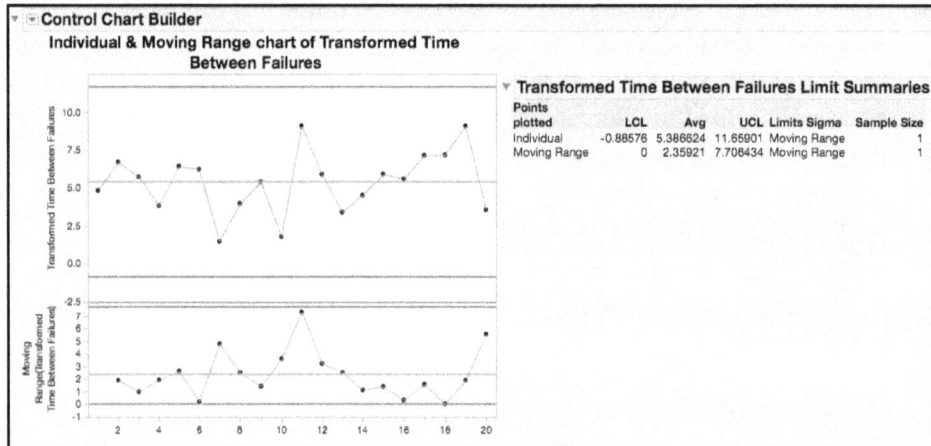

The control chart in Figure 4.46 (and ISQC Figure 7.22) shows a state of statistical control, implying that the failure rate for this valve is constant. While this is one way to deal with rare events, an alternative monitoring approach is described in the next section.

Statistical Insights

In this section, we elaborate on some of the examples provided in ISQC Chapter 7. The examples highlighted in this section include several important concepts we have encountered over our many years of applying SPC successfully to a variety of industries. For most of these examples, additional output not provided in ISQC is included to illustrate JMP functionality or further elaborate on important points.

Operating Characteristic Curve

The operating characteristic (OC) curve, see ISQC Sections 7.2.4 (Binomial) and 7.3.4 (Poisson), shows the probability, β, of not detecting a mean shift (ISQC equations 7.15 and 7.26) with the next subgroup when 3 sigma limits are used. The curve is usually shown with β on the Y axis and the mean shift, k, on the X axis. For the control chart shown in Figure 4.31, the OC curve is easily obtained in the legacy **Control Chart** platform by right-clicking on any chart point and selecting **OC Curve** from the resulting menu or by clicking on the red triangle next to **C of Number of Nonconformities** and choosing **OC Curve** from the menu (Figure 4.47).

Figure 4.47 OC Curve Selection for C Chart Data of ISQC Example 7.3

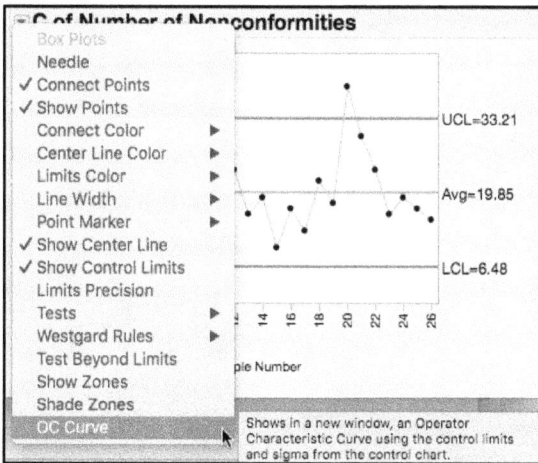

JMP produces a 2-sided OC curve, corresponding to the lower and upper control limits. At the top of the OC curve, the control chart limits LCL = 6.48, UCL = 33.21 are displayed (Figure 4.48).

Figure 4.48 OC Curve for the Number of Nonconformities C Chart

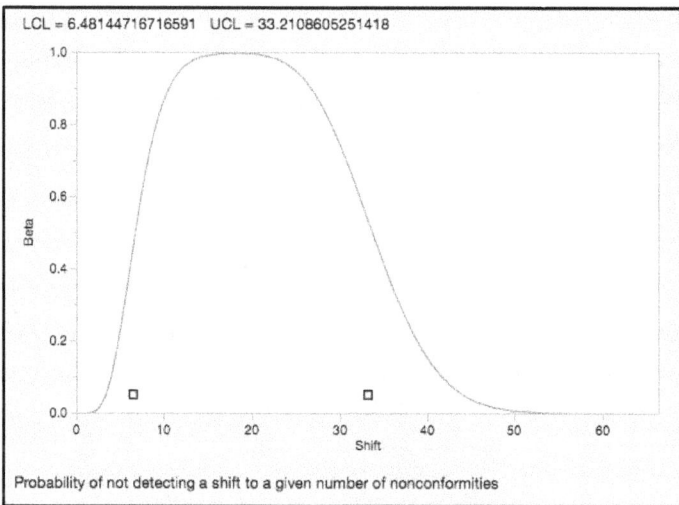

With the help of the crosshair tool, one can evaluate the Type II probability as a function of the number of defects. For example, for 40 nonconformities, the Type probability is 0.15, as shown in Figure 4.49. ISQC Table 7.13 shows the Type II probability for a given number of nonconformities c, and the value for c=40 is 0.151.

Figure 4.49 Type II Probability, β, for 40 Nonconformities.

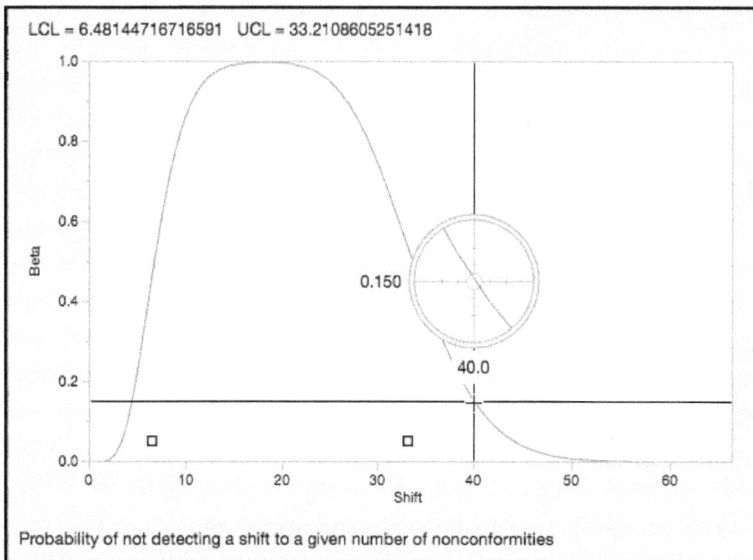

Impact of Overdispersion on P/NP and U/C Charts

The attribute charts presented in ISQC Chapter 7 use two probability distributions. The binomial distribution is used for P/NP charts (for example, fraction nonconforming) and the Poisson distribution is used for C/U charts (for example, total number of defects). As with any statistical technique, certain assumptions must be met for the effective use of the control charts. Certain types of violations of the model assumptions for attribute charts can result in higher false alarm rates, that is, there are out-of-control points when only common cause variability is present.

The most common issue we have encountered with attribute-based charts is a phenomenon referred to as *overdispersion*. This issue occurs when clustering or correlation among subgroup measurements induces a violation in the basic assumption of independent subgroup samples. This impacts the variation used in the calculation of the control chart limits. For NP charts, this means that the variance is actually $Q \times np(1-p)$ and not $np(1-p)$. For C charts, the actual variance is $Q \times \lambda$ and not λ. This issue is well-documented in the literature; see, for example, Cantell (1993) and Ramírez, et al. (1998). In ISQC Section 7.3.1, under *Alternative Probability Models for count Data*, Montgomery discusses alternative probability models and charts for count data, such as, g and h control charts based on the geometric distribution.

The most common way to accommodate overdispersion in binomial and Poisson data is to derive the control limits using a compound or mixture distribution. The beta-binomial distribution is often used for P/NP charts and the negative binomial distribution is used for C/U charts. Ramírez, et al (1998) illustrate how to monitor wafer yield data and particle data from the semiconductor industry using mixture distributions.

JMP provides several solutions to accommodate overdispersion. For P/NP charts, probability limits are calculated and used in place of the limits produced by the control chart platforms. For C/U charts, appropriate limits are derived using standard formulas in the **Control Chart Builder**.

U/C Charts (Poisson Data)

The following steps illustrate how to evaluate the C chart and accommodate overdispersion using the G chart in the **Control Chart Builder** platform. Wafer defect data from the semiconductor industry is used in this example. The Wafer Stacked.jmp data from the **Help > Sample Data Library** was used to summarize the number of defects per lot and wafer, and create the Chapter 4 Wafer Defects.jmp dataset.

1. Open Chapter 4 Wafer Defects.jmp, which has variables called *Wafer Number* and *Total Defects*. Wafer Number is the subgroup variable and Total Defects is the charting statistic.

2. Select **Analyze ▶ Quality and Process ▶ Control Chart Builder**.

3. A launch window appears. Select **Shewhart Attribute** from the first drop-down menu beneath the Columns window. Drag **Wafer Number** from the left-hand window to the **Subgroup** zone (X axis). Similarly, drag **Total Defects** from the left-hand window to the **Y** zone (Y axis).

4. Click **Warnings** and check Test Beyond. Click **Done**. The control chart is shown in Figure 4.50.

Figure 4.50 C Chart for Semiconductor Wafer Defects

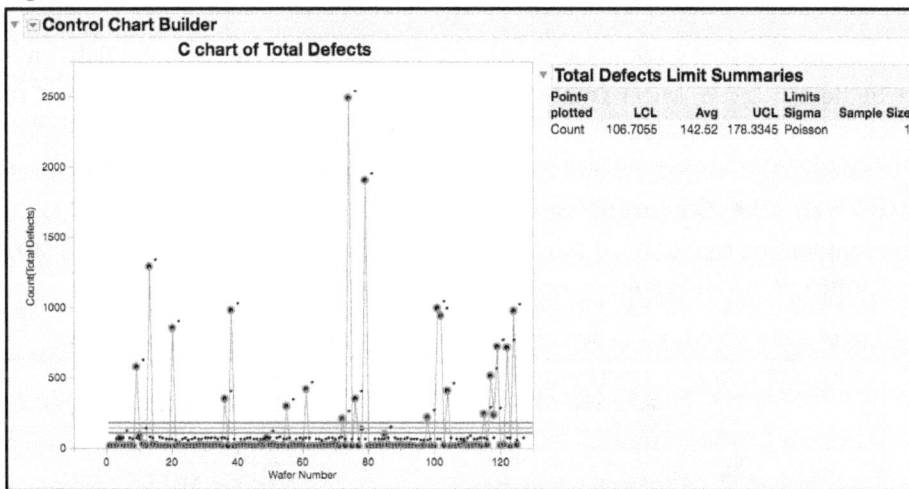

5. Make the Chapter 4 Wafer Defects.jmp table the current table by clicking on it. Select **Analyze ▶ Quality and Process ▶ Control Chart Builder** from the main menu.

6. A launch window appears. Select **Rare Event** from the first drop-down menu beneath the Columns window. Drag **Wafer Number** from the left-hand window to the **Subgroup** zone (X axis). Similarly, drag **Total Defects** from the left-hand window to the **Y** zone (Y

axis). The default **Points Statistic** is **Count** and **Limits Sigma** is **Negative Binomial (G)**. These are the correct selections for this type of data.

7. Click **Warnings** and check Test Beyond. Click **Done**. The control chart is shown in Figure 4.51.

Figure 4.51 G Chart for Semiconductor Wafer Defects

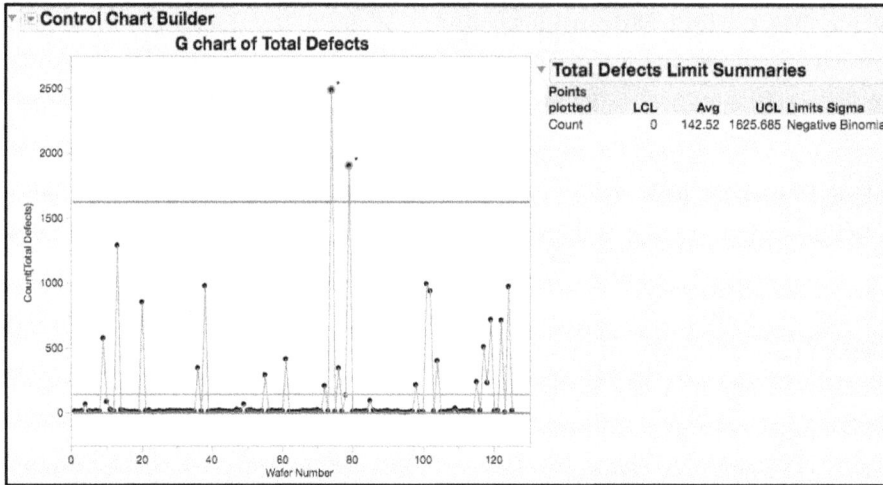

Appropriate probability limits can also be evaluated by fitting a negative binomial distribution to the data:

1. Click on Chapter 4 Wafer Defects.jmp table and select **Analyze ▶ Distribution**.

2. In the dialog box, select **Total Defects** from the Columns box and click **Y, Columns**. Then click **OK** when finished.

3. From the red triangle next to **Total Defects**, select **Discrete Fit ▶ Poisson** and then **Discrete Fit ▶ Gamma Poisson**.

4. From the red triangle next to Fitted Poisson, select **Quantiles** and enter **0.00135, 0.5** and **0.99865** in the three fields. Click **OK**. Repeat for Fitted GammaPoisson.

Figure 4.52 Poisson and Gamma Poisson Fits for Semiconductor Wafer Defects

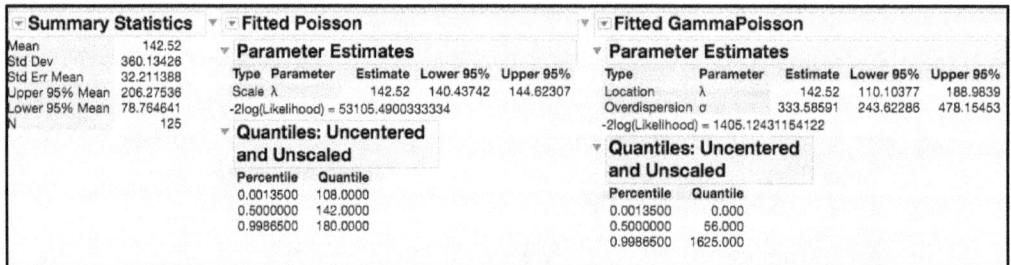

The two control charts produced for the wafer defect lead to very different interpretations regarding the stability of the process output. The C chart in Figure 4.50, which uses the Poisson distribution, has one wafer count that is within the control limits, while 124 exceed the limits. The G control chart in Figure 4.51, which uses the Negative Binomial (or Gamma Poisson) distribution, has only two wafer counts that exceed the upper control limit. Note that although wafer defects are not considered rare events, the Rare Events chart type in the Control Chart Builder is used to easily obtain a control chart using the Negative Binomial distribution.

So which distribution provides a better fit to wafer defects? The summary statistics and fitted distributions for the wafer defect data are shown in Figure 4.52. For a Poisson distribution, the mean equals the variance. One quick way to estimate overdispersion is by the ratio of the sample variance to the sample mean (that is, $Q = 360.13^2 / 142.52 = 910$). This estimate is far greater than 1, indicating the data is overdispersed. Another indication of the better fitting distribution is the smaller -2log(Likelihood) statistic for fitted distribution, which is approximately 1,405 for the Gamma Poisson and 53,105 for the Poisson.

Statistics Note 4.6: The Gamma-Poisson distribution is a mixture of a Poisson distribution where the rate parameter λ is distributed as a Gamma distribution. This distribution can be reparametrized as a negative binomial distribution.

P/NP Charts (Binomial Data)

An example of overdispersed binomial data is provided in this section. Currently, JMP does not have a control chart platform that directly uses appropriate mixture distributions to update P/NP control chart limits, so probability limits from a mixture distribution are used instead. The data comes from Table 31, Finish Defects, Galvanized Washers, of the American Society for Testing and Materials (ASTM) manual on Presentation of Data and Control Chart Analysis. There are 15 lots and each lot has 400 items, where each item is classified as good or bad. The number of bad items is tallied for each lot and charted.

1. Open Chapter 4 - ASTM Example.jmp, which has variables called *Lot Number, Lot Size,* and *Defects*. Lot Number is the subgroup variable and Defects is the charting statistic.
2. From the main menu, select **Analyze ▶ Distribution**.
3. In the launch window, select **Defects** and click **Y, Columns**. Click **OK** when finished.
4. From the red triangle next to **Defects**, select **Discrete Fit ▶ Binomial** and then **Discrete Fit ▶ Beta Binomial.**
5. From the red triangle next to **Fitted Binomial**, select **Quantiles** and enter **0.00135, 0.5**, and **0.99865** in the three fields. Click **OK**. Repeat for Fitted Beta Binomial. The output is shown in Figure 4.53.

Figure 4.53 Binomial and Beta Binomial for ASTM Defect Data

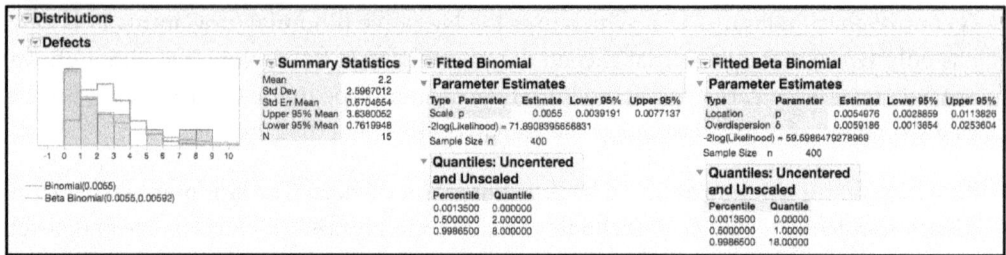

6. From the main menu, select **Analyze ▶ Quality and Process ▶ Control Chart ▶ NP**.

7. In the dialog window, select **Defects** in the Columns box and click **Process**. Similarly, select **Lot Number** and click **Sample Label**. Then enter **400** in the field for **Constant Size**. Click **OK** when finished.

Figure 4.54 NP Chart Using the Binomial for ASTM Defects

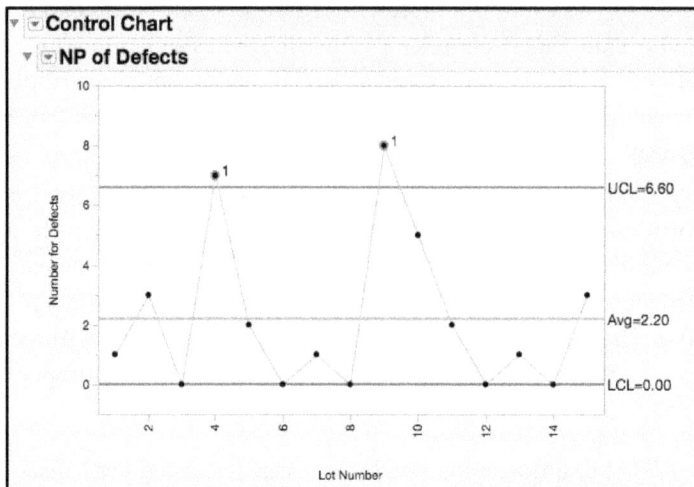

8. From the red triangle next to **Control Chart**, select **Save Limits ▶ In Column**.

9. Double-click the **Defects** label in the Chapter 4 - ASTM Example.jmp table and select **Control Limits** under the Column Properties. Change the **Avg** to 1 and the **UCL** to 18. Click **OK** when finished. These are the values for the 0.5, and 0.99865 beta binomial quantiles in Figure 4.53.

10. Re-run the NP chart for ASTM Defects (see Step 6 and Step 7). The chart is in Figure 4.55.

Figure 4.55 NP Chart Using the Beta Binomial for ASTM Defects

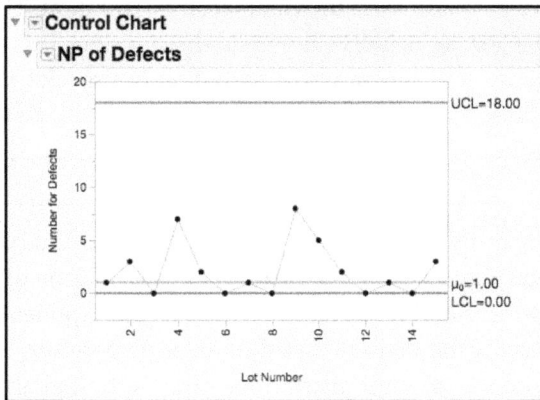

Once again, the two control charts produced for the ASTM Defects example lead to different interpretations regarding the stability of the process output. The NP chart in Figure 4.54, which uses the Binomial distribution, has two lots that exceed the UCL = 6.6. The modified NP chart in Figure 4.55, which uses the Beta Binomial distribution, has no lots that exceed the UCL = 18.

So which distribution provides a better fit to wafer defects? The summary statistics and fitted distributions for the wafer defect data are shown in Figure 4.53. For a Binomial distribution, the variance = np(1−p). One quick estimate of overdispersion is the ratio of the sample variance to the binomial variance that is, $Q = 2.6^2 / 400(0.0055)(1-0.0055) = 2.34)$ This estimate is greater than 1, suggesting that the data is overdispersed. Another indication of the better fitting distribution is the smaller -2log(Likelihood) statistic for fitted distribution, which is approximately 59.6 for the Beta Binomial and 71.9 for the Binomial.

Alternative Approaches for Rare Events

In ISQC Example 7.6, valve failures did not occur often and an approach for rare events, based on time between failures, was used. To conform to model assumptions, a transformed variable ($x = y^{0.2777}$), which was normally distributed, was charted using an XmR chart.

An alternative approach is to directly monitor the rare events. This can be done by either monitoring the number of opportunities, such as the number of days or the number of units produced between rare events or the time between rare events. For example, if we monitor rare events using the time between events, a continuous variable, then the Weibull distribution is an appropriate choice and the resulting chart is called a *T chart*. If the charting statistic is the number of opportunities between rare events, then the Negative Binomial distribution is used and the resulting chart is called a *G chart*.

> **JMP Note 4.6:** The G chart takes its name from the Geometric distribution. The control chart builder, however, uses the negative binomial distribution. This choice makes the chart more robust because the geometric distribution is a special case of the negative binomial distribution.

> **Statistics Note 4.7:** As Montgomery points out in ISQC 7.3.5, if counts occur according to a Poisson distribution then "the probability distribution of the time between events is the exponential distribution." However, the exponential distribution is a special case of the Weibull distribution and, therefore, a more robust chart is one based on the Weibull distribution. This is the T chart in the Control Chart builder.

The following steps show how to generate a T control chart for rare events data using the **Control Chart Builder** platform:

1. Open Chapter 4 - ISQC Table 7.14.jmp.
2. Select **Analyze ▶ Quality and Process ▶ Control Chart Builder** from the main menu.
3. A launch window appears. Select **Rare Event** from the first drop-down menu beneath the Columns window. Drag **Time Between Failures (hrs)** from the left-hand window to the **Y** zone (Y axis). The default **Points Statistic** is **Count** and **Limits Sigma** is **Weibull (T)**. Because we are monitoring the time between events. these are the appropriate selections for this data.
4. Click **Done** when finished.

Figure 4.56 T Chart for Valve Failures

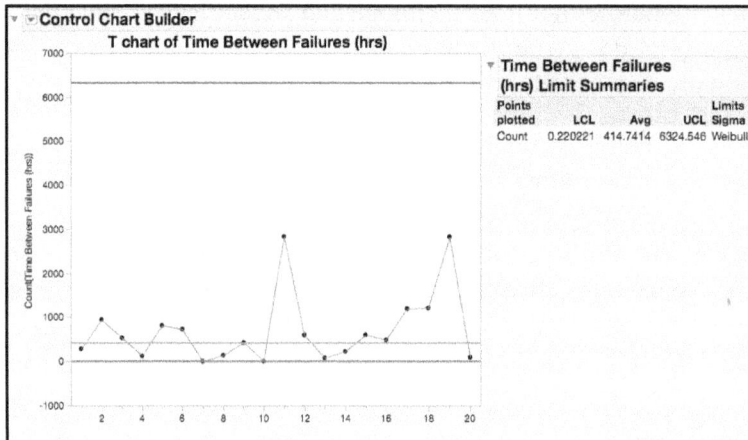

The T chart is shown in Figure 4.56. Both the charting statistic and the limits are in the units of the original data, making it easier to interpret the chart. If the time between the last and the next failure exceeds the UCL = 6,325 hrs, then the process will be out of control. Compare this to the chart in Figure 4.46 (or ISQC Figure 7.22), where the process is out of control if the transformed value is greater than UCL=11.6611. The chart in Figure 4.46 is not in units of hours but in units of $hrs^{0.2777}$. We can transform back the limits, but this introduces bias. For example, the upper control limit of 11.6611 becomes $11.661^{(1/.2777)} = 6,922.489$ hrs, which is greater than the T chart limit of 6,324.546 hrs (therefore decreasing the detection capability of the back transformed chart). Similarly, the lower limit becomes $-0.8879^{(1/.2777)} = -0.652$ hrs or a negative time value!

Statistics Note 4.8: There are times when the model assumption should not be ignored, and a more appropriate distribution should be used to calculate control limits. For attribute data, overdispersion is a primary concern because the true variance is larger than the variance modeled by the binomial and Poisson distributions.

Chapter 5: Process and Measurement System Capability Analysis

Overview

This chapter illustrates how to evaluate the capabilities of your processes and measurement systems using examples from Chapter 8, Process and Measurement System Capability Analysis, of *Introduction to Statistical Quality Control* (ISQC), and includes discussions, tips and statistical insights on alternative ways to carry out these assessments.

These assessment techniques are mostly used with data measured on a continuous scale and complement the information obtained from the control charting activities previously discussed.

Several JMP platforms are highlighted in this Chapter including **Distribution**, **Control Chart** and the **Control Chart Builder**, **Measurement System Analysis**, and **Variability / Attribute Gauge Chart**.

Process and Measurement System Capability Analysis Review

In this chapter, two types of capability analysis are explained in detail. The first type of capability relates to how well the output of a process meets the stated requirements. This scenario is referred to as *process capability* and common statistical approaches are found under the banner of process capability analysis. As Montgomery points out, conducting this analysis throughout the product cycle is useful in order to minimize variability and increase product quality.

Many metrics have been developed to quantify the capability of the process and all of them use the stated requirements, or customer requirements, in the form of specification limits and a measure of process variability. The most common process capability index is $C_p = (USL - LSL) / 6\sigma$, which measures the full width of the specification to the variation in the process. In the equation, LSL is the lower specification limit, and USL is the upper specification limit. Recall, for a normally distributed variable, 99.73% of the population values fall within XBar $\pm 3\sigma$, or equivalently, most of the spread of the distribution is represented by 6σ. However, this metric does not take into account where the process mean is centered.

Another popular process capability index is $C_{pk} = \min\{C_{pl}, C_{pu}\}$, where $C_{pl} = (\mu - LSL) / 3\sigma$ and $C_{pu} = (USL - \mu) / 3\sigma$. Unlike C_p, this metric takes into account where the process mean is centered and will reflect the performance of the mean to the closest specification limit. Note if the process mean is centered between the two-sided specification limits then $C_p = C_{pk}$. When there is only a one-sided specification limit the C_{pk} becomes C_{pl} for a LSL, or C_{pu} for a USL.

Many variations to these two basic metrics are documented in the literature. ISQC Chapter 8 describes different ways to estimate σ in the process capability equations, including the use of the sample standard deviation, an estimate from an appropriate control chart, or from quantile estimates from the raw data or a specified distribution. Confidence intervals and tests on process capability ratios provide ways to incorporate the effects of smaller sample sizes on the estimates and ways to determine if our processes are meeting minimum performance requirements, or more lofty goals, such as, 6σ process performance, $C_{pk} \geq 2$.

The other type of capability analysis described in this chapter has to do with the performance of the measurement system. Discussions and common statistical approaches are found under a number of banners, including, measurement system analysis (MSA), gauge R&R or Evaluating the Measurement Process (EMP). All of these approaches provide ways to quantify the error associated with the measurement system and further evaluate its contribution to the total variation, as well as its impact on performance indices.

The typical MSA study involves a number of operators, who measure the same parts, multiple times using the same measurement system or gauge. The repeatability of the measurement system is the variation from multiple measurements taken by the same operator measuring the same part, using the same gauge. Reproducibility is the variation from multiple measurements taken by different operators measuring the same part, using the same gauge. In other words, the different operators represent the reproducibility of the gauge and, the repeated measures of the same part by the same operator represents the repeatability of the gauge. The total variation in this type of study is typically represented using $\sigma^2_{total} = \sigma^2_{part} + \sigma^2_{gauge}$, where $\sigma^2_{gauge} = \sigma^2_{reproducibility} + \sigma^2_{repeatability}$. Variance components analysis can be used to estimate these variance components and determine their contribution to the total variation.

When the outcome of a measurement system is an attribute, such as good/bad or poor/average/excellent, an attribute gauge analysis can be used to determine the quality of the method. The study design, once again, includes different operators and a number of inspection items that need to be classified. It is helpful to predetermine the inspection items to include items from all of the ranking categories used. The analysis for this study is focused on determining the level of consistency among the operators to classify the items the same way. If the classification of the item is predetermined, then the analysis can also quantify the effectiveness of the measurement systems.

JMP Process Capability and MSA Platforms

Several platforms can be used to generate a process capability analysis in JMP, including **Distribution, Control Chart** and **Control Chart Builder**. These platforms were introduced in Chapter 2, and a detailed summary is shown in Table 5.1. In this chapter, we emphasize the **Distribution** platform for this type of analysis. However, the other platforms are illustrated for completeness. Note there is a **Process Capability** platform, found under **Analyze ▶ Quality and Process**. This platform is not discussed in detail in this chapter, since it is more useful when we want to compute capability indices for many parameters and compare the results. The **Process Screening** platform is another tool to calculate capability indices for many parameters. Both of these platforms are reviewed in Chapter 6 in this book.

Table 5.1 Overview of Features for JMP Process Capability Platforms

JMP Feature	Process Capability	Distribution	Control Chart	Control Chart Builder
Probability Distributions	Many continuous distributions are available and can be used to calculate process capability indices and limits. Can compare different distributions.	Many continuous and discrete distributions are available and can be used to calculate process capability indices and limits.	Many continuous and discrete distributions are available and can be used to calculate process capability indices and limits.	None available. Normal distribution is assumed behind the scenes.
Descriptive Statistics	Basic summary statistics are available to view. Includes the Stability Ratio.	Many summary statistics are available to view. Does not include the Stability Ratio.	Many summary statistics are available to view. Does not include the Stability Ratio.	Limited to the mean and standard deviation estimates. Includes the Stability Ratio.
Process Capability Indices (PCI)	C_p, C_{pk}, C_{pm}, C_{pl}, C_{pu}	C_p, C_{pk}, C_{pm}, C_{pl}, C_{pu}	C_p, C_{pk}, C_{pm}, C_{pl}, C_{pu}	C_p, C_{pk}, C_{pm}, C_{pl}, C_{pu}
Estimation of sigma used in PCI calculations	Within or overall sigma.	Long term, specified, or control chart estimate.	Long term or control chart estimate. A value can be specified for calculation of limits.	Long term or control chart estimate. A value can be specified for the calculation of limits.

JMP Feature	Process Capability	Distribution	Control Chart	Control Chart Builder
Process Capability Confidence Limits	For all indices and when both within sigma and overall	For all indices but only when long term sigma is used. Not available for Fitted Distributions.	For all indices but only when long term sigma is used. Not available for Fitted Distributions.	For all indices and when both within sigma and overall sigma estimates of variation are used.

Similarly, JMP has several platforms to carry out the analysis of measurement systems. The **Measurement System Analysis platform** is dedicated to gauge studies for continuous data, while the **Variability / Attribute Gauge Chart** platform can handle both continuous and attribute data. A summary of the features in these platforms is shown in Table 5.2.

Table 5.2 Overview of Features for JMP Measurement System Analysis Platforms

JMP Feature	Measurement System Analysis	Variability / Attribute Gauge Chart
Response Type	Continuous only	Continuous or Attribute
Control Charts	XBar & Range and XBar and S charts	XBar limits on a Variability Chart and S chart
Variance Components Models	EMP: Main effect, crossed, crossed with two-factor interaction, nested, crossed then nested and nested then crossed for up to 3 variables. Gauge R&R: see Variability / Attribute Gauge Chart.	Main effect, crossed, nested, crossed then nested, nested then crossed.
MSA Method	EMP & Gauge R&R	Gauge R&R
Available Statistics	EMP: Variance Components Test-retest and probable error, intraclass correlation coefficients, effective resolution and recommended measurement precision, measurement system classification. Gauge R&R: see Variability / Attribute Gauge Chart	Variability: Variance Components, AIAG output, P/T ratio, Discrimination Ratio, Misclassification Probabilities Attribute: Agreement between & within raters and across categories, Kappa statistics, Agreement counts to a standard and effectiveness to a standard. Bias and linearity.
Available Plots	XBar and Range control charts, Parallelism Plots, Bias Comparison Test-Retest Comparison, Shift Detection Profiler.	Variability Chart (control limits can be added for XBar and Std Dev), Mean Plots, Std Dev Plots.

Examples from ISQC Chapter 8

The examples from Chapter 8 of ISQC presented in this chapter are shown in Table 5.3. We show how to reproduce these examples using JMP. For some examples, additional output not provided in ISQC is shown to illustrate JMP functionality or elaborate on important points considered by the authors.

Table 5.3 Summary of Examples from ISQC Chapter 8

ISQC Example / Table Number	JMP Table Name	JMP Platform	Key Points
8.1 Bursting Strength of Glass Containers	Chapter 5 – ISQC Table 8.1	Distribution	Assess distribution fit using different tools and calculate process capability limits.
8.2 Bursting Strength of Glass Containers	Chapter 5 – ISQC Table 8.1, 8.5	Distribution Control Chart Control Chart Builder Process Capability	Calculate process capability indices using the different platforms.
8.3 Process Centering	Chapter 5 – ISQC Exercise 8.3	Distribution	Illustrate the use of C_{pm} for off centered processes.
8.4, 8.5, 8.6 Confidence Intervals for PCIs	Chapter 5 – ISQC Exercise 8.4	Distribution	Compute confidence intervals for C_p and C_{pk}.
8.7 Part Measurement Data	Chapter 5 – ISQC Table 8.6	Measurement System Analysis	Analyze data from a Gauge R&R study with a continuous response.
Table 8.7 Thermal Impedance Data	Chapter 5 – ISQC Table 8.7	Measurement System Analysis	Analyze data from a Gauge R&R study with a continuous response. Re-analyze as an EMP.
Table 8.13 Attribute Gauge Capability Analysis	Chapter 5 – ISQC Table 8.13	Variability / Attribute Gauge Chart	Analyze data from a Gauge R&R with an attribute response.
8.8, 8.9, 8.12 Setting Specification Limits	Chapter 5 – ISQC Exercise 8.8, 8.12	Distribution	Compute statistically based specification limits.

ISQC Example 8.1 Bursting Strength of Glass Containers – $\bar{x} \pm 3s$

In this example, we will show how to evaluate the Bursting Strengths for glass containers using histograms, probability plots and descriptive statistics. The data in Table 8.1 of ISQC consists of the bursting strengths for 100 glass containers, measured in pounds per square inch (psi). The following steps illustrate how to begin a process capability analysis using the **Distribution** platform in JMP.

1. Open the JMP table Chapter 5 – ISQC Table 8.1.jmp, which has variables called *Glass Container* and *Burst Strength (psi)*. In this table, Glass Container is a generic label for the ith glass container and Burst Strength (psi) is the variable of interest.

2. Select **Analyze ▶ Distribution** (Figure 5.1).

Figure 5.1 JMP Menu Selections for a Process Capability Analysis

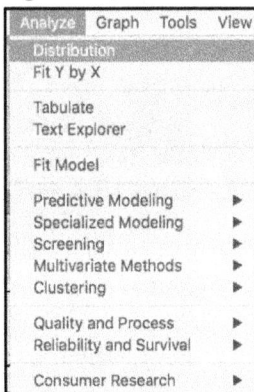

3. When the launch window appears, select **Burst Strength (psi)** as the **Y, Columns** (response) variable and click **OK** when finished.

Figure 5.2 JMP Menu Selections for Distribution Platform

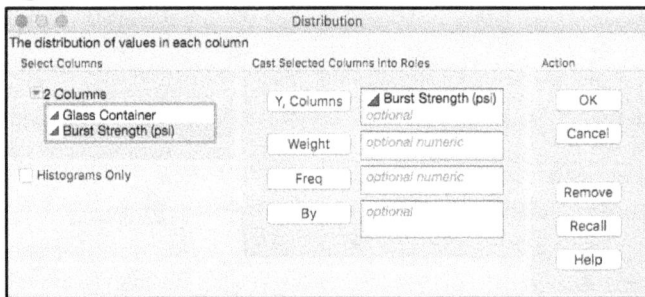

4. In order to change the default orientation of the histogram from vertical to horizontal, select the red triangle next to **Distributions** and select **Stack.** The resulting histogram and statistics is shown in Figure 5.3. Note this option can be set as default in the preferences.

Figure 5.3 Histogram and Default Output for Burst Strength

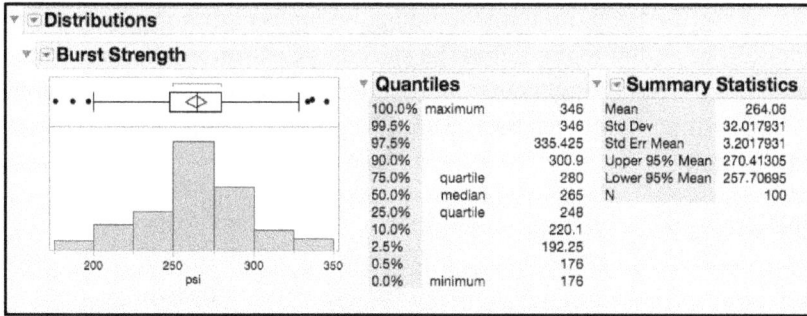

5. From the red triangle next to **Burst Strength (psi)**, select **Continuous Fit ▶ Normal**. This will add the Fitted Normal details in the right-hand side of the output window.

6. From the red triangle next to **Fitted Normal**, select **Set Spec Limits for K Sigma** and enter 3 in the **Enter K-Sigma for Capability** dialog box (Figure 5.4).

Figure 5.4 Dialog Window for K-Sigma Capability Limits

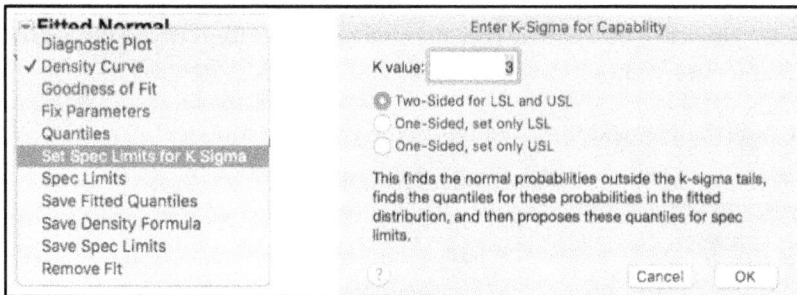

7. Click **OK** when finished.

Figure 5.5 Process Capability Limits for Burst Strength

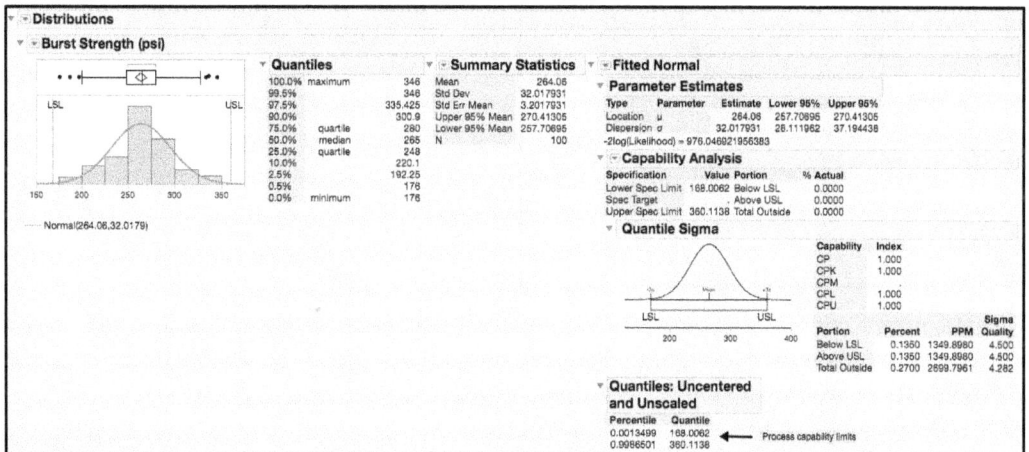

The histogram in Figure 5.5 corresponds to Figure 8.2 in ISQC. The default output in JMP includes an outlier box plot above the histogram, the quantiles, and a few summary statistics. Additional summary statistics can be added to the output by selecting **Customize Summary Statistics** under **Display Options**, from the red triangle next to the **Burst Strength (psi)** label. Also, the normal fit and all of the corresponding output can be removed by selecting **Remove Fit** from the drop-down list for the **Fitted Normal**.

The sample estimates for the mean and standard deviation are used to calculate the process capability limits given in ISQC Example 8.1. These limits are manually calculated using the mean ± 3σ, or 264.06 ± 3(32.02). This calculation can be automated in JMP by fitting a normal distribution to the data and then calculating the corresponding "K Sigma" limits. For this exercise, K = 3 and the corresponding limits are provided by the quantiles that represent the 0.00135 and 0.99865 percentiles of the normal distribution, 168 psi, and 360 psi. By design, these limits contain 99.73% of the values in a normal distribution, as shown in ISQC Figure 8.1.

In using these computations, we assume that the normal distribution appropriately represents the Burst Strength data. While the histogram provides a visual way to assess the normal fit, probability plots are also very useful to assess the fit of a distribution to the data. This plot can easily be added to the output shown in Figure 5.5.

8. From the red triangle next to **Fitted Normal**, select **Diagnostic Plot**.

Figure 5.6 Normal Probability Plot of Burst Strength

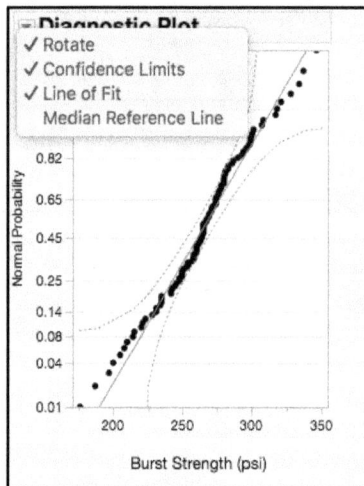

The normal probability plot in Figure 5.6 is similar to ISQC Figure 8.4. The confidence limits (red dashed lines) shown in Figure 5.6, represent 95% confidence bands for the normal probability line and can be used to visually assess the fit of the distribution to the data. These limits can be removed by deselecting **Confidence Limits** from the drop-down menu shown in Figure 5.6. As is discussed in ISQC Chapter 8, the normal probability plot can be used directly to estimate the probability of exceeding a given Burst Strength value. This is done by locating the corresponding y-axis value for a given x-axis value. For example, a Burst Strength of about 200 psi, is located at approximately 0.02 or 2% on the normal probability axis, indicating that 2% of the results are below 200 psi and 98% are above 200 psi. The **Crosshairs** tool can be used to read off the probability value of a selected Burst Strength value.

As Montgomery points out, selecting the most appropriate distribution to represent the data is important in order to proceed with process capability assessments. In addition to using probability plots, he discusses the use of additional descriptive statistics, skewness and kurtosis, and goodness-of-fit tests, such as the Shapiro-Wilk test. The following steps illustrate how to add these items to the previous JMP output.

9. Click on the red triangle next to the **Fitted Normal** and select **Goodness of Fit.** This will add the Shapiro-Wilk W Test to the **Fitted Normal** output.

Figure 5.7 Normal Distribution Goodness-of-Fit Test for Burst Strength

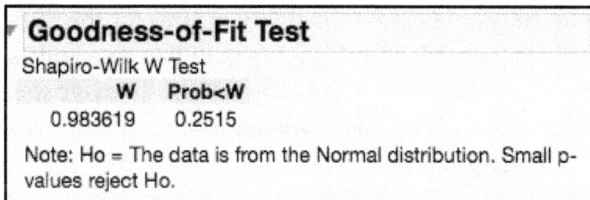

Goodness-of-Fit Test

Shapiro-Wilk W Test

W	Prob<W
0.983619	0.2515

Note: Ho = The data is from the Normal distribution. Small p-values reject Ho.

10. From the red triangle next to **Burst Strength (psi)**, select **Customize Summary Statistics** and check the boxes next to **Skewness** and **Kurtosis** and then click **OK.** You can also select this from the red triangle next to summary statistics.

Figure 5.8 Skewness and Kurtosis for Burst Strength

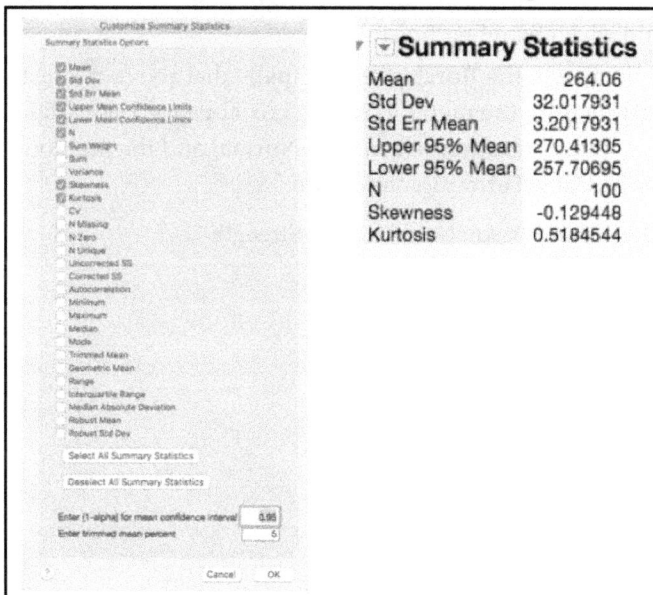

Summary Statistics

Mean	264.06
Std Dev	32.017931
Std Err Mean	3.2017931
Upper 95% Mean	270.41305
Lower 95% Mean	257.70695
N	100
Skewness	-0.129448
Kurtosis	0.5184544

The Goodness-of-fit test for the Burst Strength data is shown in Figure 5.7 and since the p-value = 0.2525 > 0.05, we assume that the normal distribution is an appropriate distribution to model this variable. The skewness = -0.129 and kurtosis = 0.518, shown in Figure 5.8, also suggest that the data are symmetrical about its mean, and that the mass in the tails is similar to that for a normal distribution (for a normal distribution the skewness and kurtosis are 0). Note these statistics are not presented for the Burst Strength data in ISQC Chapter 8. The choice of an appropriate distribution will be further discussed in the Statistical Insights section of this chapter.

> **JMP Note 5.1**: The **Distribution** platform has many tools to help evaluate the normality assumption for a given variable. Under the **Fitted** distribution report these include **Diagnostic Plot**, **Density Curve** and **Goodness of Fit**.

Example 8.2 Bursting Strength of Glass Containers – C$_{pl}$

In this example, we will show how to compute process capability indices for the Bursting Strengths for glass containers using the data in ISQC Table 8.1. Recall this data consists of the bursting strengths for 100 glass containers, measured in psi. The following lower specification limit is provided for the burst strength: LSL = 200 psi. We will use several platforms for this computation and we begin with the **Distribution** platform.

1. Open the JMP table Chapter 5 – ISQC Table 8.1.jmp, which has variables called *Glass Container* and *Burst Strength (psi)*. In this table, Glass Container is a generic label the ith glass container and Burst Strength (psi) is the variable of interest.
2. Select **Analyze ▶ Distribution**.
3. When the launch window appears, select **Burst Strength (psi)** as the **Y, Columns** (response) variable and click **OK** when finished.
4. Click on the red triangle next to the **Burst Strength (psi)** label above the histogram and select **Capability Analysis** from the menu. A dialog box appears. Enter **200** in the **Lower Spec Limit** field. Leave the distribution selection to **Normal** and the method to estimate to the standard deviation as **Long Term Sigma** (Figure 5.9).

Figure 5.9 Dialog Box for Capability Analysis for Burst Strength

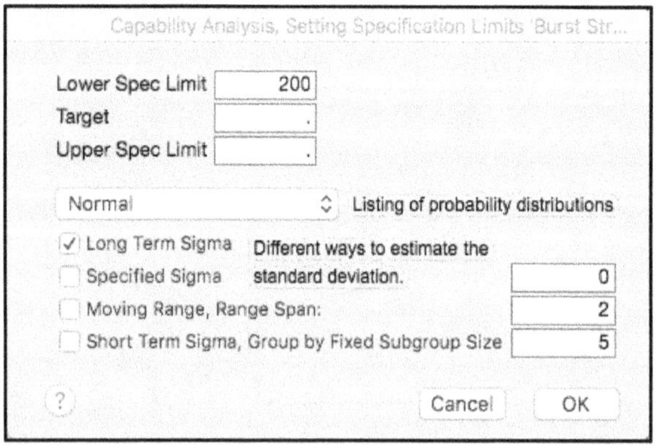

5. Click **OK** when finished (see Figure 5.10).

Figure 5.10 Process Capability Indices for Burst Strength

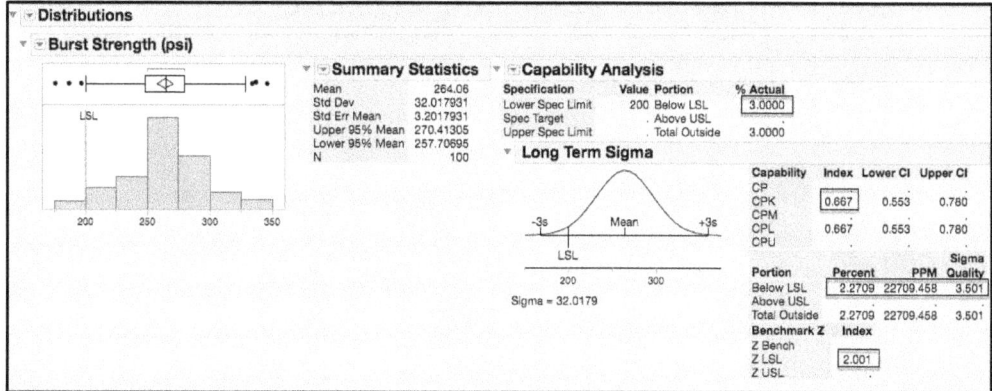

The process capability index, $C_{pl} = 0.667$, shown in Figure 5.10 matches the result in ISQC Example 8.2. Since there is only one specification limit, $C_{pk} = C_{pl} = 0.667$. The estimated fallout, or probability of being below the LSL, is 2.2709%. This calculation is done by finding the area to the left of a normal Z-score $= (\mu–LSL)/ \sigma = (264.06 – 200)/32 = 2.001$. For a normal distribution, the Z-score is also computed as $3C_{pl} = 3(0.667) = 2.001$ Note the Z-score can be added to the output by selecting **Z Bench** from the drop-down menu, when the red triangle next to **Capability Analysis** is selected. In addition, the output includes the projected PPM (parts per million defects) and the Sigma Quality level. The PPM is simply the multiplication of the probability of being below LSL and 10^6; or $0.022709 \times 10^6 = 22,709$. The Sigma Quality level is calculated as Z Bench + 1.5 = 3.501. This quantity is often used in Six Sigma programs and is used to describe the quality of the process output. Burst Strength is a "3.5 sigma" process.

> **JMP Note 5.2:** The process capability analysis output includes C_{pk}, Z-score, % Actual results out of specification and an estimate of the PPM and Sigma Quality, based on the Normal Distribution (default).

Statistics Note 5.1: The Z bench values are standardized distances to the specifications in standard deviation units. For example, the process distance to the upper specification limit, USL, is $(USL–\mu)$. If we standardize this distance by the standard deviation, σ, we get Z USL $=(USL–\mu)/ \sigma$. The Z LSL is calculated as Z LSL $=(\mu–LSL)/ \sigma$.

The **Capability Analysis** output also includes the percent of actual Burst Strength results that are below the LSL. This can be found at the top of the output, right under the Capability Analysis banner. We see that 3% of the N = 100 results are less than 200 psi. We can easily locate these results in the JMP table as follows.

6. Click on the histogram bar below the LSL in in the output window. The bar is highlighted and the results are highlighted in the JMP table (Figure 5.11).

Figure 5.11 Locating Burst Strength Results Below the LSL

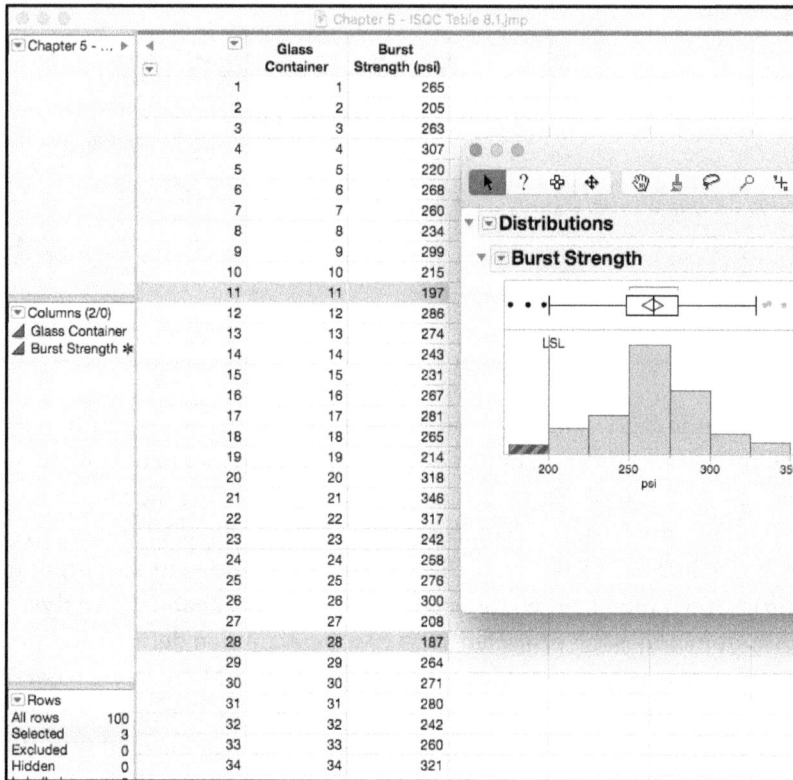

7. In order to subset the table with the selected rows, highlight the JMP table and from the main menu select **Tables ▶ Subset.** Complete appropriate selections in the dialog window and click **OK** when finished. Default settings were kept for the output in Figure 5.12.

> **JMP Note 5.3**: Double-clicking the selected histogram bar generates the subset table directly.

Figure 5.12 Subset of Burst Strength Results Less Than the LSL

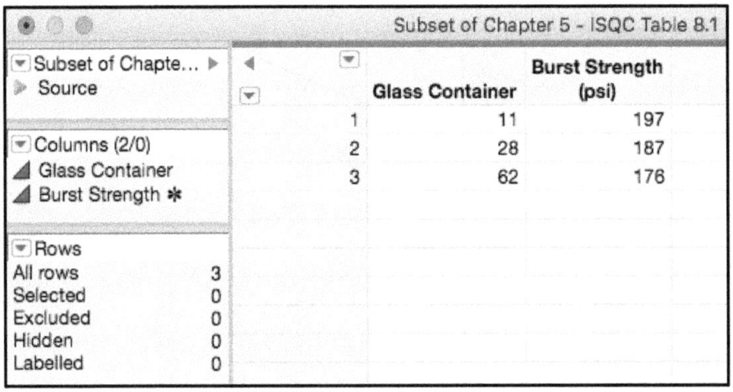

		Subset of Chapter 5 - ISQC Table 8.1	
● ○ ●			
▼ Subset of Chapte... ▶ ▷ Source	◀ ▼ ▼	Glass Container	Burst Strength (psi)
	1	11	197
▼ Columns (2/0) ◢ Glass Container ◢ Burst Strength ✳	2	28	187
	3	62	176
▼ Rows			
All rows 3			
Selected 0			
Excluded 0			
Hidden 0			
Labelled 0			

The interactivity in the **Distribution** platform makes it easy to locate data shown in the histogram or box plot. Once the rows are highlighted, the Rows information in the bottom left-hand side of the JMP table can be used to confirm the number of highlighted rows, which are 3 for this example. Two of the three values are visible in the table shown in Figure 5.11. However, we must scroll down in order to locate the third result. If the table is difficult to navigate, with many rows and/or columns, many rows are highlighted, or there is a need to do further analysis on these results, then a subset table should be created. The subset table in Figure 5.12 shows the three results that are below the LSL: 197 psi, 187 psi, and 176 psi for Glass Containers 11, 28 and 62, respectively.

Within the **Distribution** platform, there are several ways to obtain a Capability Analysis. We illustrated how to use **Capability Analysis** from the menu above the histogram. Alternatively, a Normal distribution can be fit from this same menu and from the **Fitted Normal** drop-down menu, **Spec Limits** can be selected and populated. Finally, the **Spec Limits**, located in the **Column Properties** for Burst Strength (psi) in the JMP table, can be populated. When **Distribution** is launched, the **Capability Analysis** will be included as part of the default output.

> **JMP Note 5.4**: In order to facilitate a process capability analysis in the different platforms, it is helpful to use the **Spec Limits Column Property** in the JMP table. This information is saved as metadata for the column and it will be used automatically every time a capability analysis is requested.

The **Process Capability** platform, one of the new generation quality tools within JMP, can also be used to calculate process capability indices, especially if we have several responses to analyze. Below we illustrate how to generate the process capability indices in Figure 5.10 using the **Process Capability** Platform.

1. Open, or highlight, the JMP table Chapter 5 – ISQC Table 8.1.jmp.

2. Choose **Analyze ▶ Quality and Process ▶ Process Capability** as shown in Figure 5.13.

Figure 5.13 Selecting the Process Capability Platform

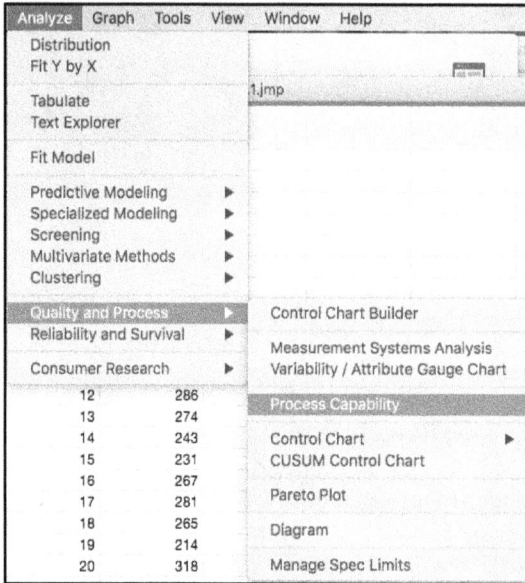

3. When the dialog window appears, select **Burst Strength (psi)** as the **Y, Process**. Open the **Distribution Options** to reveal the **Normal** distribution as the default (Figure 5.14). Note the continuous distribution options, including the Best Fit. Click **OK** when finished.

Figure 5.14 Process Capability Platform Dialog Window

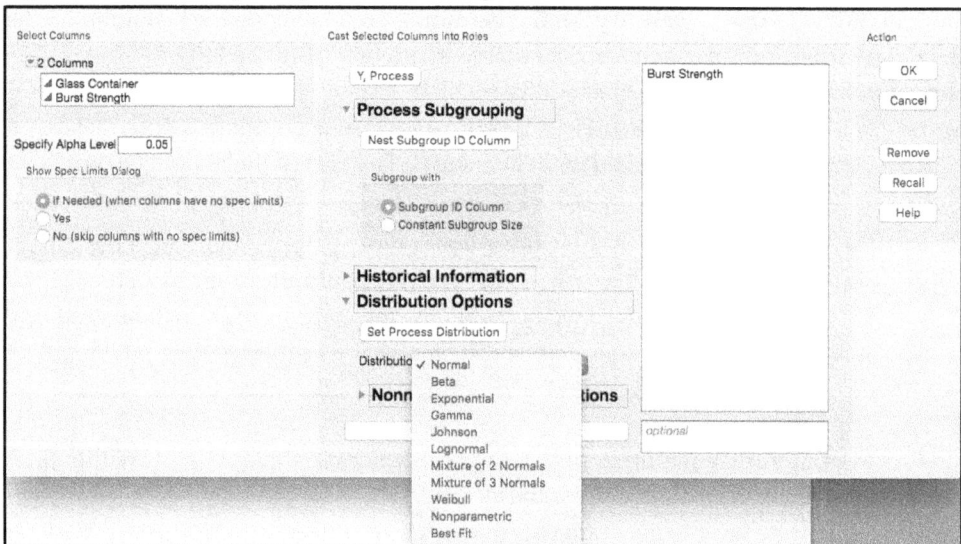

4. By default, if the column has no **Spec Limits** property, the **Spec Limits** window comes up (Figure 5.15). Enter the LSL=200 in the **LSL** field. Click **OK** when done.

Figure 5.15 Spec Limits Window

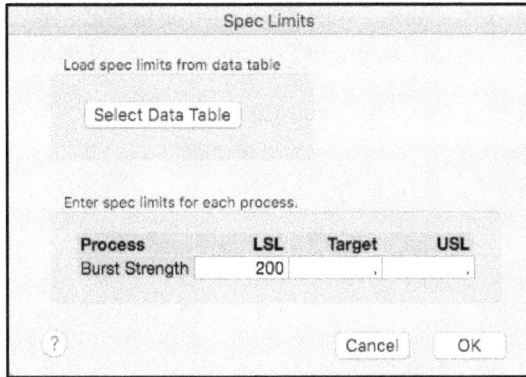

5. By default, the **Process Capability** platform does not show a summary with the process capability indices (in Chapter 6 we illustrate some of the output in this platform). To generate a report with the capability indices, click on the red triangle next to **Process Capability**, and select **Individual Detail Reports** as shown in Figure 5.16.

Figure 5.16 Selecting Individual Details Reports in the Process Capability Report

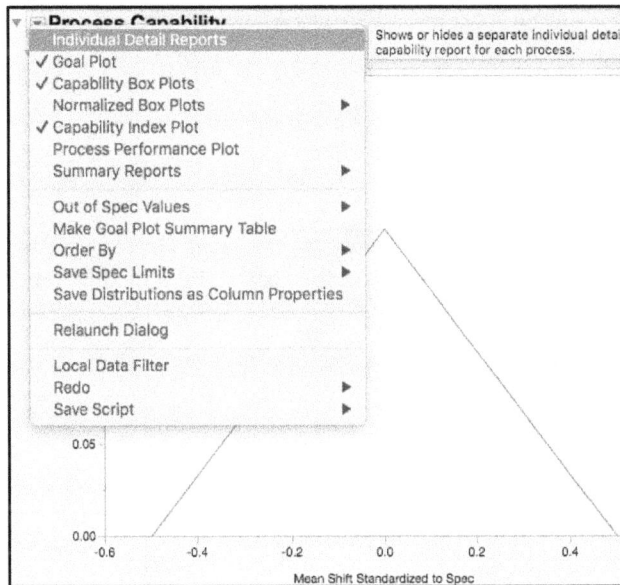

The **Individual Detail Reports** (Figure 5.17) shows the LSL and histogram of the burst strength data, overlaid with two normal densities that are generated using the **Within** and **Overall Sigma**. The **Within Sigma** is estimated using a moving range of size 2, and the **Overall Sigma** is estimated using all the data. The **Process Summary** shows the LSL, the sample size, the sample mean, the overall sigma, the within sigma, and the **Stability Ratio**. (See Chapter 6 for a detailed explanation and examples of the Stability Ratio.) Two capability reports are included, the **Within Sigma Capability** and the **Overall Sigma Capability**. These reports show the Cpk and Cpl, along with 95% confidence limits, calculated using the within sigma standard deviation, and the overall sigma standard deviation. Finally, a **Nonconformance** report details the actual percent below the LSL, and the expected percent, within and overall, estimated using the normal distribution.

Figure 5.17 Individual Details Reports in the Process Capability Report

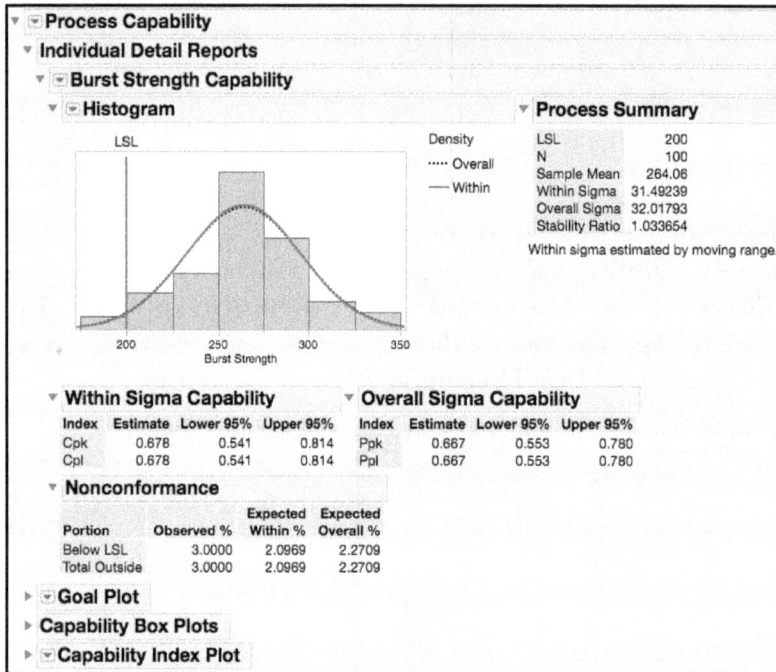

6. To identify the out of spec observations, click on the red triangle next to **Process Capability**, and select **Out of Spec Values ▶ Select Out of Spec Values** (Figure 5.18).

Figure 5.18 Selecting Out of Spec Values

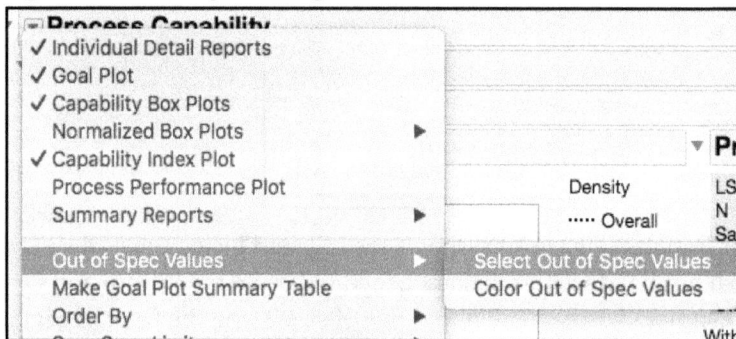

7. The selected out of spec values are shown in the histogram (Figure 5.19), similar to the histogram in Figure 5.11. Double-clicking the selected histogram bar allows one to generate a subset table like the one shown in Figure 5.11.

Figure 5.19 Selected Out of Spec Values

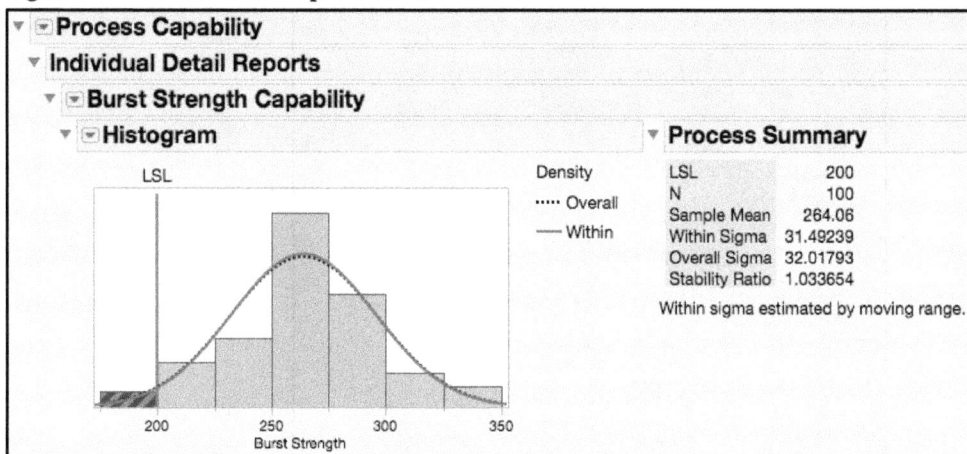

JMP has several additional platforms that compute process capability indices. Before moving on to the next ISQC example, we will show how to use the **Control Chart** and **Control Chart Builder** platforms to calculate process capability indices. The use of these two platforms requires the identification of a control chart type that is appropriate for the data structure. Although no data structure is provided for ISQC Table 8.1, in the description of ISQC Table 8.5, there are 20 subgroups of size n = 5 and an XBar and Range chart was constructed.

1. Open Chapter 5 – ISQC Table 8.5.jmp. Right-click on **Burst Strength (psi)** and select **Column Properties ▶ Spec Limits**. A dialog box will appear. Enter **200** in the **Lower Spec Limit** field (Figure 5.20) and then click **OK** when finished. An asterisk will appear next to **Burst Strength (psi)** under Columns, in the left-hand side of the table, indicating the presence of a column property.

Figure 5.20 Burst Strength Column Properties Window

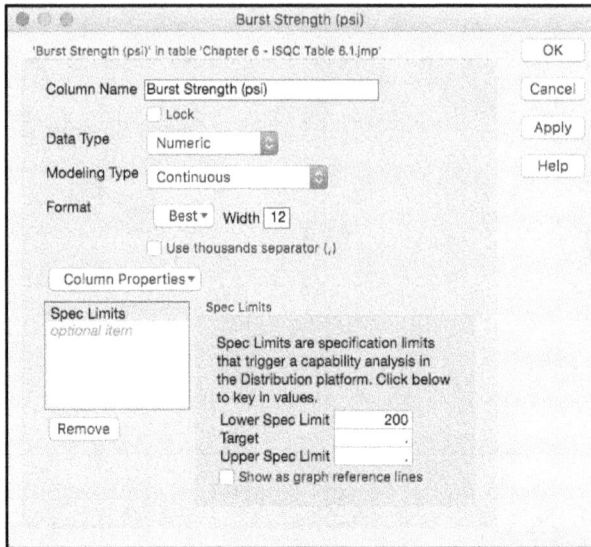

2. Select **Analyze ▶ Quality and Process ▶ Control Chart ▶ XBar**.
3. A launch window will appear. Highlight **Burst Strength (psi)** from the left hand window and click **Process** and then highlight **Sample** from the left hand window and click **Sample Label**. Finally, click the button next to **Capability**, in the lower right-hand side of the dialog window (Figure 5.21).

Figure 5.21 Launch Window for XBar Chart for Burst Strength

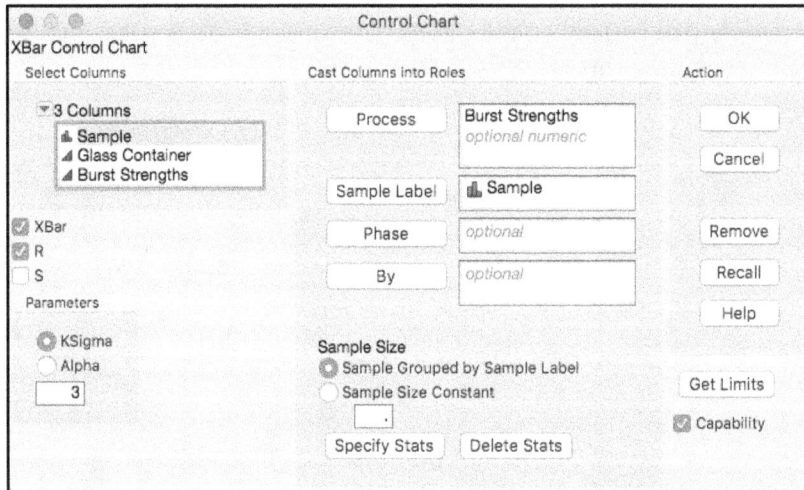

4. Click **OK** when finished. The output is shown in Figure 5.22.

Figure 5.22 Control Chart Capability Analysis for Burst Strength

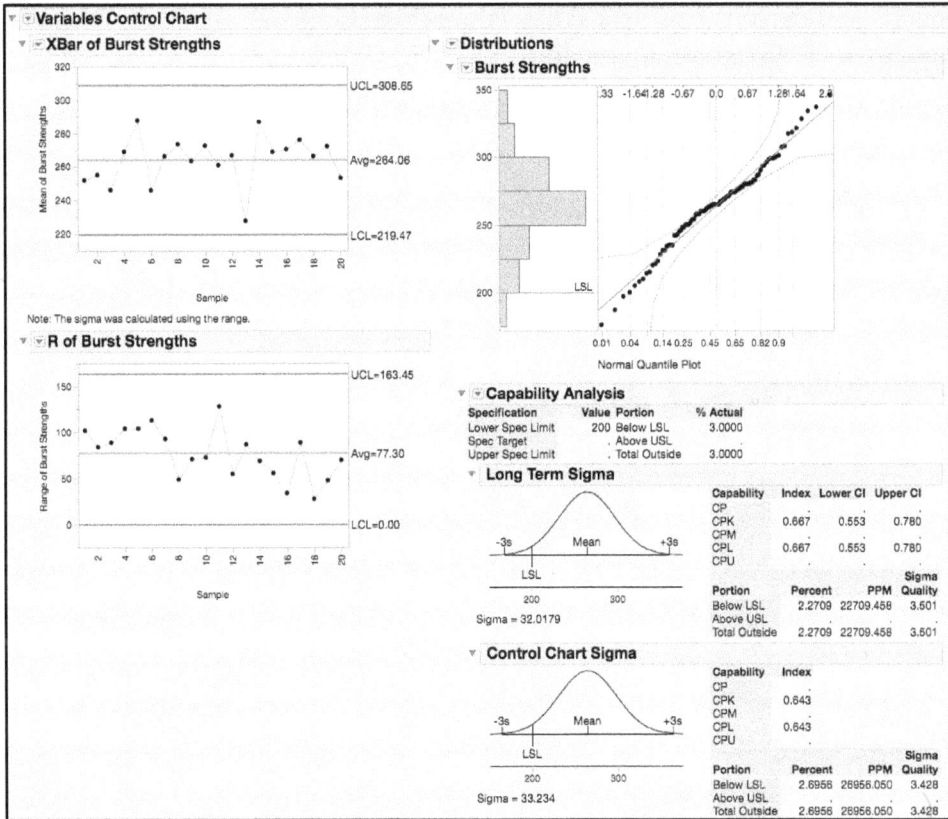

The default output is shown in Figure 5.22. The XBar and Range control charts are shown on the left-hand side of the output window, while the Capability Analysis is shown on the right-hand side. The drop-down menu next to the **Burst Strength** label above the histogram provides the same options as the similar drop-down menu in the **Distribution** platform. For example, summary statistics can be added, different distributions can be fit to the data, and statistical intervals can be computed.

The Capability Analysis under **Long Term Sigma** includes the same output as was shown for the Distribution platform in Figure 5.10. Note the Z-scores can be added to the output in Figure 5.22 by selecting **Z Bench** from **the Capability Analysis** drop-down menu. The C_{pl} is 0.667, using the long-term sigma estimate. The long-term sigma uses the sample standard deviation, s = 32.0179, to estimate the standard deviation used in the denominator of C_{pl}.

The **Control Chart Sigma** output provides statistics that use the same standard deviation used in the calculation of the XBar control limits: Average Range / d2 = 77.30 / 2.326 = 33.234. When the process is in control, these two estimates will be similar and so will the two estimates of C_{pl}. However, when the process is out-of-control, the long term sigma will be larger than the control chart sigma and its corresponding estimate of C_{pl} will be less than the Control Chart Sigma C_{pl}.

This issue is discussed further in Chapter 6 of this book. In this example, the two estimates of the standard deviations are similar, and the control chart C_{pl} = 0.643. The control charts in Figure 5.22 are shown in ISQC Figure 8.12, and the C_{pl} is in ISQC Section 8.4.

The following steps illustrate how to produce the Capability Analysis using the **Control Chart Builder** platform.

1. Highlight Chapter 5 – ISQC Table 8.5.jmp by clicking on it. Note this table should include the specification limits in the column properties for **Burst Strength** (see Step 1 from previous example).
2. Select **Analyze ▶ Quality and Process ▶ Control Chart Builder**.
3. When the dialog window appears, select **Sample** and drag it onto the **Subgroup** zone (X axis) and then select **Burst Strength (psi)** and drag it onto the **Y** zone (Y axis).

Figure 5.23 Control Chart Builder Capability Analysis for Burst Strength

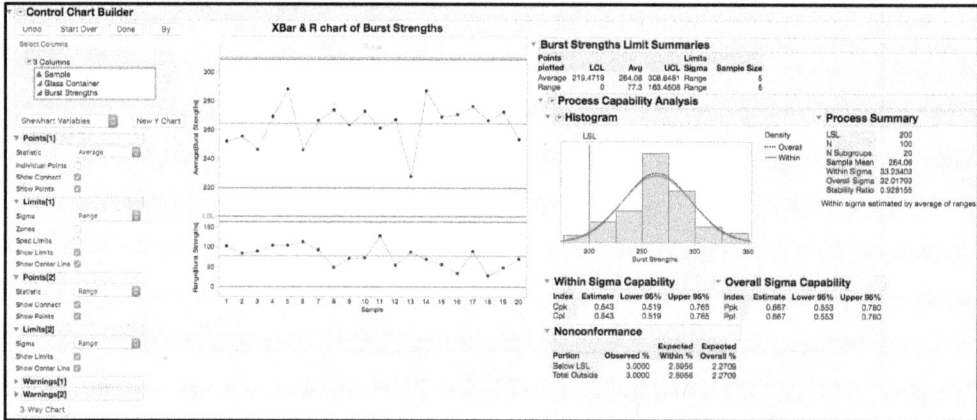

The default output from **Control Chart Builder** is shown in Figure 5.23. Once again, the XBar and Range control charts are shown in the left-hand side of the output, while the Capability Analysis is on the right-hand side. The output differs slightly from what was presented in Figure 5.22 for the **Control Chart** platform. For example, the options in the drop-down menus are more limited. The red triangle next to **Process Capability Analysis** includes options to add or remove the histogram, process summary, nonconformance information, and process capability indices. Note the Z-scores (**Z Benchmark**) can be added to the output using this menu. The **Histogram** drop-down menu allows us to manipulate the features of the histogram, such as, showing the density curves and specification limits. Also, the process capability labels are slightly different in the **Control Chart Builder**. The **Overall Sigma Capability** metrics are the same as **Long Term Sigma** metrics in Figure 5.22 and the **Within Sigma Capability** section is the same as the **Control Chart Sigma** section in the same figure.

The **Process Summary** report includes estimates of the two different standard deviations used in the two calculations. This is the same output produced by the **Process Capability** platform. The

Overall Sigma is 32.02, while the Within Sigma is 33.23. The Stability Ratio, which was introduced in in JMP version 13, is the ratio of these two variances, that is, $32.02^2 / 33.23^2 = 0.93$. As was previously stated, when the process is in control the two standard deviations, or equivalently, variances, estimates should be similar and the Stability Ratio will be close to 1. On the other hand, when the process is out-of-control Overall Sigma > Within Sigma and the Stability Ratio will be greater than 1. This metric will be discussed in detail in Chapter 6 of this book.

ISQC Example 8.3 Process Centering

In this example, we calculate another process capability index called C_{pm}. This index more effectively incorporates where a process is centered relative to its specification limits. A parameter must have a 2-sided specification in order to calculate it. A specific data set to illustrate this example is not available in ISQC Example 8.3, so a simulated data set was created based on the processes in ISQC Figure 8.11 The simulated data includes 100 observations for two responses, where Response 1 ~ $N(50, 5^2)$ and Response 2 ~ $N(57.5, 2.5^2)$. The specification limits for this example are LSL = 35 and USL = 65 and the Target value = 50. A description of the C_{pm} index and equations can be found in ISQC Section 8.3.4.

The following steps illustrate how to compute C_{pm} using the **Distribution** platform.

1. Open Chapter 5 – ISQC Example 8.3.jmp, which has variables called *Sample Number*, *Response (center=50)* and *Response (center=57.5)*.
2. From the main menu select **Analyze ▶ Distribution**.
3. When the dialog box appears, select **Response (center=50)** and **Response (center=57.5)** as the **Y, Columns** (response) variable and click **OK** when finished.
4. From the red triangle next to **Distributions** select **Stack**. From the red triangle next to **Response (center=50)**, deselect **Display Options ▶ Quantiles and Outlier Box Plot** and select **Continuous Fit ▶ Normal**. Repeat for **Response (center=57.5)**.

Figure 5.24 C$_{pm}$ Index for Centered and Off-Centered Processes

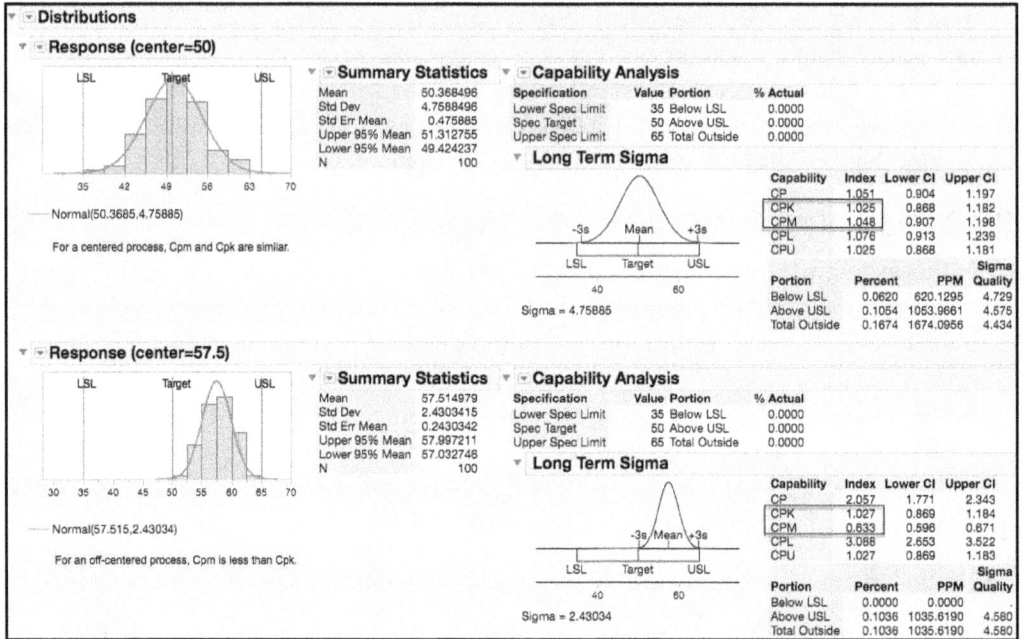

The JMP output for the centered and off-centered processes is shown in Figure 5.24. The C$_{pm}$ statistics are similar as those shown in ISQC Example 8.3. For the centered process C$_{pm}$ is approximately 1, while for the off-centered process C$_{pm}$ is 0.63. As Montgomery points out in ISQC Section 8.3.4, since both of these processes have a C$_{pk}$ ≈ 1, this index is not able to discern the differences in the performance of the two processes within the specification limits. However, the C$_{pm}$ index better reflects the off-target performance of the second process. Not only does the off-centered process have a smaller C$_{pm}$ relative to the C$_{pm}$ of the centered process, but it is also smaller than its C$_{pk}$ index. Where keeping a process on target is desired, C$_{pm}$ should be added to the arsenal of summary statistics used to monitor the quality of a process.

ISQC Example 8.4, 8.5 & 8.6 Confidence Intervals for C$_P$ & C$_{pk}$

In these examples, confidence intervals and tests for process capability indices are shown. Since no specific data are provided in ISQC Chapter 8, data were simulated using parameters that closely match the output in ISQC Example 8.4, Example 8.5, and Example 8.6. Specific details can be found in the **Formula** and **Column Properties** for the response of interest in the corresponding JMP tables.

The following steps illustrate how to construct confidence intervals for process capability indices using the **Distribution** platform. We begin with ISQC Example 8.4.

1. Open **Chapter 5 – ISQC Example 8.4.jmp**, which has variables called *Sample Number* and *Response*.
2. From the main menu, select **Analyze ▶ Distribution**.

3. When the launch window appears, select **Response** as **Y, Columns** and click **OK** when finished.

4. From the red triangle next to **Distributions** select **Stack**. From the red triangle next to **Response**, deselect **Display Options ▶ Quantiles** and **Outlier Box Plot**.

Figure 5.25 Confidence Interval for C_P for ISQC Example 8.4

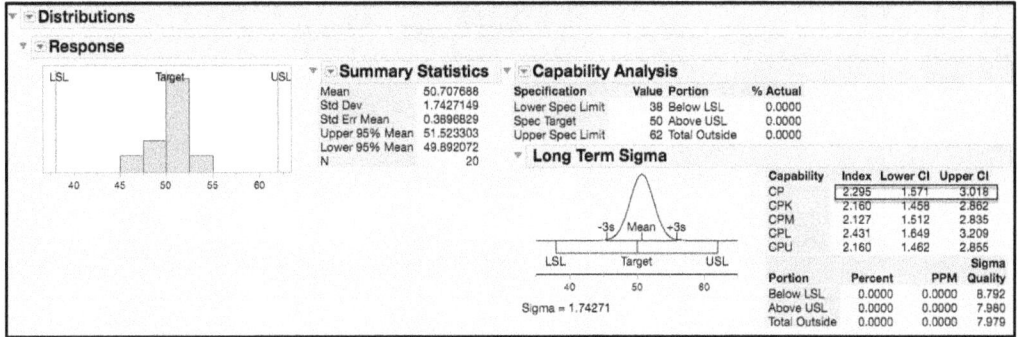

The output in Figure 5.25 shows C_P = 2.295 for a process with a mean = 50.7, standard deviation = 1.74 and LSL = 38 and USL = 62. With n = 20 observations, the 95% confidence interval for the estimate of C_P is (1.571, 3.018). The confidence interval takes into account the sampling error and sample size. Even though the estimated C_P = 2.295, it can be as low as 1.571 or as high as 3.018, with 95% confidence. These simulated data results are very similar to the estimated C_P and 95% confidence interval in ISQC Example 8.4.

The C_P confidence intervals calculated in JMP agree with the formulas shown in ISQC equation 8.19 and are, by default, 95% confidence intervals. We now illustrate ISQC Example 8.5.

1. Open Chapter 5 – ISQC Example 8.5.jmp, which has variables called *Sample Number* and *Response*.

2. From the main menu, select **Analyze ▶ Distribution**.

3. When the launch window appears, select **Response** as **Y, Columns** and click **OK** when finished.

4. From the red triangle next to **Distributions,** select **Stack**. From the red triangle next to **Response,** deselect **Display Options ▶ Quantiles and Outlier Box Plot**. The results are shown in Figure 5.26.

Figure 5.26 Confidence Interval for C_{pk} for ISQC Example 8.5

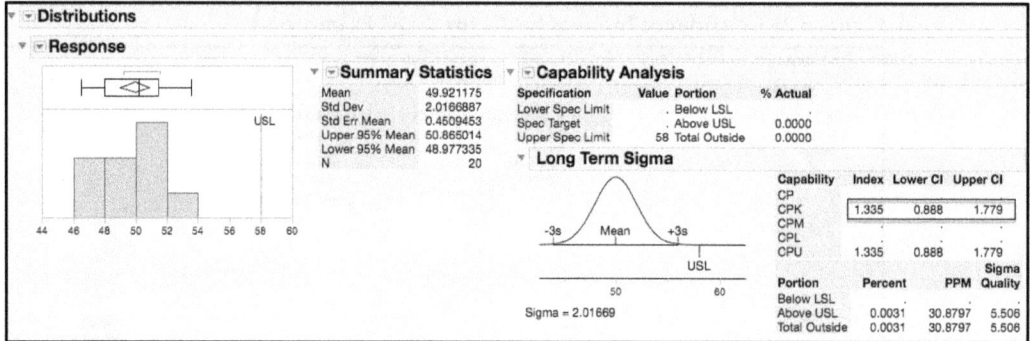

The output in Figure 5.26 shows C_{pk} = 1.335 for a process with a mean = 49.92, standard deviation = 2.02 and USL = 58. With n = 20 observations, the 95% confidence interval for the estimate of C_{pk} is (0.888, 1.779). The interval takes into account the sample size and the sampling error. Even though the estimated C_{pk} = 1.335, it can be as low as 0.888 or as high as 1.779, with 95% confidence. These simulated data results are very similar to the estimated C_{pk} and 95% confidence interval in ISQC Example 8.5.

The C_{pk} confidence intervals calculated in JMP agree with the formulas shown in ISQC equation 8.21 and are, by default, 95% confidence intervals. We now illustrate ISQC Example 8.6.

 Statistics Note 5.2: The C_{pk} can be thought of as the standardized distance, in units of standard deviation, to the nearest specification.

---

1. Open Chapter 5 – ISQC Example 8.6.jmp, which has variables called *Sample Number, Response 1* and *Response 2*.

2. From the main menu select **Analyze ▶ Distribution**.

3. When the launch window appears, select **Response 1** and **Response 2** as the **Y, Columns** (response) variable and click **OK** when finished.

4. From the red triangle next to **Distributions,** select **Stack**. While holding down the control key (command on a Mac), from the red triangle next to **Response 1** deselect **Display Options ▶ Quantiles** and **Outlier Box Plot** and select **Continuous Fit ▶ Normal**. The output is shown in Figure 5.27.

Figure 5.27 Hypothesis Test for H₁: C$_P$ > 1.33

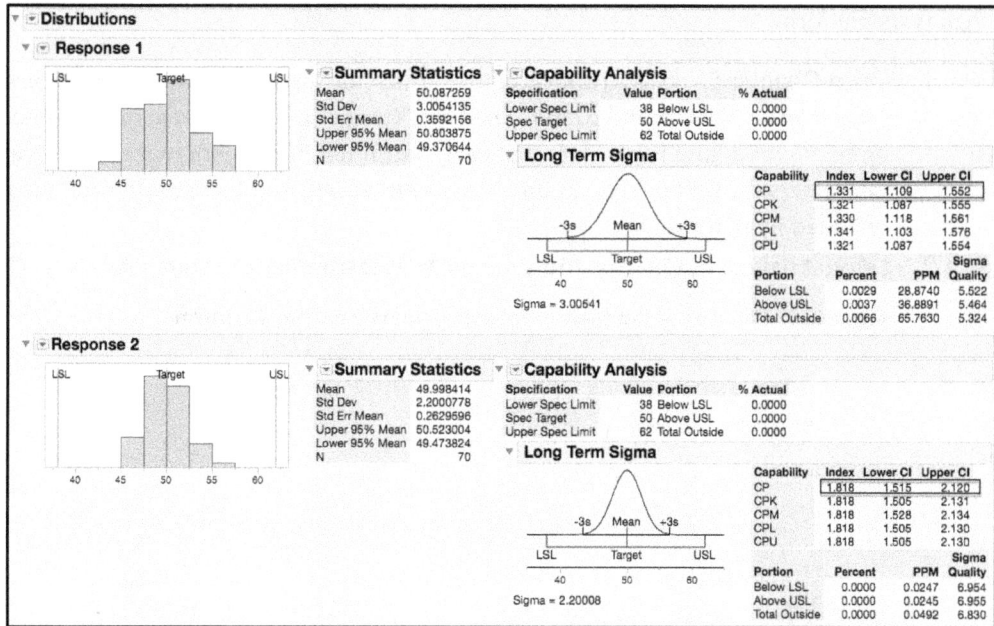

▼ Distributions

▼ Response 1

Summary Statistics

Mean	50.087259
Std Dev	3.0054135
Std Err Mean	0.3592156
Upper 95% Mean	50.803875
Lower 95% Mean	49.370644
N	70

Capability Analysis

Specification	Value	Portion	% Actual
Lower Spec Limit	38	Below LSL	0.0000
Spec Target	50	Above USL	0.0000
Upper Spec Limit	62	Total Outside	0.0000

▼ Long Term Sigma

Sigma = 3.00541

Capability	Index	Lower CI	Upper CI
CP	1.331	1.109	1.552
CPK	1.321	1.087	1.555
CPM	1.330	1.118	1.561
CPL	1.341	1.103	1.578
CPU	1.321	1.087	1.554

Portion	Percent	PPM	Sigma Quality
Below LSL	0.0029	28.8740	5.522
Above USL	0.0037	36.8891	5.464
Total Outside	0.0066	65.7630	5.324

▼ Response 2

Summary Statistics

Mean	49.998414
Std Dev	2.2000778
Std Err Mean	0.2629596
Upper 95% Mean	50.523004
Lower 95% Mean	49.473824
N	70

Capability Analysis

Specification	Value	Portion	% Actual
Lower Spec Limit	38	Below LSL	0.0000
Spec Target	50	Above USL	0.0000
Upper Spec Limit	62	Total Outside	0.0000

▼ Long Term Sigma

Sigma = 2.20008

Capability	Index	Lower CI	Upper CI
CP	1.818	1.515	2.120
CPK	1.818	1.505	2.131
CPM	1.818	1.528	2.134
CPL	1.818	1.505	2.130
CPU	1.818	1.505	2.130

Portion	Percent	PPM	Sigma Quality
Below LSL	0.0000	0.0247	6.954
Above USL	0.0000	0.0245	6.955
Total Outside	0.0000	0.0492	6.830

The intent of ISQC Example 8.6 is to determine if a Supplier meets the qualification criteria of having C$_P$ of at least 1.33, for a given parameter. The test C$_P$ ≥ 1.33 is a one-sided test of hypothesis, and we reject the null hypothesis if the estimated C$_P$ exceeds the criterion for n=70 in ISQC Table 8.4, of 1.33x1.10=1.45. The output shown in Figure 5.27 includes process capability indices C$_P$, and their 95% confidence intervals, for two different responses. For Response 1, C$_P$ = 1.331 and the 95% confidence interval is (1.109, 1.552). Based on the criterion listed above, the Supplier would not be qualified since the C$_P$ is less than 1.45. As Montgomery points out, even though C$_P$ = 1.331, in order to demonstrate that the capability is at least 1.33, the estimated C$_P$ "will have to exceed 1.33 by a considerable amount." We can also see that the lower confidence limit for Response 1 is 1.109, which is substantially below 1.33.

The results for Response 2, in Figure 5.27, show C$_P$ = 1.818 and the 95% confidence interval is (1.515, 2.120). Since the estimate for C$_P$ > 1.46, we have demonstrated that C$_P$ is at least 1.33. Therefore, the supplier met the qualification criterion. In addition, the lower confidence limit is 1.515, > 1.33, which suggests that, with 95% confidence, the process capability for Response 2 is at least 1.515.

ISQC Example 8.7 Measuring Gauge Capability

In this example, we will show how to conduct an analysis of data from a Gauge R&R study. The study design included twenty units of a product, where one process operator measured each unit two times using one instrument. The data consists of a total of 40 observations and it is organized in a vertical manner, where all 40 observations are included in one column and identified appropriately.

The following steps illustrate how to analyze this data set using the **Measurement Systems Analysis** platform.

1. Open Chapter 5 – ISQC Table 8.6.jmp, which has variables called *Part Number, Measurement Number* and *Result*. The Part Number indicates an actual part and has a Nominal modeling type. Note that each part number appears twice. The Measurement Number is the first or second time it was measured, and the Result is the measurement obtained from the gauge.

2. Select **Analyze ▶ Quality and Process ▶ Measurement Systems Analysis** (Figure 5.28).

   Figure 5.28 Launching the Measurement Systems Analysis Platform

   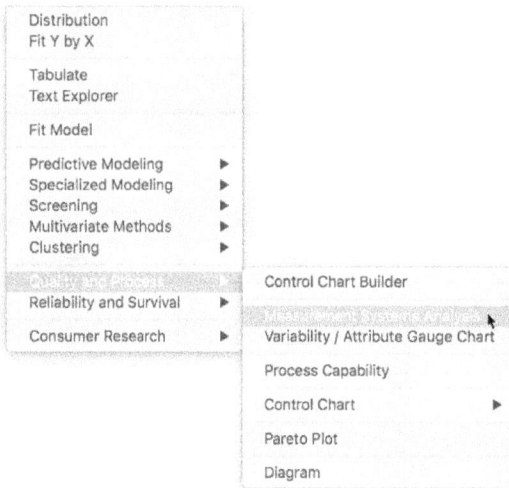

3. The **EMP Measurement System Analysis** launch window appears (Figure 5.29). Select **Result** and click **Y, Response,** and then select **Part Number** and click **Part, Sample ID**. Select **EMP** for **MSA Method** and **Range** for **Chart Dispersion Type** and **Main** for **Model Type**.

Figure 5.29 Launch Window for EMP Measurement Systems Analysis

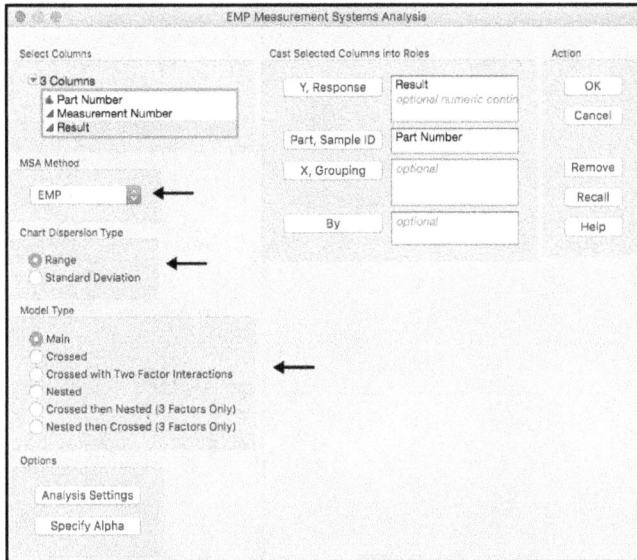

A few of the fields in the interface are further defined:

○ **MSA Method:** two options are provided, EMP (Evaluating the Measurement Process) and Gauge R&R. The two different options produce different ways of summarizing the results, different statistical tests, and different graphics. Note Gauge R&R produces the same output as the **Variability / Attribute Gauge Chart** platform. See JMP help for more details.

○ **Chart Dispersion Type:** designates the type of chart for showing variation. If EMP is selected for MSA, then XBar and Range or XBar and S control charts are automatically displayed. If Gauge R&R is selected for MSA, only S charts are available.

○ **Model Type:** specifies the type of variance components model fit to the results, in order to estimate the different sources of variation in the study.

4. Click **OK** when done. The output is shown in Figure 5.30.

Figure 5.30 Default Output for MSA EMP Analysis

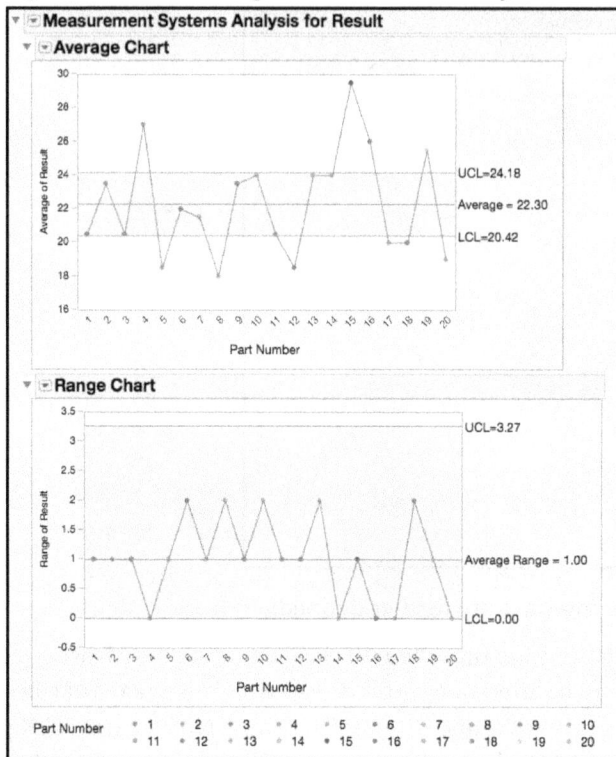

The default output shown in Figure 5.30 matches the control charts in ISQC Figure 8.14. As Montgomery points out, in a measurement study we expect the XBar chart to have many points that exceed the control limits. Recall, the XBar limits are calculated using an estimate of variation from the Range chart. In this study design, the Range chart represents the measurement error, so the XBar chart is a visual representation of the measurement error (control limits) and the part-to-part variation. An estimate of measurement error can be obtained from the Range chart using the average Range and the control chart constant d_2. In ISQC Example 8.7, $\sigma_{gauge} = 1.0 / 1.128 = 0.887$.

Further analyses of this study are presented in the text following ISQC Example 8.7, including a description of P/T ratio (precision to tolerance), variance components and the discrimination ratio (DR). We will continue this example by switching to the Gauge R&R MSA method in this platform. Note additional output produced by the EMP MSA method is discussed in the Statistical Insights section, later in this chapter.

5. Re-launch the **Measurement Systems Analysis** by selecting **Redo ▶ Relaunch Analysis** from the red triangle next to **Measurement Systems Analysis for Result** in Figure 5.30. Alternatively, activate the JMP table and repeat Step 2 above.

6. When the launch window appears, select **Result** and click **Y, Response,** and then select **Part Number** and click **Part, Sample ID**. Select **Gauge R&R** for **MSA Method** and **Main Effect** for **Model Type**.

Figure 5.31 Launch Window using Gauge R&R MSA Method

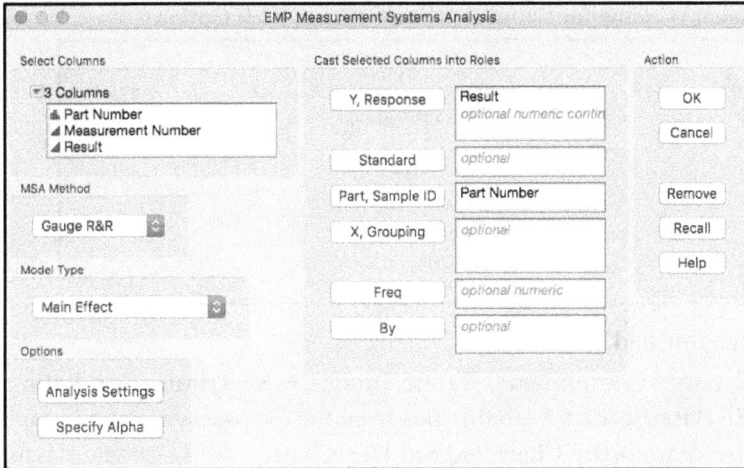

7. Click **OK** when finished. A number of options are available using the red triangle next to **Variability Gauge** as shown in Figure 5.32.

Figure 5.32 Options for Gauge R&R MSA Method

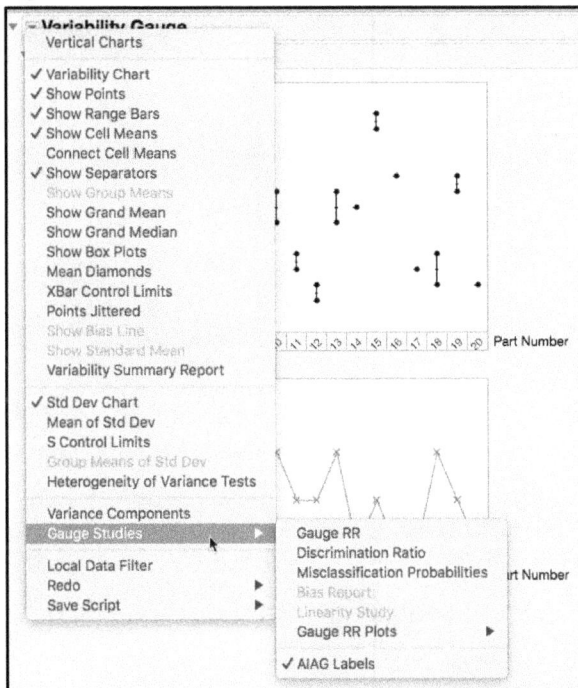

8. Select **Gauge Studies** ▶ **Gauge RR** from the drop-down menu in Figure 5.32. A dialog box will appear. Leave the **tolerance entry method** as **LSL and/or USL**. Type **6** into the field for **K, Sigma Multiplier** and ensure that **LSL = 5** and **USL = 60**. Note the specification limits were imported from the Result's Column Properties in the JMP table.

Figure 5.33 Dialog Window for Gauge Studies ▶ Gauge R&R Option

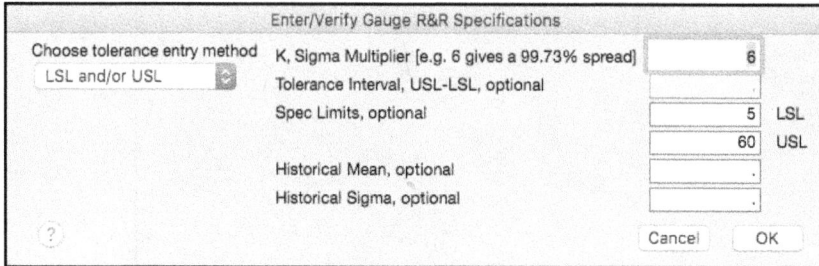

Enter/Verify Gauge R&R Specifications	
Choose tolerance entry method	K, Sigma Multiplier [e.g. 6 gives a 99.73% spread] 6
LSL and/or USL	Tolerance Interval, USL-LSL, optional
	Spec Limits, optional 5 LSL
	60 USL
	Historical Mean, optional
	Historical Sigma, optional
?	Cancel OK

9. Click **OK** when finished.

10. Next select **Variance Components, Gauge Studies** ▶ **Discrimination Ratio** and **Gauge Studies** ▶ **Misclassification Probabilities** from the drop-down menu in Figure 5.32. Finally, deselect **Variability Chart and Std Dev Chart**.

Figure 5.34 Selected Output for Gauge R&R MSA Method

▼ Variability Gauge
▼ Variability Chart for Result
▼ Analysis of Variance

Source	DF	SS	Mean Square	F Ratio	Prob > F
Part Number	19	377.4	19.8632	26.4842	<.0001*
Within	20	15	0.75		
Total	39	392.4	10.0615		

▼ Variance Components

Component	Var Component	% of Total	20 40 60 80	Sqrt(Var Comp)
Part Number	9.555579	92.7		3.0914
Within	0.750000	7.3		0.8660
Total	10.306579	100.0		3.2104

▼ Gauge R&R

Measurement Source		Variation (6*StdDev)	% of Tolerance		which is 6*sqrt of
Repeatability	(EV)	5.196152	9.45	Equipment Variation	V(Within)
Reproducibility	(AV)	0.000000	0.00	Appraiser Variation	
Gauge R&R	(RR)	5.196152	9.45	Measurement Variation	V(Total)-V(Part Number)
Part Variation	(PV)	18.548230	33.72	Part Variation	V(Part Number)
Total Variation	(TV)	19.262317	35.02	Total Variation	V(Total)

    6 k
26.9757 % Gauge R&R = 100*(RR/TV)
0.28014 Precision to Part Variation = RR/PV
    5 Number of Distinct Categories = 1.41(PV/RR)
    5 LSL
   60 USL
   55 Tolerance = USL-LSL
0.09448 Precision/Tolerance Ratio = RR/(USL-LSL)
Using last column 'Part Number' for Part.

▼ Variance Components for Gauge R&R

Component	Var Component	% of Total	20 40 60 80
Gauge R&R	0.7500000	7.28	
Repeatability	0.7500000	7.28	
Reproducibility	0.0000000	0.00	
Part-to-Part	9.5565789	92.72	

▼ Discrimination Ratio

Source	Ratio
Part Number	5.14628123

▼ Misclassification Probabilities

Description	Probability
P(Good part is falsely rejected)	0.0000
P(Bad part is falsely accepted)	0.3065
P(Part is good and is rejected)	0.0000
P(Part is bad and is accepted)	0.0000
P(Part is good)	1.0000

Much of the available output for a Gauge R&R analysis is shown in Figure 5.34. The Precision / Tolerance Ratio = 0.09448 is in the last row of the output under the **Gauge R&R** banner. This number implies that the gauge error is 9.448% of the tolerance (USL – LSL). Montgomery points out that a P/T ratio of 0.1 or less indicates adequate gauge capability. Note the JMP P/T result is slightly different from the P/T of 0.09673 (= 5.32/55) shown in ISQC Section 8.7.1. This is due to the two different methods of estimating the Gauge R&R error: Rbar/d2 (1.0/1.128 = 0.8865248) in ISQC and EMS (expected mean square) estimate (0.8660254) in JMP.

The variance component estimates are found under the **Variance Components** banner and under the **Variance Components for Gauge R&R** banner. The variance component estimates for part variation and gauge error are identical in both sections of the output; however, the first banner also provides an estimate of the total variation and the square root of the variance component. The variance component break down, $\sigma^2_{total} = \sigma^2_{part} + \sigma^2_{gauge}$, shows that 10.3065 = 9.5566 + 0.7500. These estimates are slightly different from what is presented in ISQC Section 8.7.1, 10.05 = 9.26 + 0.79, due to the different estimation methods and roundoff error.

Estimates for the two measures shown in ISQC equations 8.26 and 8.27 are also found under the **Variance Components** or **Variance Components for Gauge R&R** banners. Both of these measures are ratios (or equivalently % = 100*ratio) of the specific variance component to the total variation and can be found in the **% of Total** columns in both tables. (For example, ρ_{part} = 92.7% = 9.5566 / 10.3065 and ρ_{gauge} = 7.3% = 0.7500/ 10.3065.) These numbers are expressed as ratios in ISQC Section 8.7, ρ_{gauge} = 0.0786 and ρ_{part} = 1 - 0.0786 = 0.9214, and differ slightly from the JMP output.

The **Discrimination Ratio (DR)** is provided in the JMP output. It provides a measure of the relative usefulness of a given measurement for a specific product. Guidelines suggest that when the DR is less than 2, the measurement system cannot adequately detect production variation and the measurement process needs improvement, and if DR is greater than 4 then the measurement system adequately detects production variation. Figure 5.34 shows a Discrimination Ratio of 5.15, while ISQC Section 8.7.1 shows DR = 24.45. It should be noted that two different equations were used for these calculations. For ISQC see equation 8.29 and for JMP see Wheeler (2005).

The **Misclassification Probabilities** are included at the bottom of Figure 5.34. Although these were not presented for ISQC Example 8.7, the topic is addressed in ISQC Section 8.7.4. We also find these calculations to provide useful information, when the study includes representative part-to-part variation. These calculations show the probability for a good part being rejected or a bad part being accepted, which can happen whenever measurement error is greater than zero. These two corresponding probabilities are found next to the labels **P(Good part is falsely rejected)** and **P(Bad part is falsely accepted)**. The formulas for these two misclassification probabilities are given in ISQC equations 8.36 and 8.37.

Finally, the output under **Variance Components for Gauge R&R** comes from the Auto Industry Action Group (AIAG) of the American Society for Quality. It is included here for completeness. An interesting discussion of this analysis can be found in Wheeler (2009).

> **JMP Note 5.5**: The JMP Measurement System Analysis (MSA) offers three options for computing variance components and, by default, selects the best analysis option for the data. The three options are Expected Mean Squares (EMS), Restricted Maximum Likelihood (REML), and Bayesian.

ISQC Table 8.7 Thermal Impedance Data for Gauge R&R Experiment

In this example, we show how to conduct an analysis of data from another Gauge R&R study. The study design included ten units of a product, where three inspectors measured each unit three times using one instrument. The data consists of a total of 90 observations and it is organized in a vertical manner, where all 90 observations are included in one column and identified appropriately.

The following steps illustrate how to analyze this data set using the **Measurement System Analysis** platform. Note this example is further discussed in the Statistical Insights section of this chapter.

1. Open Chapter 5 – ISQC Table 8.7.jmp, which has variables called *Part Number, Operator, Test Number Thermal Impedance*. The Part Number indicates an actual part and has a Nominal modeling type. Note that each part number is repeated nine times. The operator represents the three inspectors in the study and it also has a Nominal modeling type. The test number represents the first, second, or third time it was measured, and the thermal impedance is the measurement obtained from the gauge.
2. Select **Analyze ▶ Quality and Process ▶ Measurement System Analysis**.
3. When the launch window appears, select **Thermal Impedance** and click **Y, Response** and then select **Part Number** and click **Part, Sample ID** and select **Operator** and click **X, Grouping**. Select **Gauge R&R** for **MSA Method** and **Crossed** for **Model Type** (Figure 5.35). When finished click **OK**.

Figure 5.35 Launch Window for MSA with Part and Operator

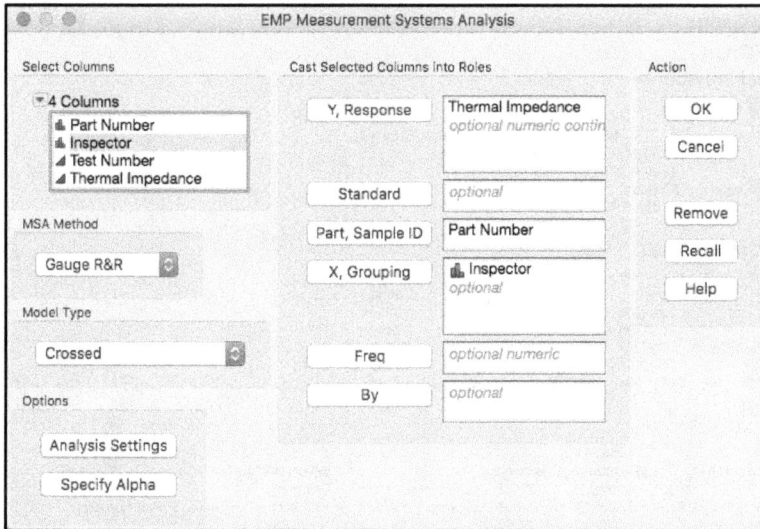

4. From the red triangle next to **Variability Gauge** select **Gauge Studies ▶ Gauge RR** from the drop-down menu. A dialog box will appear. Leave the **tolerance entry method** as **LSL and/or USL**. Type **6** into the field for **K, Sigma Multiplier** and ensure that **LSL = 18** and **USL = 58**. Note the specification limits were imported from the Thermal Impedance's Column Properties in the JMP table.

5. Next, from the same menu, select **Variance Components** and deselect **Variability Chart and Std Dev Chart**. Click **OK** when finished.

Figure 5.36 Analysis Output for Thermal Impedance Gauge R&R

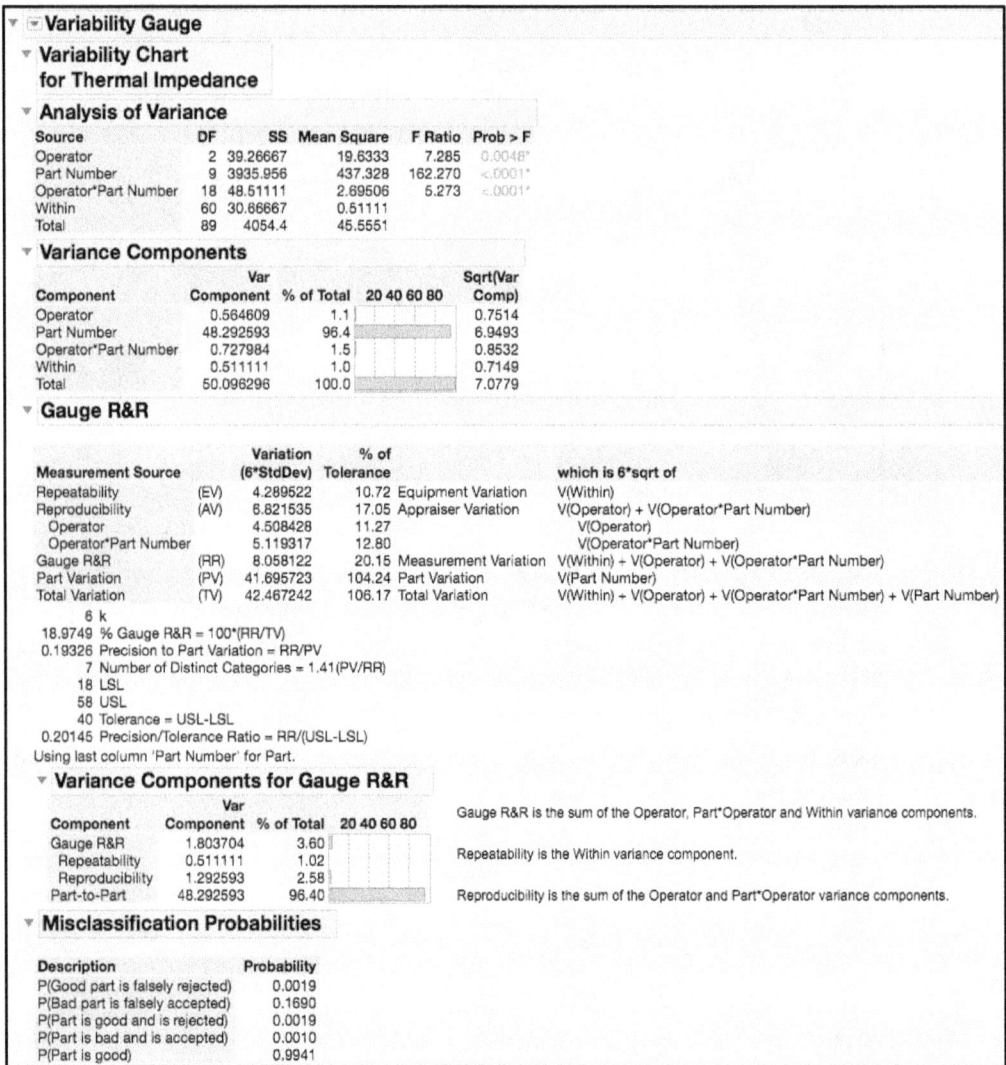

▼ ⊟ **Variability Gauge**

　▼ **Variability Chart**
　　for Thermal Impedance

　▼ **Analysis of Variance**

Source	DF	SS	Mean Square	F Ratio	Prob > F
Operator	2	39.26667	19.6333	7.285	0.0048*
Part Number	9	3935.956	437.328	162.270	<.0001*
Operator*Part Number	18	48.51111	2.69506	5.273	<.0001*
Within	60	30.66667	0.51111		
Total	89	4054.4	45.5551		

　▼ **Variance Components**

Component	Var Component	% of Total	20 40 60 80	Sqrt(Var Comp)
Operator	0.564609	1.1		0.7514
Part Number	48.292593	96.4		6.9493
Operator*Part Number	0.727984	1.5		0.8532
Within	0.511111	1.0		0.7149
Total	50.096296	100.0		7.0779

　▼ **Gauge R&R**

Measurement Source		Variation (6*StdDev)	% of Tolerance		which is 6*sqrt of
Repeatability	(EV)	4.289522	10.72	Equipment Variation	V(Within)
Reproducibility	(AV)	6.821535	17.05	Appraiser Variation	V(Operator) + V(Operator*Part Number)
Operator		4.508428	11.27		V(Operator)
Operator*Part Number		5.119317	12.80		V(Operator*Part Number)
Gauge R&R	(RR)	8.058122	20.15	Measurement Variation	V(Within) + V(Operator) + V(Operator*Part Number)
Part Variation	(PV)	41.695723	104.24	Part Variation	V(Part Number)
Total Variation	(TV)	42.467242	106.17	Total Variation	V(Within) + V(Operator) + V(Operator*Part Number) + V(Part Number)

　　　6 k
　18.9749 % Gauge R&R = 100*(RR/TV)
　0.19326 Precision to Part Variation = RR/PV
　　　　7 Number of Distinct Categories = 1.41(PV/RR)
　　　18 LSL
　　　58 USL
　　　40 Tolerance = USL-LSL
　0.20145 Precision/Tolerance Ratio = RR/(USL-LSL)
Using last column 'Part Number' for Part.

　▼ **Variance Components for Gauge R&R**

Component	Var Component	% of Total	20 40 60 80	
Gauge R&R	1.803704	3.60		Gauge R&R is the sum of the Operator, Part*Operator and Within variance components.
Repeatability	0.511111	1.02		Repeatability is the Within variance component.
Reproducibility	1.292593	2.58		
Part-to-Part	48.292593	96.40		Reproducibility is the sum of the Operator and Part*Operator variance components.

　▼ **Misclassification Probabilities**

Description	Probability
P(Good part is falsely rejected)	0.0019
P(Bad part is falsely accepted)	0.1690
P(Part is good and is rejected)	0.0019
P(Part is bad and is accepted)	0.0010
P(Part is good)	0.9941

ISQC Section 8.7.2 uses the Analysis of Variance (ANOVA) method to analyze the data from ISQC Table 8.7. We selected a **Crossed** model type in Figure 5.35 because every operator measured the same 10 parts three times, resulting in Operator, Part and Operator*Part terms. The ANOVA table is provided at the top of the output shown in Figure 5.36 and matches ISQC Table 8.8. Note the Expected Mean Squares are not provided in this platform. The Part Number exhibits the largest variation and results in the largest F Ratio = 162.270.

The variance component estimates are found under the **Variance Components** banner and under the **Variance Components for Gauge R&R** banner. The variance component estimates for part variation and gauge error are identical in both sections of the output; however, the first banner breaks out the Reproducibility components (Operator and Operator*Part Number) also provides an estimate of the total variation and the square root of the variance components. The variance component break down, $\sigma^2_{total} = \sigma^2_{part} + \sigma^2_{gauge}$, shows that 50.0963 = 48.2926 + 1.8037. These estimates agree with what is presented in ISQC Section 8.7.2, 48.29 + 1.80 = 50.09. The % of Total columns, indicate that the part-to-part variation contributes 96.4% to the total variation, while the gauge error contributes 3.60%. All components that contribute to the Gauge R&R appear to contribute in a similar manner, with Operator = 1.1%, Operator*Part Number = 1.5% and Within (repeatability) = 1.0%.

The Precision / Tolerance Ratio = 0.20145 is in the last row of the output under the **Gauge R&R** banner. This number implies that the gauge error is 20.145% of the tolerance (USL – LSL). Montgomery points out that a P/T ratio of 0.1 or less indicates adequate gauge capability. The JMP output shows P/T = 0.20145, in agreement with the P/T equation

$6\hat{\sigma}_{Gauge} / (USL - LSL) = 6\sqrt{1.8037} / (58 - 18) = 0.20145$. Note that the reported P/T = 0.27 shown in ISQC Section 8.7.2 is incorrect. This is the value one would obtain if one uses the gauge variance 1.8037 in the formula instead of the gauge standard deviation. In other words, the P/T was calculated as $6 \times 1.8037 / (58 - 18) = 0.27$.

Once again, the **Misclassification Probabilities** are included at the bottom of Figure 5.36. These calculations show the probability for a good part being rejected or a bad part being accepted, which can happen whenever measurement error is greater than zero. These two corresponding probabilities are found next to the labels **P(Good part is falsely rejected)** and **P(Bad part is falsely accepted)**.

ISQC Table 8.13 Attribute Gauge Capability Analysis

In this example, we will show how to analyze data generated from a Gauge R&R study, where the response of interest is attribute in nature, such as, good/bad or class a / class b / class c. The study design in ISQC Table 8.13 shows a study for a bank that uses underwriting to analyze mortgage loan applications. Each underwriter must analyze the relevant information and classify the application into one of four categories: Decline (decline the applicant of funding), Fund – 1 (fund, but higher risk applicant), Fund-2 (fund, lower risk applicant) and Fund-3 (fund, lowest risk applicant). Three underwriters, John, Sue, and Fred, were asked to classify 30 applications twice.

The JMP table has a total of 60 rows, with columns for the application number (Application), the correct classification of the application (Classification), the first or second review of each application (Review Number) and a column each for Sue's, Fred's, and John's classification of the application (Sue, Fred, John). The following steps illustrate how to analyze this data using the **Variability / Attribute Gauge Chart** platform.

1. Open Chapter 5 – ISQC Table 8.13.jmp, which has variables called *Application, Classification, Review Number, Sue, Fred,* and *John*. Note that Classification, Sue, Fred, and John have a Nominal modeling type.

2. Select **Analyze ▶ Quality and Process ▶ Variability / Attribute Gauge Chart**.

3. When the launch window appears, first select **Attribute** for **Chart Type**. Next select **Sue**, **Fred**, and **John** and click **Y, Response.** Select **Classification** and click **Standard**. Then select **Application** and click **X, Grouping** (Figure 5.37).

Figure 5.37 Launch Window for Attribute Gauge Study

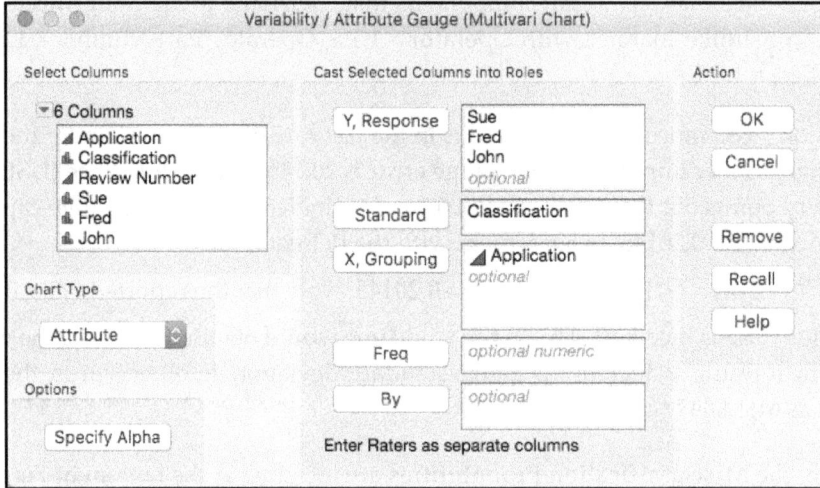

4. Click **OK** when finished.

The output for this analysis is too large to capture in one figure and therefore, will be presented in sections. Figure 5.38a shows a plot of the % **Agreement** for each of the thirty applications and then for each underwriter. The blue lines in the plots show the agreement rates for the underwriters for each application and the red lines in the plots show the effectiveness rates, or agreement to the standard, for the underwrites for each application. In order to locate the applications with high or low rates, click on the points in the first figure to highlight them in the JMP table then use **Tables ▶ Subset**. The data in Figure 5.38b show the raw data associated with the highest and lowest effectiveness. For example, for Application 7, all underwriters classified it as 'Fund-3', which also matches the standard classification of 'Fund-3'. For Application 1, three of the 6 attempts classified it as 'Fund-3', 2 as 'Fund-2' and 1 as 'Fund-1'; and, the standard classification is 'Fund-1'.

Figure 5.38a Default Output for Attribute Gauge Study

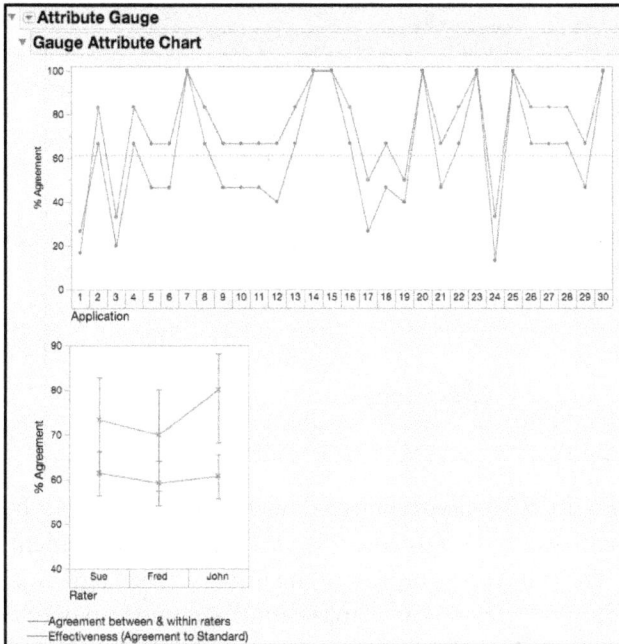

Figure 5.38b Data Subset for High Effectiveness

	Application	Classification	Review Number	Sue	Fred	John
1	7	Fund-3	1	Fund-3	Fund-3	Fund-3
2	7	Fund-3	2	Fund-3	Fund-3	Fund-3
3	14	Fund-2	1	Fund-2	Fund-2	Fund-2
4	14	Fund-2	2	Fund-2	Fund-2	Fund-2
5	15	Fund-1	1	Fund-1	Fund-1	Fund-1
6	15	Fund-1	2	Fund-1	Fund-1	Fund-1
7	20	Fund-1	1	Fund-1	Fund-1	Fund-1
8	20	Fund-1	2	Fund-1	Fund-1	Fund-1
9	22	Fund-2	1	Fund-1	Fund-2	Fund-2
10	22	Fund-2	2	Fund-2	Fund-2	Fund-2
11	25	Fund-3	1	Fund-3	Fund-3	Fund-3
12	25	Fund-3	2	Fund-3	Fund-3	Fund-3
13	29	Decline	1	Decline	Fund-3	Decline
14	29	Decline	2	Decline	Decline	Fund-3

High Effectiveness

Columns (6/0)
- Application
- Classification
- Review Number
- Sue
- Fred
- John

Rows
All rows	14
Selected	0
Excluded	0
Hidden	0
Labelled	0

Figure 5.38b Data Subset for Low Effectiveness

	Application	Classification	Review Number	Sue	Fred	John
1	1	Fund-1	1	Fund-3	Fund-2	Fund-1
2	1	Fund-1	2	Fund-3	Fund-2	Fund-3
3	3	Fund-1	1	Fund-3	Fund-2	Fund-1
4	3	Fund-1	2	Fund-3	Fund-2	Fund-1
5	17	Fund-3	1	Decline	Fund-1	Fund-3
6	17	Fund-3	2	Fund-3	Fund-1	Fund-3
7	19	Decline	1	Fund-3	Fund-3	Decline
8	19	Decline	2	Fund-3	Decline	Decline
9	24	Fund-3	1	Decline	Fund-1	Fund-3
10	24	Fund-3	2	Fund-3	Fund-2	Fund-1

Subset of Chapter 5 – ISQC Table 8.13

Columns (6/0): Application, Classification, Review Number, Sue, Fred, John

Rows: All rows 10, Selected 0, Excluded 0, Hidden 0, Labelled 0

Low Effectiveness

The information in the bottom graph in Figure 5.38a is summarized next in the JMP output in an **Agreement Report**. As is shown in Figure 5.38c, Sue's agreement is 61.48%, Fred's agreement is 59.26%, and John's agreement is 60.74%. These numbers represent the underwriter's agreement with himself or herself and the other underwriters for a given application, summed up over all of the applications. Sue, Fred, and John have similar performance, with approximately a 60% agreement. The overall agreement is 23.33%, with 7 out of 30 matches; and with 95% confidence, this rate goes from 11.79% to 40.93%.

Figure 5.38c Agreement Report for Attribute Gauge Study

Agreement Report

Rater	% Agreement	95% Lower CI	95% Upper CI
Sue	61.4815	56.4141	66.3115
Fred	59.2593	54.1539	64.1722
John	60.7407	55.7438	65.5225

Number Inspected	Number Matched	% Agreement	95% Lower CI	95% Upper CI
30	7	23.333	11.792	40.928

The next piece of output, shown in the first part of Figure 5.38d, looks at agreement between all of the pairwise comparisons of the underwriters. It uses a Kappa statistic to quantify the level of agreement. The Kappa statistic can take on values between 0 and 1. Kappa values closer to 1 indicate agreement and Kappa values closer to 0 suggest less agreement. The Kappa statistics do not seem to support a high level of agreement between any two underwriters, with all values less than 0.5. The bottom part of the output in Figure 5.38d shows agreement for each underwriter with the standard. Since John has the highest Kappa statistic (0.73), he has the highest agreement between his classifications and the standard. Sue has the next highest Kappa (0.63) followed by Fred (0.58).

Figure 5.38d Agreement Comparison for Attribute Gauge Study

Rater	Compared with Rater	Kappa	.2	.4	.6	.8	Standard Error
Sue	Fred	0.4000					0.0840
Sue	John	0.4586					0.0842
Fred	John	0.3529					0.0865

Rater	Compared with Standard	Kappa	.2	.4	.6	.8	Standard Error
Sue	Classification	0.6339					0.0775
Fred	Classification	0.5791					0.0806
John	Classification	0.7285					0.0695

The output in Figure 5.38e appears next in the JMP output. The first table shown provides information about the agreement level for each underwriter with themselves. For example, Sue's two classifications for a given application matched each other for 23 of the 30 applications, which results in a Rater Score = 23 / 30 = 75.67%. John had the lowest Rater Score of 60% because he only matched 18 of the 30 applications. These estimates agree with the estimates presented at the top of ISQC Table 8.14. However, the 95% confidence intervals are slightly different because JMP confidence intervals are score confidence intervals, while the ones in Table 8.14 are based of the F distribution. The bottom table in Figure 5.38e, **Agreement across Categories**, shows Kappa statistics for the agreement in classification over that which would be expected by chance. All the kappa statistics are < 0.51 indicating that there is no strong agreement between the raters.

Figure 5.38e Agreement Comparison for Attribute Gauge Study

Agreement within Raters

Rater	Number Inspected	Number Matched	Rater Score	95% Lower CI	95% Upper CI
Sue	30	23	76.6667	59.0717	88.2076
Fred	30	21	70.0000	52.1242	83.3353
John	30	18	60.0000	42.3204	75.4094

Agreement across Categories

Category	Kappa	.2	.4	.6	.8	Standard Error
Decline	0.5062					0.0471
Fund-1	0.3401					0.0471
Fund-2	0.5873					0.0471
Fund-3	0.4401					0.0471
Overall	0.4628					0.0289

The final output is shown in Figure 5.38f. The **Effectiveness Report** is only available when a standard is used and provides information about the consistency between the classification and the standard. The first table shows the correct and incorrect classifications for each category and underwriter. Note these counts include the replicates for each application, using all 60 results for each underwriter. Therefore, the standard included 8 (4x2) Decline, 14 (7x2) Fund-1, 20 (10x2) Fund-2 and 18 (9x2) Fund-3. For example, Sue correctly classified 6 'Decline', 8 'Fund-1', 15 'Fund-2' and 15 'Fund-3', for a total of 44 correct classifications. She incorrectly classified 2 'Decline', 6 'Fund-1', 5 'Fund-2' and 3 'Fund-3', for a total of 16 incorrect classifications.

Figure 5.38f Effectiveness Report for Attribute Gauge Study

▼ Effectiveness Report

▼ Agreement Counts

Rater	Correct(Decline)	Correct(Fund-1)	Correct(Fund-2)	Correct(Fund-3)	Total Correct	Incorrect(Decline)	Incorrect(Fund-1)	Incorrect(Fund-2)	Incorrect(Fund-3)	Grand Total
Sue	6	8	15	15	44	2	6	5	3	60
Fred	3	9	18	12	42	5	5	2	6	60
John	7	13	15	13	48	1	1	5	5	60

▼ Effectiveness

Rater	Effectiveness	95% Lower CI	95% Upper CI	Error rate
Sue	73.3333	60.9913	82.8674	0.2667
Fred	70.0000	57.4913	80.1018	0.3000
John	80.0000	68.2182	88.1715	0.2000
Overall	74.4444	67.6080	80.2593	0.2556

▼ Misclassifications

Standard Level	Decline	Fund-1	Fund-2	Fund-3
Decline	.	0	0	2
Fund-1	0	.	11	11
Fund-2	0	5	.	1
Fund-3	8	7	1	.
Other	0	0	0	0

In the second table in Figure 5.38f, Sue's Effectiveness score is 44 / 60 = 73.33%. Based on all of the attempts, John has the highest Effectiveness rate of 48 / 60 = 80% and Fred's is the lowest, 42 / 60 = 70%. The last table in Figure 5.38f shows the misclassification rates, where the rows represent the standard and the columns contain the levels given by the underwriters. For the three underwriters, there are a total of 46 incorrect classifications, with 11 'Fund-1' incorrectly classified as 'Fund-2' and 11 'Fund-1' incorrectly classified as 'Fund-3'. I this case, a standard result of 'Fund-1' had the most misclassifications (11+11=22).

Some of the output presented in Figures 5.38a through Figure 5.38f can be found in the output shown in ISQC Table 8.14. For example, the **Within Appraisers** and **Between Appraiser** information in ISQC is also shown in Figure 5.38c and Figure 5.38e. We agree with Montgomery's conclusion, that "there is not great agreement in this study."

ISQC Example 8.8 Meeting Customer Specifications

In this example, we provide a data set to illustrate the concepts in ISQC Example 8.8. The specification limits for the length of the Final Dimension is 12 ± 0.10. The following steps illustrate how to analyze this data set using the **Distribution** platform.

1. Open Chapter 5 – ISQC Example 8.8.jmp, which has variables called *Assembly Number, Part 1, Part 2, Part 3, Part 4,* and *Final Dimension.*
2. From the main menu, select **Analyze ▶ Distribution**.
3. When the launch window appears, highlight **Part1**, **Part2**, **Part3**, **Part4**, and **Final Dimension** and click Y, **Columns**. When finished, click **OK**.
4. From the red triangle next to **Part 1**, hold the control key (command on Mac) and deselect **Display Options ▶ Quantiles**. Next, while holding the control key (command on Mac), select **Display Options ▶ Customize Summary Statistics**. When the dialog box comes up, click **Variance** and then click **OK**. The output is shown in Figure 5.39.

Figure 5.39 Output for ISQC Example 8.8

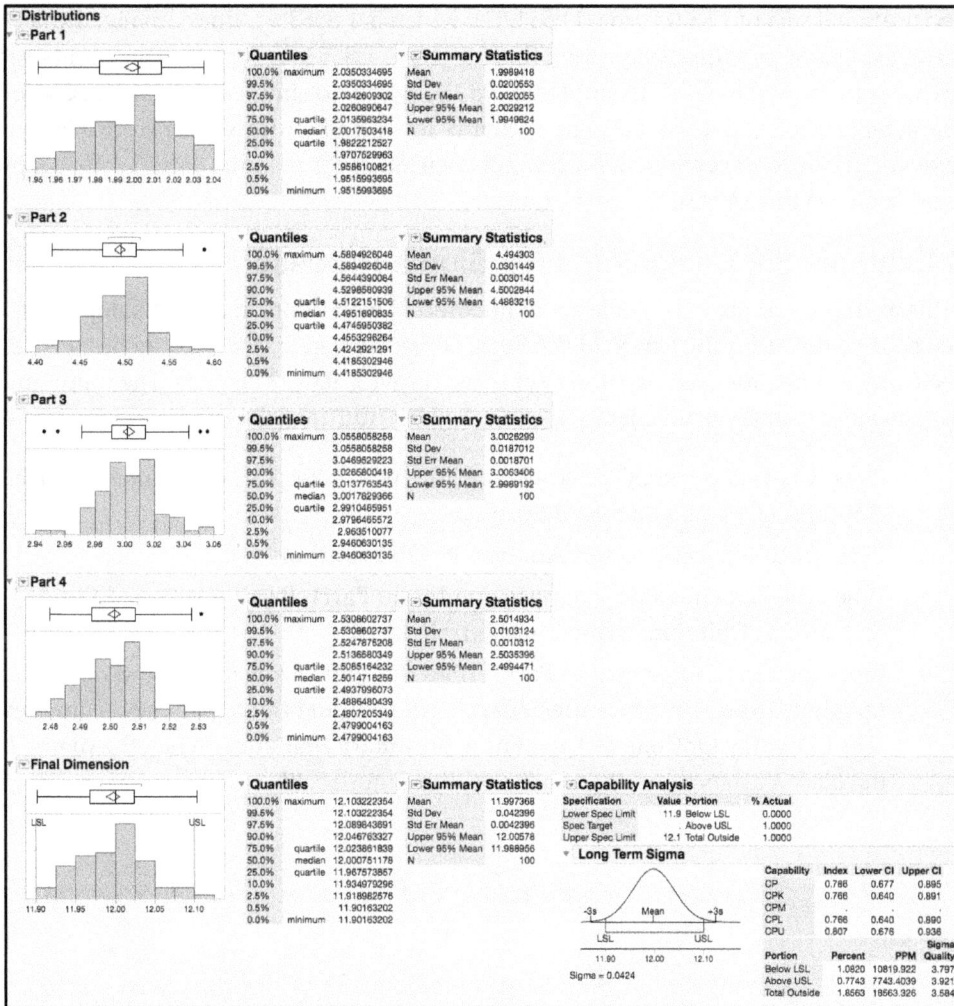

The output in Figure 5.39 includes summary statistics for Part 1 through Part 4, along with a capability analysis for the Final Dimension. This example illustrates the impact of tolerance stack-up problems when the performance of the individual components is not well controlled or specified in an optimal manner. As was discussed in ISQC Example 8.8, when four parts are assembled, the mean of the Final Dimension is the sum of the means of the four parts and the variance of the Final Dimension is the sum of the variances of the four parts. In Figure 5.39, we see that the means for Parts 1 through 4 are 1.9989, 4.4943, 3.0026 and 2.5015, respectively. The Final Dimension mean is 11.9974, as was expected. Similarly, the variance of Part 1 through Part 4 is 0.0004, 0.00009, 0.0003 and 0.0001, respectively. The Final Dimension variance is 0.0018, as is expected.

The **Capability Analysis** provides an estimate of the percent of assemblies that will be out of specification limits of 12 ± 0.10, or (11.9, 12.1), which is 1.8563%. Alternatively, the assembled parts that will all be within the specification limits is 100 – 1.8653, or 98.14%. This number is slightly off to what is presented in ISQC Example 8.8, 98.173%, due to simulated data. The C_{pk} for this parameter is 0.766 and it has '3.8 sigma' quality level. This process is far from a '6 sigma' process. However, it can be improved by reducing the variation for the four parts, while keeping them on target; which is the topic of the next example.

ISQC Example 8.9 Designing a Six Sigma Process

In this example, we provide a data set to illustrate the concepts in ISQC Example 8.9. The specification limits for the length of the Final Dimension are 5.00 ± 0.06 inches. Process capability specification limits for each part were set to its Target ± 0.01732 inches. The following steps illustrate how to analyze this data set using the **Distribution** platform.

1. Open Chapter 5 – ISQC Example 8.9.jmp, which has variables called *Assembly Number, Part1, Part2, Part3,* and *Final Dimension*.
2. From the main menu, select **Analyze ▶ Distribution**.
3. When the launch window appears, highlight **Part1, Part2, Part3**, and **Final Dimension** and click Y, **Columns**. When finished, click **OK**.
4. From the red triangle next to Part 1, hold the control key (command on Mac) and deselect **Display Options ▶ Quantiles**. Next, while holding the control key (command on Mac), select **Display Options ▶ Customize Summary Statistics**. When the dialog box comes up, click **Variance** and then click **OK**. See Figure 5.40.

Figure 5.40 Output for ISQC Example 8.9

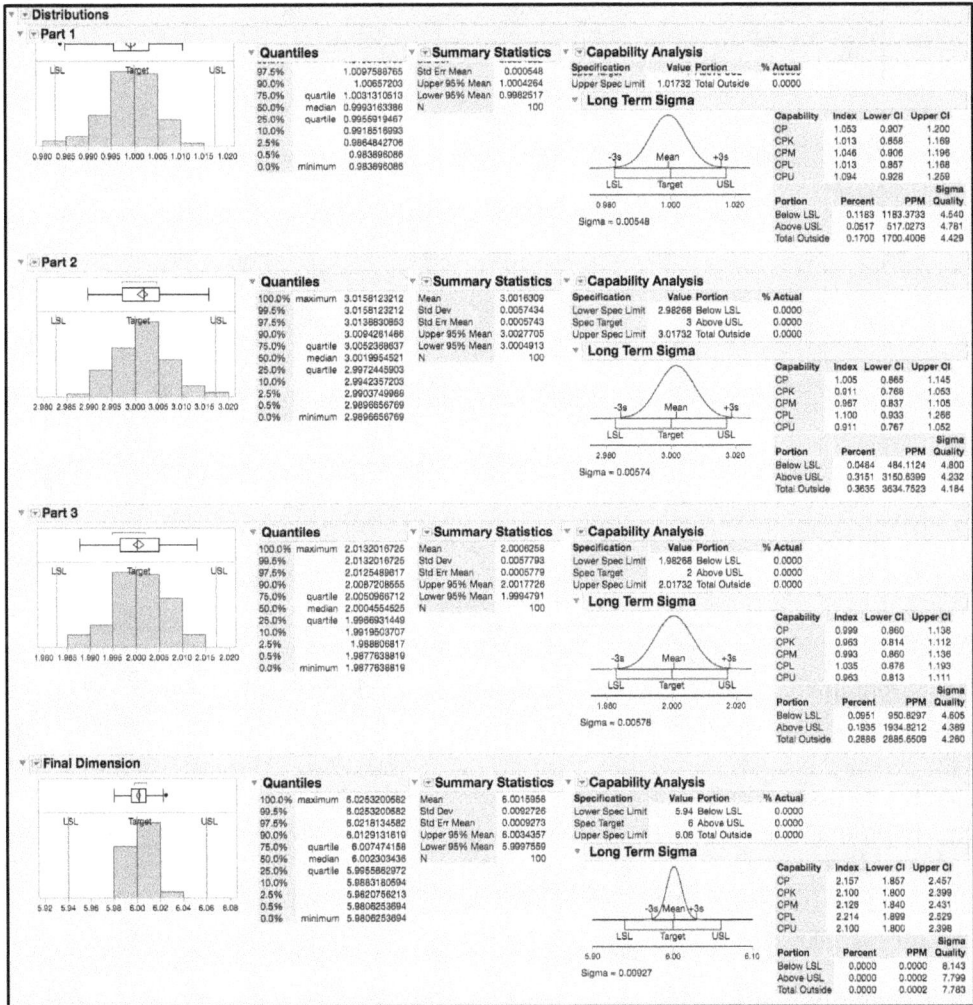

5. Under the **Final Dimension** output, from the red triangle next to **Capability Analysis** select **Capability Animation**. The Capability Animation window will appear. Shift the process mean by dragging the square on top of the blue normal distribution to the right, until it says '**Mean Shift 2.5σ**'. Note the LSL and USL can also be moved in the same manner.

Figure 5.41 Capability Animation for Final Dimension

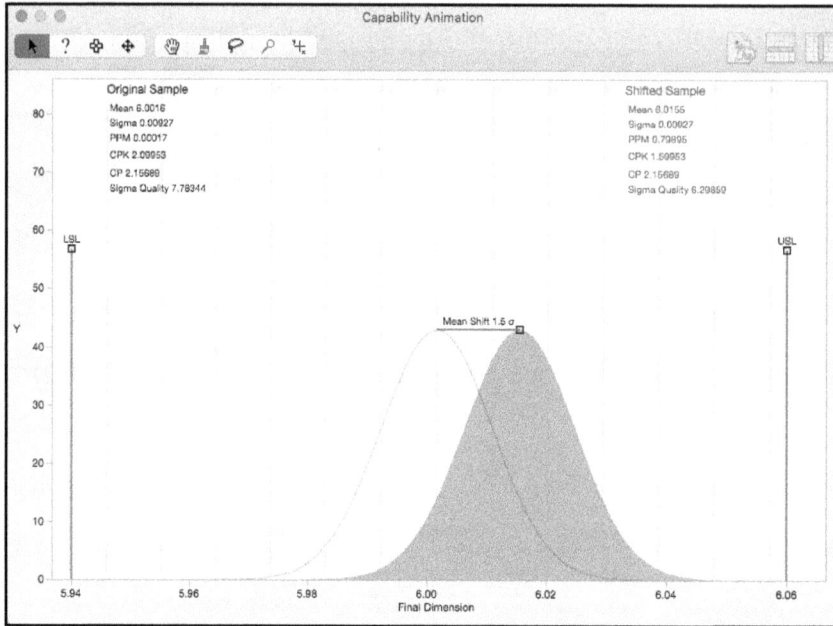

The output in Figure 5.40 shows the summary statistics for Part 1, Part 2, Part 3, and the Final Dimension, along with their Capability Analysis. In ISQC Example 8.9, it was shown how to determine the maximum value for the variance of each part that is required to achieve a very high quality process for the final assembly. It was determined that if $\sigma^2_{final} = 0.010^2$ is required for a high quality product then the variance for each of the three components (σ^2_{part}) can be no greater than 0.000033 ($\sigma = 0.00574$). The summary statistics show that the standard deviations for Part 1, Part 2, and Part 3 are 0.00548, 0.00574 and 0.00578, respectively. The standard deviation for the final assembly is 0.00927.

The **Capability Analysis** also shows that C_p is close to 1 for all three parts. For the Final Dimension, C_p is 2.157, which is at better than a '6 sigma' level of quality. **Capability Animation** is a very nice tool in JMP that allows the mean and / or the specification limits to be reset in order to see the impact on the performance indices. The graph in Figure 5.41 shows the impact of shifting the mean of the Final Dimension from 5.00 to 5.025, which is a 2.5σ shift toward the USL. In practice, it is not unusual to see a drift in the mean over time. Even with this shift, the process is still operating at a $C_{pk} = 1.6$, which is a '6.3 sigma' level of quality. This example illustrates the importance of tolerance stack up and the need to control the quality of all of the individual components to achieve a '6 sigma' level of quality.

ISQC Example 8.12 Constructing a Tolerance Interval

In this example, we provide a data set to illustrate the concepts in ISQC Example 8.12. The following steps illustrate how to analyze this data set using the **Distribution** platform.

1. Open Chapter 5 – ISQC Example 8.12.jmp, which has variables called *Sample Number* and *Response*.
2. From the main menu, select **Analyze ▶ Distribution**.
3. When the launch window appears, select **Response** and click on **Y, Columns** and click **OK**.
4. Click on the red triangle next to **Response** and select **Tolerance Interval** from the drop-down menu. In the Tolerance Intervals dialog box, enter **0.99** in the field next to **Specify Proportion to cover**. Click on radio button for **Two-sided** (Figure 5.42).

Figure 5.42 Dialog Box for Tolerance Intervals

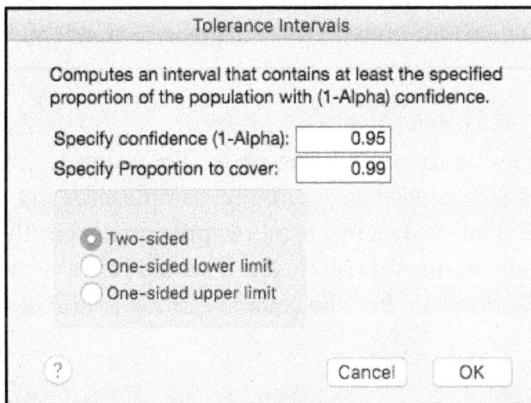

5. Click **OK** when finished.

Figure 5.43 Tolerance Interval for Solid-Fuel Rocket Propellant

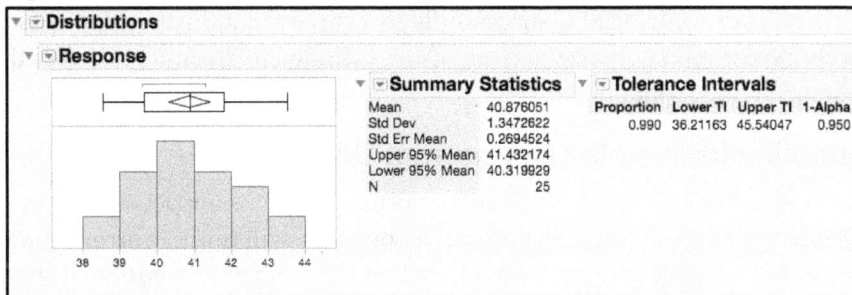

In Figure 5.43 we see that the mean and standard deviation for this data are 40.876 and 1.347, respectively. The tolerance interval is (35.212, 45.540), with 95% confidence (1-Alpha) and 99% coverage (Proportion). We expect 99% of future results to fall between 35.212 and 45.540. The "k" multiplier used in this calculation is 3.46. The value of k=2.972 used in ISQC Example 8.12 is the one corresponding to 99% confidence and 95% coverage. For 95% confidence and 99% coverage, the value from ISQC Appendix VII is 3.457, in close agreement with the k value in JMP.

Statistics Note 5.3: Howe (1969) gives an approximation for the two-sided tolerance interval multiplier k when the data can be approximated by a normal distribution. The k constant for a $(1-\alpha)\%$ confidence and $\gamma\%$ coverage, for a sample of size n, can be approximated by

$$\sqrt{\frac{(n-1)\left(1+\dfrac{1}{n}\right)z_{\frac{1-\gamma}{2}}^2}{\chi_{1-\alpha,\,n-1}^2}}$$

Where $z_{1-\gamma/2}$ is the $1-\gamma/2$ quantile of the normal distribution, and $\chi_{1-\alpha,\,(n-1)}^2$ is the $1-\alpha$ quantile of the Chi-Square distribution with (n-1) degrees-of-freedom.

In industry, statistical tolerance intervals are often used to set process capability based specification limits. The sample size can influence the coverage used. For example, for smaller sample sizes (n ≤ 30), a coverage value of 99% is used and for larger sample sizes (n ≥ 75), a coverage value of 99.73% coverage, equivalent to 3 sigma, is more appropriate. For this example, setting the specifications to the 99% tolerance interval (36.21, 45.54) would result in an initial $C_{pk} = k/3 = 3.46/3 = 1.153$. If a larger initial C_{pk} is desired, then the coverage of the tolerance interval should be increased.

Statistical Insights

In this section, we elaborate upon some of the examples provided in ISQC Chapter 8. The examples highlighted in this section include several important concepts we have encountered over our many years of applying SPC successfully to a variety of industries. For most of these examples, additional output not provided in ISQC is included to illustrate JMP functionality or further elaborate on important points.

Process Capability Indices for Nonnormal Data

There is a discussion in ISQC Section 8.3.3 about normality and its impact on the interpretation of C_p and C_{pk}. More specifically, a large departure in normality will result in errors in the interpretation of the expected process fallout based on the normal distribution. It should be noted that published tables of the expected process fall out for different values of C_p or C_{pk} are most likely based on the normal distribution and therefore, assume that this is an appropriate distribution for the data at hand. In JMP, this information is found in the Capability Analysis section, which is labeled: **Portion Below LSL**, **Above USL** and **Total Outside**.

There are several suggestions provided in ISQC Section 8.3.3 for computing process capability indices for nonnormal data. These include computing PCIs in a transformed scale that achieve normality, using Luceños's C_{pc} index (see ISQC equation 8.10) or using quantile-based formulas using a more appropriate distribution for the data (see ISQC equation 8.11). We will explore some of these options in this section using the **Distribution** platform in JMP. The data shown in ISQC

Table 6.8 in Example 6.6 will be illustrated first. The data set includes Resistivity readings for 25 silicon wafers and the natural log of each result.

1. Open Chapter 5 – ISQC Table 6.8.jmp, which has variables called *Sample, Resistivity* and *log(Resistivity)*.
2. From the main menu, select **Analyze ▶ Distribution**.
3. When the launch window appears, select **Resistivity** and **log(Resistivity)** in the left hand box and then click **Y, Columns**. Click **OK** when finished.
4. From the red triangle next to the **Resistivity** label, hold the control key (command on Mac) and deselect **Display Options ▶ Quantiles** and then deselect **Outlier Box Plot**.
5. From the red triangle next to **Resistivity** select **Continuous Fit ▶ Normal** (Figure 5.44a) and then select **Continuous Fit ▶ LogNormal** (Figure 5.44b). From the red triangle next to **log(Resistivity)** select **Continuous Fit ▶ Normal** (Figure 5.44c).

Figure 5.44a Process Capability Results for Resistivity Normal Distribution

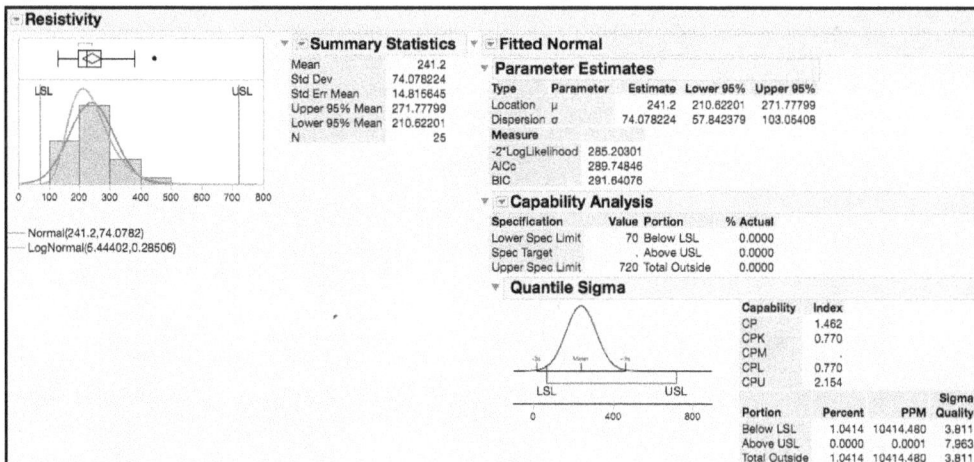

Figure 5.44b Process Capability Results for Resistivity Lognormal Distribution

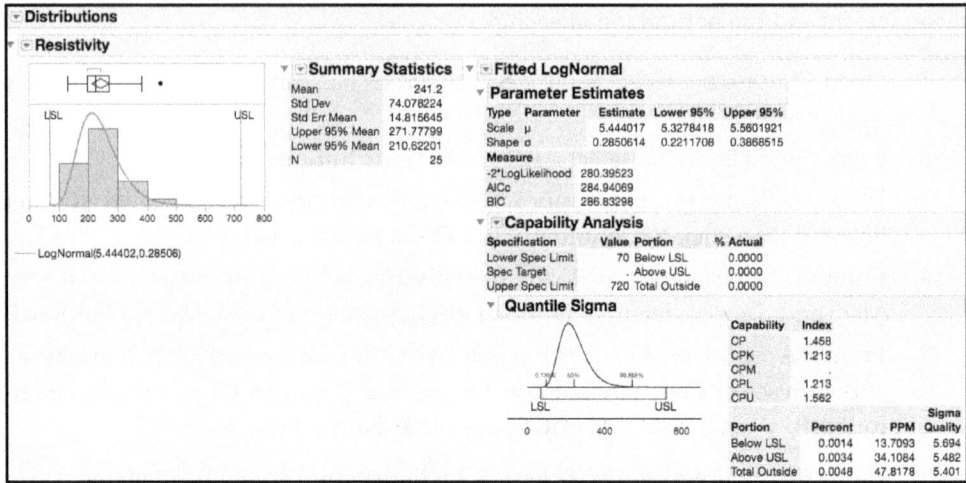

Figure 5.44c Process Capability Results for log(Resistivity) Normal Distribution

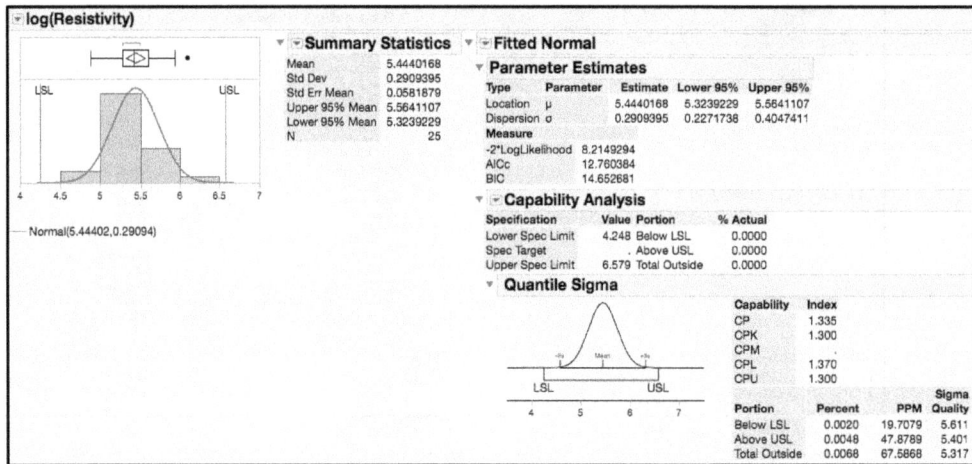

In order to carry out a capability analysis for this data, we arbitrarily set LSL = 70 and USL = 720. The first part of the output shown in Figure 5.44a and Figure 5.44b show the data and capability analyses for the original Resistivity results, while Figure 5.44c shows the analysis for the natural log transformed results.

Both a normal (Figure 5.44a) and lognormal distribution (Figure 5.44b) was fitted to the Resistivity results. Since the specification values were set in the Column Properties, the capability analysis is automatically produced by JMP. The capability indices shown are calculated using the quantile-based approach described in ISQC equation 8.11. The normally derived C_p = 1.462 and C_{pk} = 0.770, while the lognormal C_p = 1.458 and C_{pk} = 1.213. The portion outside of the specification limits for the normal and lognormal distributions is 1.04% and 0.0048%, respectively. Finally, the

Sigma Quality for each distribution is '3.811 sigma' and '5.401 sigma', for the normal and lognormal, respectively.

Statistics Note 5.4: To compute process capability indices for nonnormal data, one can use the quantiles from a scientifically justified probability distribution that approximates the data well.

These two analyses for silicon wafer Resistivity result in two very different conclusions regarding the state of the process. The normal based analysis concludes that the process is significantly off-center and results in an unacceptable capability at the LSL, while the lognormal analysis suggests that capability is greater than 1 and is performing closer to a '6 sigma' level of quality.

The analysis for the natural log transformed data in Figure 5.44c assumes that the transformation resulted in normally distributed data. The capability estimates are $C_p = 1.335$ and $C_{pk} = 1.300$. For this data, these results are similar to the capability indices obtained using the lognormal quantiles. However, this might not always be the case and there might be differences in the two estimates. We prefer, whenever possible, to select an appropriate distribution that approximates the data and makes sense, scientifically, and use the fitted quantiles to estimate the capability indices. The **Distribution** platform in JMP calculates capability indices for the continuous distributions (for example, normal, lognormal, gamma, weibull, and exponential) as well as the discrete distributions (Poisson, Gamma Poisson, Binomial, and Beta Binomial).

JMP Note 5.6: The **Process Capability** platform can be used to compute capability indices for nonnormal data. One can specify the distribution as well as the type of computation for capability indices, either quantile based, or z-score based.

We can also use the **Process Capability** platform to perform a capability analysis for Resistivity using the lognormal distribution.

1. Open Chapter 5 – ISQC Table 6.8.jmp, which has variables called *Sample, Resistivity* and *log(Resistivity)*.
2. Select **Analyze ▶ Quality and Reliability ▶ Process Capability**.
3. Select **Resistivity** as the **Y, Process** variable. From the **Distributions Options** select **Lognormal** under **Distribution** as shown in Figure 5.45. In the **Distribution Options**, select **Resistivity** and click the **Set Process Distribution** button. The dialog window now shows Resistivy&Dist(Lognormal) to indicate that the lognormal distribution is being applied to Resistivity (Figure 5.45). Leave the default value of **Percentiles** under **Nonnormal Capability Indices Method**. Click **OK**.

Figure 5.45 Process Capability Platform Dialog Window for Resistivity

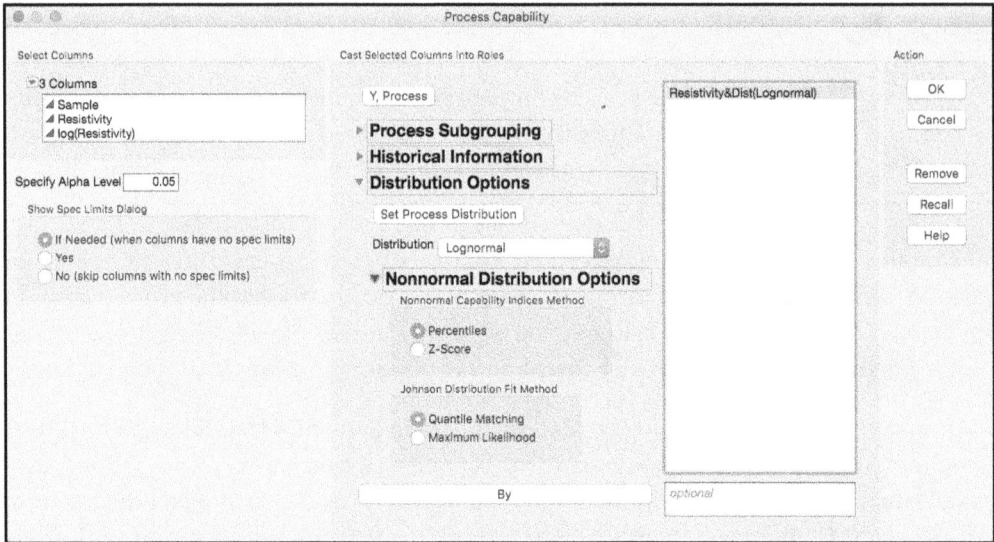

4. To generate a report with the capability indices, click on the red triangle next to **Process Capability** and select **Individual Detail Reports.** The output is shown in Figure 5.46. These results are the same as the ones in Figure 5.44b, and similar to those in Figure 5.44c.

Figure 5.46 Process Capability Output for Resistivity with Lognormal Distribution

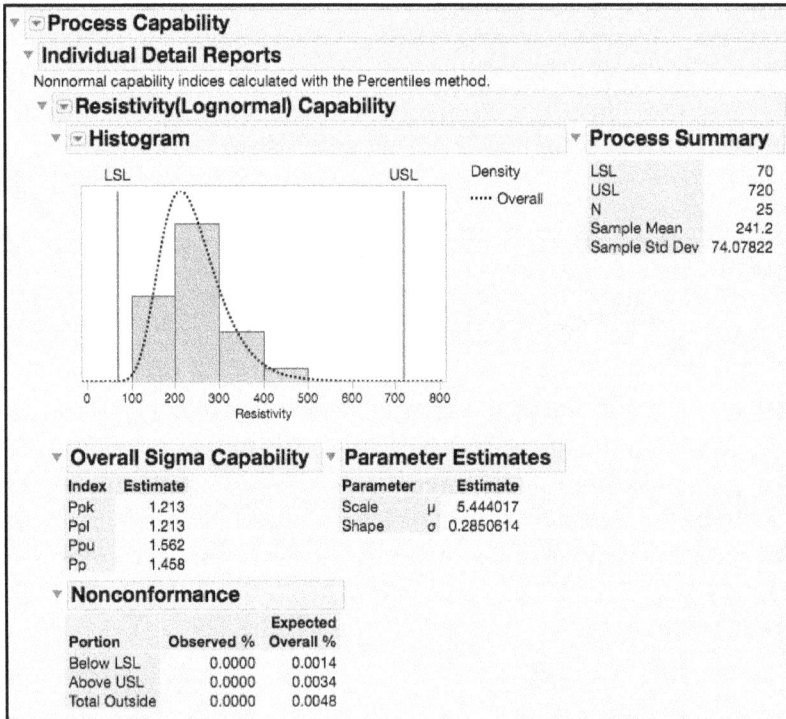

JMP Note 5.7: The distribution for the **Process Capability** platform can be pre-specified as a column property.

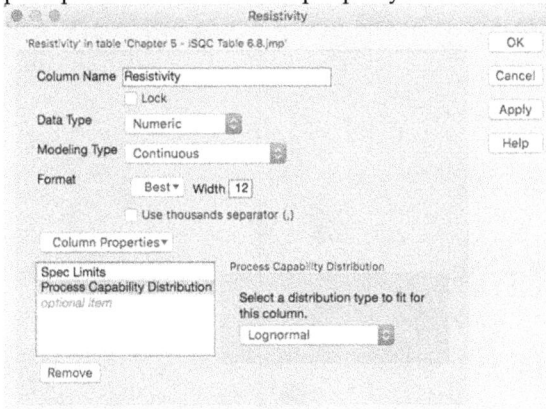

Process Capability Indices and Sample Size

Confidence intervals for C_p and C_{pk} are presented in ISQC Section 8.3.5. Since these are point estimates, similar to a sample mean or sample standard deviation, they contain sampling error. Therefore, these estimates should take into account the sampling error, as well as the sample size. Equations 8.20 and 8.21 in ISQC provide $1-\alpha$ confidence intervals for C_p and C_{pk}, respectively. Both of these equations incorporate the effect of sample size, n, directly in the calculations, as well as through the selection of the chi-square quantiles in equation 8.20.

Figure 5.47 shows the width of the 95% confidence interval as a function of sample size, for a $C_{pk} = 1.33$, using ISQC equation 8.21. For a sample size of 20, the 95% confidence interval is [0.88, 1.78] (ISQC Example 8.5), with a width of ±0.45. As the sample size increases the width of the interval decreases. However, even for a sample size of 100 the 95% confidence interval for a $C_{pk} = 1.33$ is [1.13, 1.53], with a width of ±0.195. This means that with 95% confidence the true C_{pk} can be as low as 1.13, not as good of a situation, or as large as 1.53. We need a sample size of 400 to have an interval width of ±0.1, or a 95% interval of [1.23, 1.43].

Figure 5.47 95% Confidence Interval Width for $C_{pk} = 1.33$ versus Sample Size

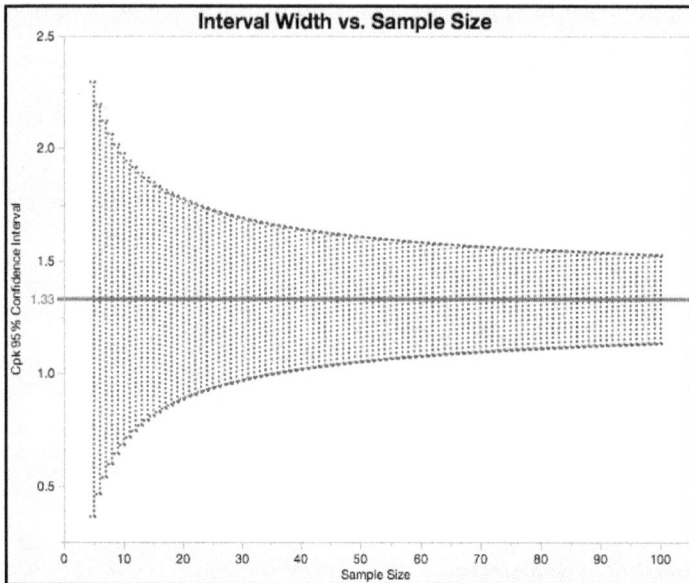

EMP for Measurement System Analysis

Earlier in this chapter, we analyzed the data in ISQC Table 8.7 using the Gauge R&R MSA Method in the **Measurement System Analysis** platform. In this section, we would like to illustrate some of the functionality available in JMP using the EMP MSA Method.

1. Open Chapter 5 – ISQC Table 8.7.jmp, which has variables called *Part Number, Operator, Test Number* and *Thermal Impedance*.
2. Select **Analyze ▶ Quality and Process ▶ Measurement Systems Analysis**.
3. When the launch window appears, select **Thermal Impedance** and click **Y, Response** and then select **Part Number** and click **Part, Sample ID** and select **Operator** and click **X, Grouping**. Select **EMP** for **MSA Method** and **Crossed** for **Model Type**.
4. When finished, click **OK**.

Figure 5.48 Default Output for EMP MSA Method

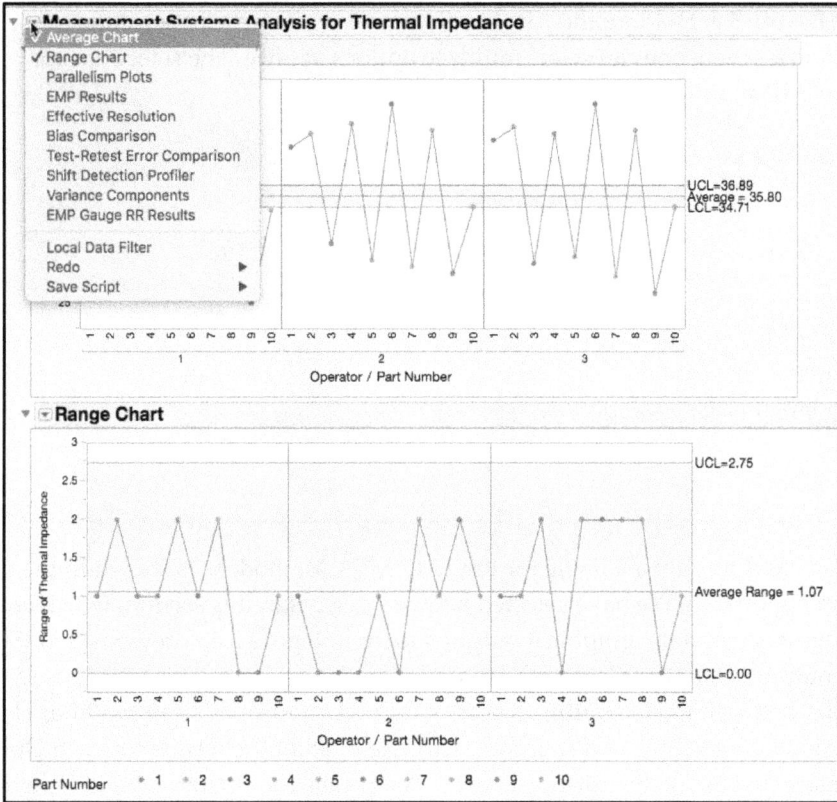

5. Click on the red triangle next to **Measurement System Analysis for Thermal Impedance** (Figure 5.48). Select **Bias Comparison**, **Test-Retest Comparison**, **EMP Results**, and **Effective Resolution**. The output is shown in Figure 5.49a through Figure 5.49d.

JMP Note 5.8: Holding the Alt (Mac Option key) launches a selection window where one can select multiple options at once. The selections for the EMP analysis in step 5 are shown below.

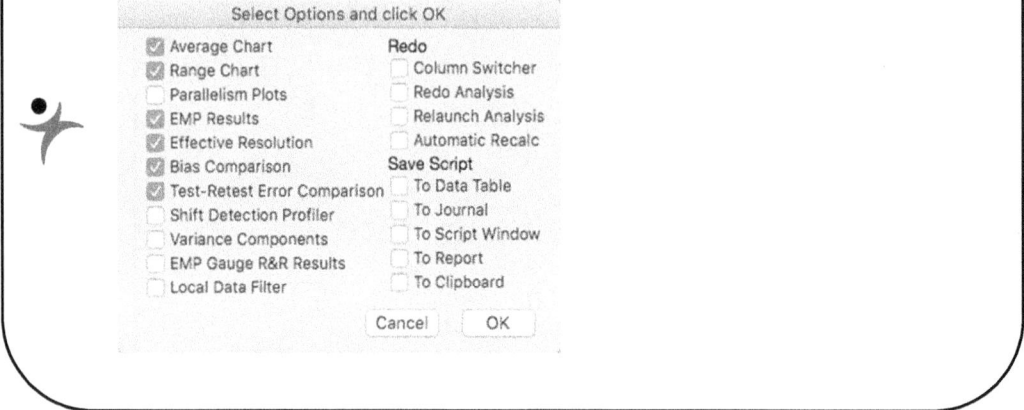

Select Options and click OK

☑ Average Chart	Redo
☑ Range Chart	☐ Column Switcher
☐ Parallelism Plots	☐ Redo Analysis
☑ EMP Results	☐ Relaunch Analysis
☑ Effective Resolution	☐ Automatic Recalc
☑ Bias Comparison	Save Script
☑ Test-Retest Error Comparison	☐ To Data Table
☐ Shift Detection Profiler	☐ To Journal
☐ Variance Components	☐ To Script Window
☐ EMP Gauge R&R Results	☐ To Report
☐ Local Data Filter	☐ To Clipboard

Cancel   OK

Many additional outputs are available for the EMP MSA Method, as is shown in the menu highlighted in Figure 5.48. We have selected a few to discuss in this section, which we find very useful. Note we discussed the graphical output shown in Figure 5.48 previously in ISQC Example 8.7. The first output is the **Bias Comparison**, shown in Figure 5.49a. This plot is used to determine if there is a bias between the Operators. It consolidates the information in the XBar chart in Figure 5.48. It uses Analysis of Means and charts the average of all of the results for each Operator. The limits are constructed using the test-retest error. The means for Operator 1 and Operator 2 plot outside of the decision limits, indicating that their mean difference is significantly larger than the test-retest error. Note in this analysis, Operator is treated as a fixed effect, where emphasis is placed on understanding and removing operator differences and therefore, improving the measurement process.

Figure 5.49a Bias Comparison for Thermal Impedance

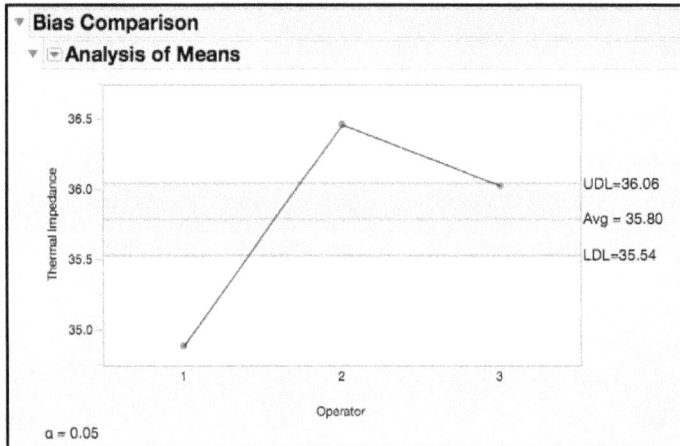

Statistics Note 5.5: An analysis of means is a graphical decision tool for comparing a set of averages or variances with respect to their overall average. You can think of it as a control chart but with decision limits instead of control limits, or as an alternative to an analysis of variance (ANOVA). Prof. Ellis Ott introduced the analysis of means in 1967 as a logical extension of the Shewhart control chart.

The plot in Figure 5.49b is used to determine if the test-retest error is similar among the operators. It consolidates the information found in the Range Chart in Figure 5.48. It also uses Analysis of Means to determine if the average Range for any of the operators is statistically different. Since all of the points are within the control limits, we can assume that the test-rest error for each operator is similar.

Figure 5.49b Test-Retest Error Comparison for Thermal Impedance

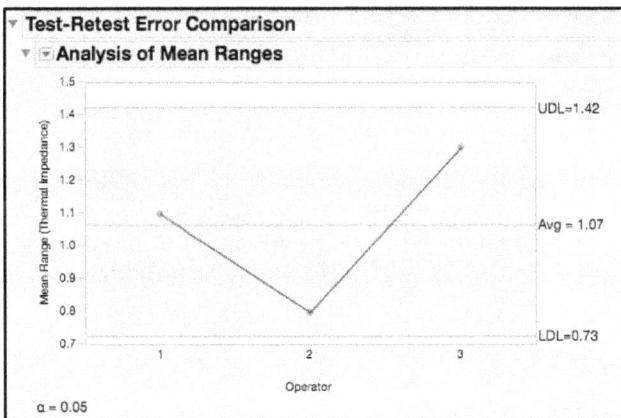

The EMP Results section is shown in Figure 5.49c. As was described above, Operator and Part*Operator are not included in the measurement error. The measurement error in this study is estimated using the repeat measurements, where the same operator measured the same part multiple times. In Figure 5.49c, an estimate of the test-retest error is 0.6302 and has 54.7 degrees of freedom. The probable error is 0.67σ test-retest = 0.67(0.6302) = 0.4251. This is used to describe the error associated with a single measurement; for example, the thermal impedance reading is 35.0 ± 0.4251. The intraclass correlation (no bias) is the ratio of part-to-part variation over the total variation, using all of the data and the intraclass correlation (with bias) is the same ratio, but with operator bias removed The bias impact is the difference in the two intraclass correlation estimates. Finally, the EMP MSA method provides a measurement system classification, which is based on the intraclass correlation coefficient. The thermal impedance measurement system has a 'First Class' classification, which is the best performance. Definitions for the four classes are shown under the **Monitor Classification Legend**.

Figure 5.49c EMP Results for Thermal Impedance

▼ **EMP Results**

EMP Test	Results	Description
Test-Retest Error	0.6302	Within Error
Degrees of Freedom	54.7	Amount of information used to estimate within error
Probable Error	0.4251	Median error for a single measurement
Intraclass Correlation (no bias)	0.9874	Proportion of variation attributed to part variation without including bias factors
Intraclass Correlation (with bias)	0.9673	Proportion of variation attributed to part variation with bias factors
Bias Impact	0.0201	Amount by which the bias factors reduce the intraclass correlation

System	Classification
Current (with bias)	First Class
Potential (no bias)	First Class

▼ **Monitor Classification Legend**

Classification	Intraclass Correlation	Attenuation of Process Signal	Probability of Warning, Test 1 Only*	Probability of Warning, Tests 1-4*
First Class	0.80 - 1.00	Less than 11%	0.99 - 1.00	1.00
Second Class	0.50 - 0.80	11% - 29%	0.88 - 0.99	1.00
Third Class	0.20 - 0.50	29% - 55%	0.40 - 0.88	0.92 - 1.00
Fourth Class	0.00 - 0.20	More than 55%	0.03 - 0.40	0.08 - 0.92

* Probability of warning for a 3 standard error shift within 10 subgroups using Wheeler's tests, which correspond to Nelson's tests 1, 2, 5, and 6.

The last piece of output shown is the **Effective Resolution** of the measurement system. This information is used to determine the recorded precision for a measurement. The current measurement increment shown in Figure 5.49d is 1 and can be verified in ISQC Table 8.7. The smallest and largest effective increments, 0.0935 and 0.9351, provide a bound for measurement precision. For example, there is no additional value in recording data below 0.0935. If we round this to 0.1, then the data should not be recorded past the first decimal place. The largest effective increment is 0.9351, which suggests that the recorded data precision should be increased.

Figure 5.49d Effective Resolution Results for Thermal Impedance

▼ **Effective Resolution**

Source		Value	Description
Probable Error	(PE)	0.4251	Median error for a single measurement
Lower Bound Increment	(0.1*PE)	0.0425	Measurement increment should not be below this value
Smallest Effective Increment	(0.22*PE)	0.0935	Measurement increment is more effective above this value
Largest Effective Increment	(2.2*PE)	0.9351	Measurement increment is more effective below this value
Current Measurement Increment	(MI)	1	Measurement increment estimated from data (in tenths)

Action: Consider adding a digit
Reason: The measurement Increment of 1 is greater than the largest effective measurement increment and might need to be adjusted to record more digits.

Statistics Note 5.6: There are many different ways to analyze data from a measurement system study. Some analyses view 'operator' effects as part of the noise of the measurement system, while others, treat it as an experimental factor that can be studied and improved upon.

Chapter 6: Process Health Assessment

Overview

This chapter illustrates how to assess the health of a process by evaluating the stability and capability of the process parameters. The concepts will be illustrated using the **Process Screening** platform in JMP, which was introduced in version 13, and examples based on a composite data set from a number of sources. The techniques presented in this chapter apply to measurements using a continuous scale.

Process Health Assessment Review

In this chapter, we illustrate techniques to carry out a *process health assessment* or PHA. The inspiration for this chapter is the addition of the **Process Screening** platform, introduced in JMP version 13. There is not a one-to-one connection between this chapter and a specific chapter in ISQC 7th edition, like there is for the other chapters in this book. However, many ideas found in this chapter were discussed in ISQC chapters. For example, SPC for variables data was presented in ISQC Chapter 6 and Process Capability was discussed in ISQC Chapter 7.

There are two key dimensions to assessing the health of a process which include, evaluating the capability and the stability of the process parameters. In order to evaluate the process capability, a parameter must have a specification or performance limits. As was previously discussed in Chapter 5 in this book (ISQC Chapter 8), process capability indices, such as P_p or P_{pk} are often used to determine how well a parameter meets its specification limits. A capable process is one that has a high probability of meeting the specification and produces no, or very little defective units. ISQC Chapter 8 defines a capable process as one with $P_{pk} > 1$.

Statistics Note 6.1: Process capability indices have two notations depending on how the standard deviation, sigma, is estimated. If sigma is estimated using a long-term estimate of variation, then the index uses P (performance), giving P_p or P_{pk}. The formula for P_p is

$$P_p = \frac{\text{USL-LSL}}{6s}, \text{ where } s \text{ is the sample standard deviation.}$$

If sigma is estimated using a short-term variation estimate, the index uses C (capability), giving C_p or C_{pk}. The formula for C_p is

$$C_p = \frac{\text{USL-LSL}}{6\overline{R}/d_2}, \text{ where } \overline{R}/d_2 \text{ is the short-term estimate from a XBar and R}$$

chart. For an IR the short-term is estimated using $\overline{MR}/d_2$.

See also Section 8.3 in ISQC Chapter 8.

Process stability is traditionally assessed by looking for violations of runs tests, such as, Western Electric Rules, on a Shewhart control chart. If runs tests violations are present, then the process is deemed to be unstable, otherwise the process is stable. While this approach has a long history of use and is straight forward, it has several limitations. For example, examining the stability of many parameters would be difficult since it would require someone to view and examine many control charts and summarize their outcome. An even bigger challenge relates to the false alarms associated with the use of these tests. Britt, Ramírez, and Mistretta (2016) quantify the false signaling rates for Nelson rules 1 through 4, when n = 30 to n = 100 subgroups are plotted on an XmR chart. They point out that the false signaling rate is as high as 50% for n=100.

Process stability metrics were proposed by Ramírez and Runger (2006) to quantify the stability of a parameter. One such metric is the Stability Ratio (SR); which is a measure of the long-term to short-term variation, $SR = \sigma^2_{\text{long-term}} / \sigma^2_{\text{short-term}}$. The long-term variation can be estimated using the overall sample variation, s^2, using all the data, while the short-term variation is estimated from the within-subgroup variation that is used to construct the control limits in a Shewhart chart. For example, for an XmR chart, the estimate of short-term variation is given by $\overline{MR}/d_2$ and the stability ratio $SR = s^2/\overline{MR}/d_2$. If the process is stable, operating with common cause variation, then these two estimates of variation will be similar and SR will be close to 1. However, if the process is unstable, operating with special cause variation, then the long-term variation will be larger than the short-term variation and SR > 1. Ramírez and Runger use an approximate test of significance based on the F-distribution to classify the performance of a parameter as *stable* or *unstable*. The performance of the SR Test is documented in Britt, Ramirez, and Mistretta for random noise and step shifts in the mean using an XmR chart for a different number of subgroups from n = 30 to n = 100. As of JMP version 13, the SR is classified, by default, using a value of 1.5 for all sample sizes. If

SR < 1.5 then the parameter is stable; otherwise it is unstable. Section 6.4 of ISQC Chapter 6 also discusses the impact of process instability on estimates for sigma.

For a given parameter, the Stability Ratio can be combined with P_{pk} to determine the health of the process. The outcome of the two assessments results in four distinct process states. In theory, a continuous improvement strategy would accompany each of the four process states. An ideal situation occurs when the parameter is classified as capable (P_{pk} > 1.33) and stable (SR <1.5). Since the process is meeting customer requirements in a predictable manner, the focus is on maintaining this state of performance. In contrast, the least ideal situation occurs when the parameter is incapable (P_{pk} < 1.33) and unstable (SR >1.5). Some suggest that the process should first be brought into a state of control and then re-engineered to meet specifications. Another scenario occurs when the parameter is incapable (P_{pk} < 1.33) and stable (SR <1.5). This might result in a potential yield issue. In this scenario, the process is performing to its potential but might require extra resources to meet customer demand. Finally, the parameter might be capable (P_{pk} > 1.33) and unstable (SR >1.5). Unfortunately, this scenario is often ignored, since the product is within specification. However, in this situation, an effort should be made to reduce variation and increase the predictability of the process. In ISQC Section 7.5, Montgomery provides actions to taken to improve the process for the four process states.

JMP Platforms for Process Health Assessments

The **Process Screening** platform is used to assess the process capability and process stability of multiple process parameters. In addition, the Stability Ratio is available in the **Process Capability** and the **Control Chart Builder** platforms. These platforms were introduced in Chapter 2. In this chapter, we focus on the use of these platforms for continuous data. Table 6.1 provides a summary of the features we find most useful in each one.

Table 6.1 Overview of JMP Features for Process Health Assessment Platforms

JMP Feature	Process Screening	Process Capability	Control Chart Builder
Process Stability Metrics	Stability Ratio, Control Chart Alarm Rates and Alarm Counts, Latest Alarm.	Stability Ratio	Stability Ratio
Process Capability Metrics	P_{pk}, C_{pk}, Out of Spec count and rate, Spec Centered Mean, Spec Scaled Standard Deviation	C_{pk}, C_{pl}, C_{pu}, C_p, C_{pm}, P_{pk}, P_{pl}, P_{pu}, P_p, Non-conformance (actual and estimated)	C_{pk}, C_{pl}, C_{pu}, C_p, C_{pm}, P_{pk}, P_{pl}, P_{pu}, P_p, Non-conformance (actual and estimated).
Process Performance Graph	Yes	Yes (called Process Performance Plot)	No

JMP Feature	Process Screening	Process Capability	Control Chart Builder
Control Charts	Thumb nail of charts in platform or link to **Control Chart Builder**.	N/A	XmR, Xbar & Range, Xbar & Std Dev, 3-Way Chart.
Process Capability Graphs	Goal Plot in platform or link to **Process Capability**.	Goal Plot, Capability Box Plots, Capability Index Plot, Normalized Box Plots, Histograms overlaid with specifications and normal distribution.	Histograms overlaid with specifications and normal distribution.
Number of Parameters	One or many	One or many	One
Links to other platforms	Process Capability & Control Chart Builder	N/A	N/A
Column Properties	Requires Spec Limits.	Spec Limits are preferred but not required.	Requires Spec Limits.
Process shifts and drifts	Metrics and graphs provided to detect a step change or drift in the process mean, using other techniques.	N/A	N/A

Examples for Chapter 6

As was noted previously, there is not a one-to-one connection between this chapter and a corresponding chapter in ISQC. Therefore, the data presented here is a composite from data in Ramirez (2016), exercise data sets from ISQC Chapter 10 and simulated data. We loosely refer to this JMP table name as Semiconductor. The specification limits for each process variable were set to illustrate a given process state. The same JMP table will be used in all examples presented in this chapter. However, the specific parameters were selected to illustrate different concepts and highlight important features of the **Process Screening** platform.

Table 6.2 Overview of Process Health Assessment Examples

JMP Table Name	Column Name	JMP Platform	Key Points
Chapter 6 – Semiconductor 6.1	E	Control Chart Builder	Perform PHA for a stable and capable parameter
Chapter 6 – Semiconductor 6.1	D	Control Chart Builder	Perform PHA for an unstable and capable parameter
Chapter 6 – Semiconductor 6.1	A	Control Chart Builder	Perform PHA for a stable and incapable parameter

JMP Table Name	Column Name	JMP Platform	Key Points
Chapter 6 – Semiconductor 6.1	J	Control Chart Builder	Perform PHA for an unstable and incapable parameter
Chapter 6 – Semiconductor 6.1	A through J	Process Screening	Perform PHA for ten parameters; graphical and statistical output.
Chapter 6 – Semiconductor 6.1	A through J	Process Capability	Perform PHA for ten parameters and show key features of this platform.

Process Health Assessment Using the Control Chart Builder

In this section, we will show how to carry out a process health assessment for a single parameter using the **Control Chart Builder**. The JMP data table contains 10 parameters to assess the overall health of this process. Each parameter has a specification that represents an internal release limit. The data includes one hundred measurements and is assumed to represent 4 months of manufacturing output. For each parameter, a subgroup size of n = 1 is appropriate and therefore, an XmR chart is used for monitoring purposes. We will illustrate the analysis for four parameters (E, D, A and J) in Chapter 6 – Semiconductor 6.1 and discuss their outcomes.

1. Open the JMP table Chapter 6 – Semiconductor 6.1.jmp, which has variables called *Sample ID, A, B, C, D, E, F, G, H, I,* and *J.* In this table, Sample ID is the subgroup variable, and A through J are the process parameters.

2. Double click on column E. The **Column Information** window will appear. Select **Column Properties ▶ Spec Limits**. Ensure that the specification limits are included: LSL = 0.5, Target = 1, USL = 1.5; otherwise enter them in the appropriate fields and then click **OK**. (see Figure 6.1).

Figure 6.1 Column Information for Parameter E

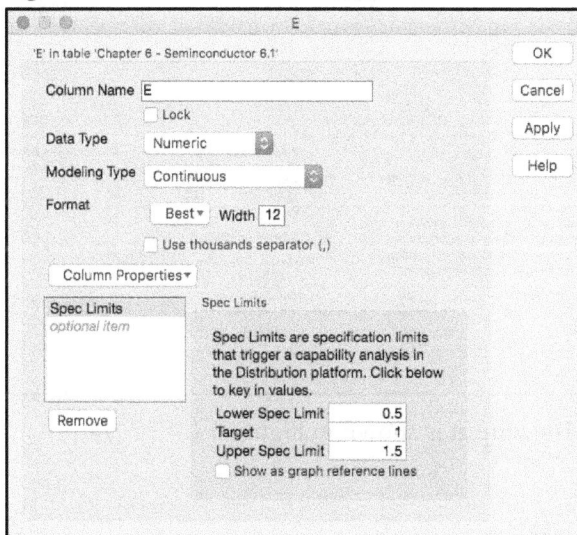

3. Select **Analyze ▶ Quality and Process ▶ Control Chart Builder** (Figure 6.2).

    Figure 6.2 Launching the Control Chart Builder

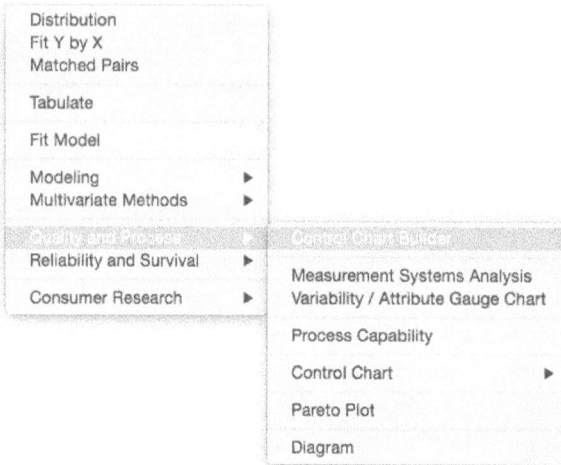

4. From the Columns Window on the left-hand side, drag parameter **E** to the **Y** zone. Click **Warnings(1)** and select **Tests 1** and **2** (Figure 6.3).

    Figure 6.3 Control Chart Builder Output for Parameter E

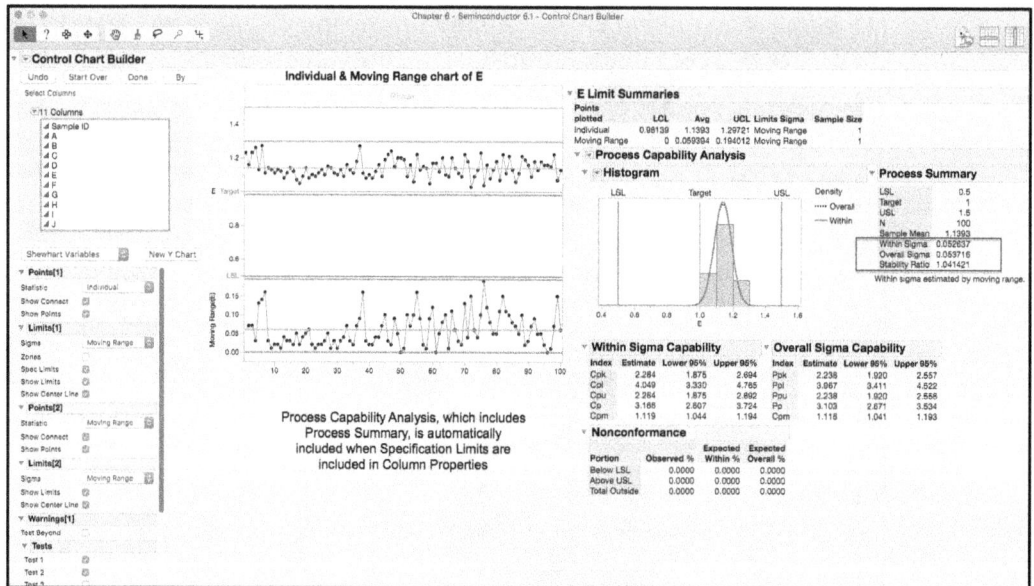

5. Click **Done** when finished. The output is shown in Figure 6.4.

Figure 6.4 PHA for Parameter E

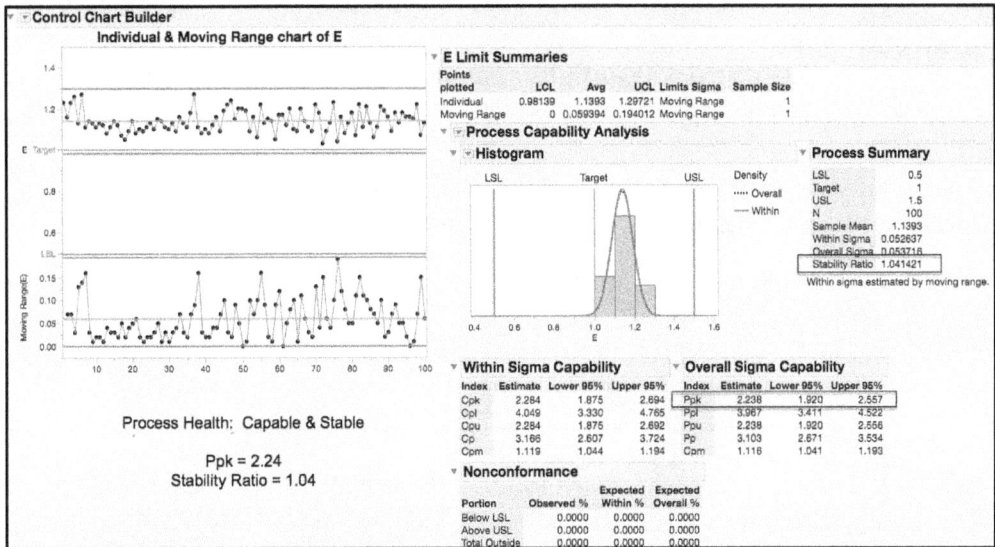

The output shown in Figure 6.4 can be used to assess the health for parameter E in terms of its capability and stability. The process capability index P_{pk} is used to determine if the process is capable. In JMP, by default, a process is capable if its $P_{pk} \geq 1.33$, but this value can be changed to accommodate different levels of capability. Since $P_{pk} = 2.24$, Parameter E can easily meet its specification limits, LSL = 0.5 and USL = 1.5. The histogram in Figure 6.4 indicates that the data is well within the specification, but slightly off target, toward the USL. If the process can be centered to the target = 1, then $P_{pk} = P_p = 3.1$. The **Nonconformance** table indicates that none of the results exceed the specification limits and none are expected to, based on probability calculations using the normal distribution approximation.

Process stability is assessed using the Stability Ratio. Recall, the Stability Ratio measures the long-term variation (overall sigma) to the short-term variation (within sigma). The long-term variation is estimated using the sample variance, or "overall sigma" in Figure 6.4: $s^2 = 0.053716^2$. The short-term sigma is estimated from the moving range control chart using $\overline{MR}/d_2$, or "within sigma" in Figure 6.4, $s^2_{MR} = 0.052637^2$. Finally, SR = $s^2 / s^2_{MR} = 0.053716^2 / 0.052637^2 = 1.04$.

For a stable process operating with common cause variation, SR = 1 and, SR > 1 for an unstable process, operating with special cause variation. JMP proposes a default critical value for SR of 1.5 to determine if a process is stable or not. A more accurate critical value can be obtained by obtaining the critical value from an F-distribution using the degrees of freedom described in Ramírez and Runger (2006). For this parameter, which uses an XmR chart and N = 100, the F-critical is the quantile associated with $F_{0.95, (100-1), 0.62(100-1)} = F_{0.95, 99, 61.38} = 1.48$. Since the Stability Ratio for parameter E is 1.04 < 1.48, the parameter is stable. The control chart also indicates that the behavior is stable over time, with most points randomly bouncing around the mean, and no violations in Test 1 or Test 2.

Statistics Note 6.2: The critical value for the Stability Ratio for a parameter monitored using an XmR control chart can be estimated using the 95% quantile associated with an F-distribution with $n-1$ and $0.62(n-1)$ numerator and denominator degrees of freedom, respectively.

The overall assessment for parameter E is Capable and Stable. This is an ideal situation and the focus should be on maintaining this level of performance over time.

6. Repeat Step 3 and from the Columns Window on the left-hand side, drag parameter **D** to the **Y** zone. Click **Warnings(1)** and select **Tests 1** and **2**. Click **Done** when finished. The output is in Figure 6.5.

Figure 6.5 PHA for Parameter D

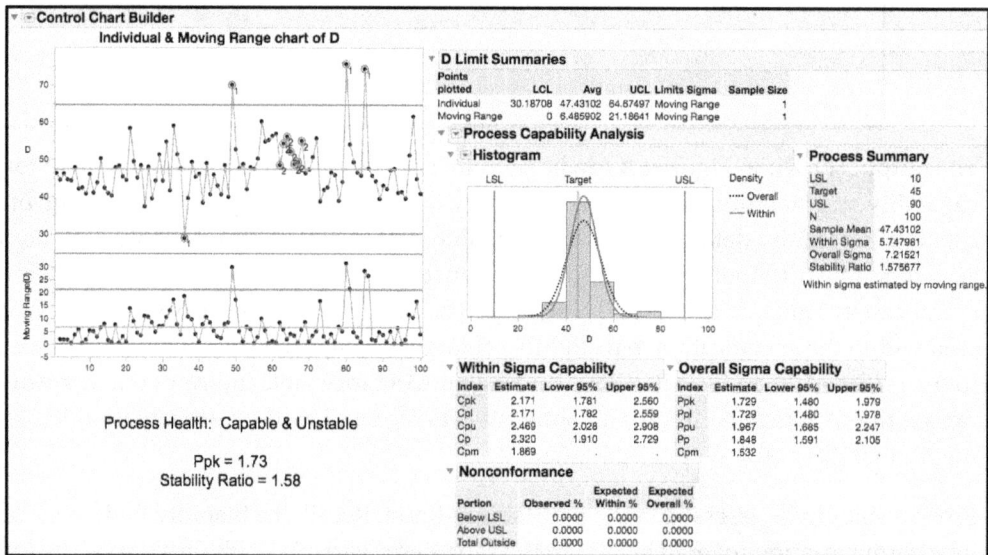

This output shown in Figure 6.5 can be used to assess the health for parameter D, in terms of its capability and stability. Since $P_{pk} = 1.73 \geq 1.33$, Parameter D can meet its specification limits, LSL = 10 and USL = 90. The histogram in Figure 6.5 shows that the data is well within the specifications and is on target. The **Nonconformance** table indicates that none of the results exceed the specification limits and none are expected to, based on probability calculations using the normal distribution approximation.

For process stability, the long-term variation is estimated using the sample variance, or "overall sigma" in Figure 6.5: $s^2 = 7.21521^2$. The short-term sigma is estimated from the moving range control chart using $\overline{MR}/d_2$, or "within sigma" in Figure 6.5, $s^2_{MR} = 5.7478^2$. Finally, SR = s^2 / s^2_{MR} = $7.21521^2 / 5.7478^2 = 1.58$. Since the sample size is also N = 100 for this parameter, the default critical value of 1.5 can be replaced by the more accurate value of 1.48. The Stability Ratio for parameter D is 1.58, which is greater than 1.48, so the parameter is unstable. The control chart also indicates that the behavior is unstable over time, with several points outside of the control limits and a run of 15 points above the centerline.

The overall assessment for parameter D is Capable and Unstable. Although the results are within the specification limits, the spurious data points might be indicative of extraneous variation in the process, which could eventually result in an out-of-specification point. The run of fifteen points above the centerline is also suspicious.

7. Repeat Step 3 and from the Columns Window on the left-hand side, drag parameter **A** to the **Y** zone. Click **Warnings(1)** and select **Tests 1** and **2**. Click **Done** when finished. The output is shown in Figure 6.6.

Figure 6.6 PHA for Parameter A

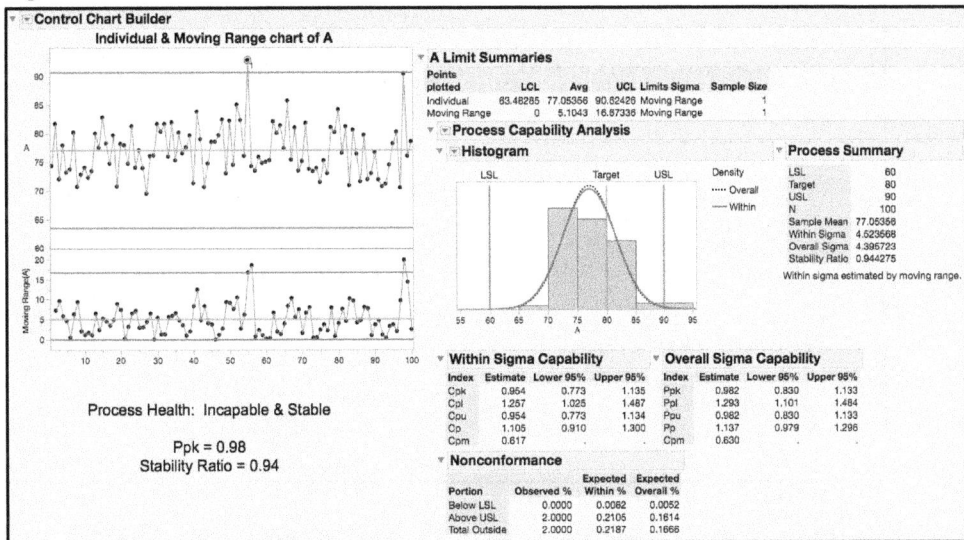

This output shown in Figure 6.6 can be used to assess the health for parameter A, in terms of its capability and stability. Since $P_{pk} = 0.98 < 1.33$, Parameter A is incapable of meeting its specification limits, LSL = 60 and USL = 90. The histogram in Figure 6.6 shows that the data is close to the upper specification, and slightly below target. The **Nonconformance** table indicates that 2% (n=2) of the results exceeded the specification limits and 0.1666% are expected to, based on probability calculations using the normal distribution approximation.

For process stability, the long-term variation is estimated using the sample variance, or "overall sigma" in Figure 6.6: $s^2 = 4.3957^2$. The short-term sigma is estimated from the moving range control chart using $\overline{MR}/d_2$, or "within sigma" in Figure 6.6, $s^2_{MR} = 4.5236^2$. Finally, SR = s^2 / s^2_{MR} = $4.3957^2/4.5236^2 = 0.94$. Since the sample size is also N = 100 for this parameter, the default critical value of 1.5 can be replaced by the more accurate value of 1.48. The Stability Ratio for parameter A is 0.94, which is less than 1.48, so the parameter is stable and operating with common cause variation.

Note that the one point outside of the control limit might be a false signal or indicative of a smaller shift in the mean.

The overall assessment for parameter A is Incapable and Stable. This means that the process is doing its best, as designed. Since the parameter is closer to target, the capability can be more significantly improved by reducing the overall variation.

8. Repeat Step 3 and from the Columns Window on the left-hand side, drag parameter **J** to the **Y** zone. Click **Warnings(1)** and select **Tests 1** and **2**. Click **Done** when finished. The output is in Figure 6.7.

Figure 6.7 PHA for Parameter J

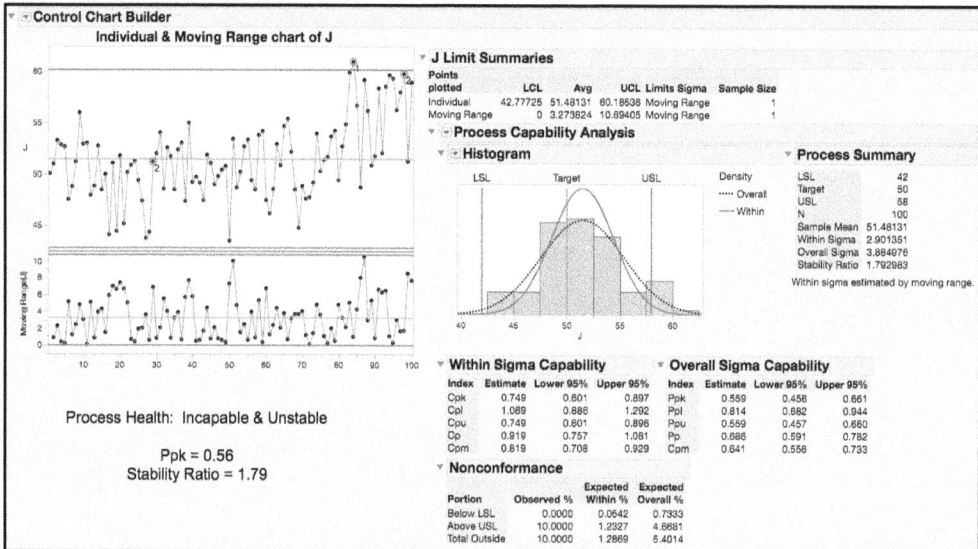

The output shown in Figure 6.7 can be used to assess the health for parameter J, in terms of its capability and stability. Since $P_{pk} = 0.58 < 1.33$, Parameter J is incapable of meeting its specification limits, LSL = 42 and USL = 58. The histogram in Figure 6.7 shows that the data is close to, or outside of, the specifications, and slightly above target. The **Nonconformance** table indicates that 10% (n = 10) of the results exceeded the specification limits and 5.4% are expected to, based on probability calculations using the normal distribution approximation.

For process stability, the long-term variation is estimated using the sample variance, or "overall sigma" in Figure 6.7: $s^2 = 3.88497^2$. The short-term sigma is estimated from the moving range control chart using $\overline{MR}/d_2$, or "within sigma" in Figure 6.7, $s^2_{MR} = 2.90135^2$. Finally, SR = s^2 / s^2_{MR} = $3.88497^2/ 2.90135^2 = 1.79$. Since the sample size is also N = 100 for this parameter, the default critical value of 1.5 can be replaced by the more accurate value of 1.48. The Stability Ratio for parameter J is 1.79, which is greater than 1.48, so the parameter is unstable and operating with special cause variation. The control chart shows several runs tests violations and, a closer examination, shows a potential shift in the mean at the end of the series.

The overall assessment for parameter J is Incapable and Unstable. This is the most undesirable situation. Some experts suggest trying to first bring the process into a state of control and then trying to re-center the process.

> **JMP Note 6.1**: The specification limits must be specified in the Spec Limits Column Property to produce the process capability and process stability metrics using the Control Chart Builder.

Process Health Assessment Using the Process Screening Platform

In this section, we will show how to carry out a process health assessment for many parameters using the **Process Screening** platform. We will use the same JMP data table, Chapter 6 – Semiconductor 6.1.jmp, which contains 10 parameters to assess the overall health of this process. Recall, each parameter has a specification that represents an internal release limit. The data includes one hundred measurements and it is assumed to represent 4 months of manufacturing output. For each parameter, a subgroup size of n = 1 is appropriate and therefore, an XmR chart is used for monitoring purposes.

The following steps illustrate how to use the **Process Screening** platform.

1. Open Chapter 6 – Semiconductor 6.1.jmp, which has variables called *Sample ID, A, B, C, D, E, F, G, H, I,* and *J*. In this table, Sample ID is the subgroup variable, and A through J are the PHA parameters.

2. Make sure that all parameters have their specification limits included in the **Spec Limits** Column Properties in the JMP table. This can be verified by double clicking on the column label (for example, **A** and selecting **Column Properties ▶ Spec Limits)**. Alternatively, we can click on the * symbol next to each column name in the Columns panel in the left-hand side of the JMP table. A list of the column properties will pop up. We can select Spec Limits to view the limits entered (Figure 6.8).

Figure 6.8 Identifying Column Properties

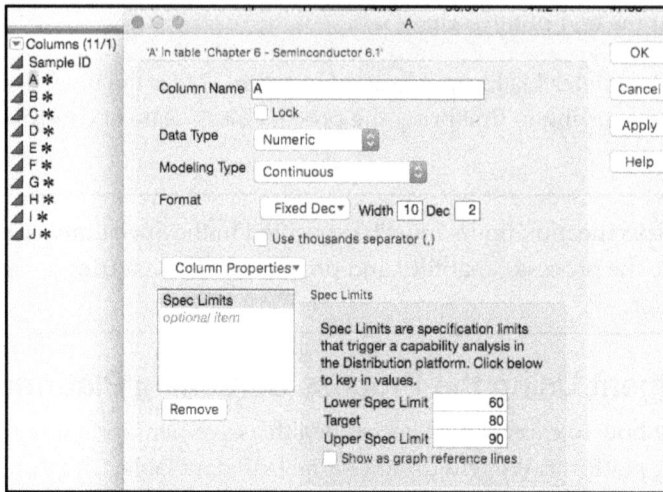

3. From the main menu, select **Analyze ▶ Screening ▶ Process Screening** (Figure 6.9).

Figure 6.9 Launching the Process Screening Platform

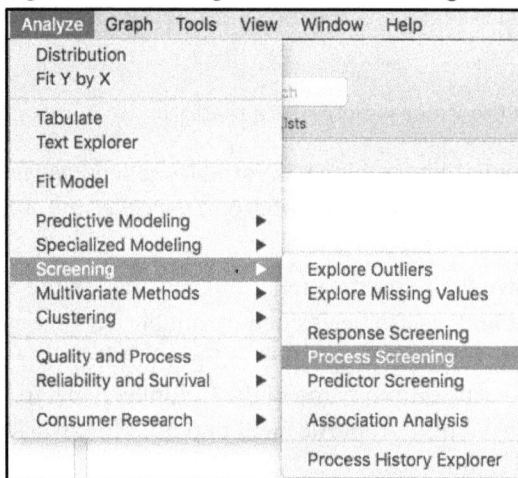

4. A **launch** window will appear. In the left-hand window, click **A** and hold the shift key and select **J**, so they all parameters are highlighted and then click **Process Variables.** Make sure that **Control Chart Type** is set to **Indiv and MR**. Leave the default settings for the remaining fields. Some of these will be discussed later in this section.

Figure 6.10 Launch Window for Process Screening

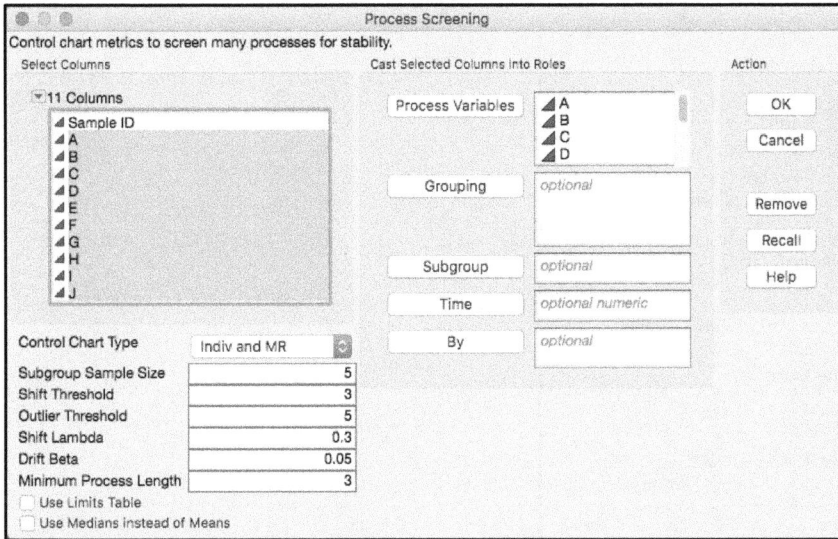

5. Click **OK** when finished. The default output contains the Stability Ratio, the within and overall estimates of variability, alarm rates for runs tests, and capability information (Figure 6.11).

Figure 6.11 Process Screening Default Output

Column	Stability Ratio	Variability Within Sigma	Overall Sigma	Summary Mean	Count	Control Chart Alarms Alarm Rate	Test1	Latest Alarm	Out of Spec Count	Capability Out of Spec Rate	Latest Out of Spec	Cpk	Ppk
I	8.71	5.09357	15.0336	506.52	100	0.39000	39	1	0	0	.	2.845	0.964
H	3.74	3.22264	6.23204	200.01	100	0.16000	16	1	19	0.19	4	0.518	0.268
J	1.79	2.90135	3.88498	51.4813	100	0.01000	1	1	10	0.1	1	0.749	0.559
C	1.71	4.68673	6.12121	42.193	100	0.02000	2	28	3	0.03	29	0.911	0.697
D	1.58	5.74798	7.21521	47.431	100	0.04000	4	15	0	0	.	2.171	1.729
F	1.06	0.05962	0.06127	1.2154	100	0.01000	1	15	0	0	.	2.150	2.092
E	1.04	0.05264	0.05372	1.1393	100	0.00000	0	51	0	0	.	2.284	2.238
B	0.97	5.80158	5.71084	86.7321	100	0.02000	2	23	1	0.01	24	1.624	1.650
A	0.94	4.52357	4.39572	77.0536	100	0.01000	1	3	2	0.02	3	0.954	0.982
G	0.90	0.06427	0.06092	1.216	100	0.00000	0	32	10	0.1	2	0.436	0.460

6. Click on the red triangle next to the **Process Screening** banner at the top of the window to bring up a list of options. A brief description of some of the options that we will illustrate in this chapter is shown below. To launch the online help, click on the **?** at the top of the window and then click anywhere in the default output shown in Figure 6.11.

Figure 6.12 Process Screening Platform Options

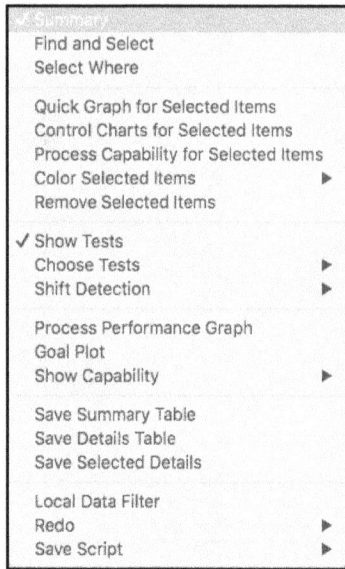

a. **Quick Graph for Selected Items**: produces thumbnail size graphs for all parameters selected in the Summary table. The plots contain the control limits and specification limits, if the y-axis scale permits them.

b. **Control Charts for Selected Items**: launches the **Control Chart Builder**, which contains default output of control chart with runs tests engaged, and the process capability analysis and stability ratio metrics.

c. **Process Capability for Selected Items**: launches the **Process Capability** platform, which contains default output of process capability indices, Stability Ratio information, Goal Plot, Capability Box Plots, and Capability Index Plot.

d. **Show Tests:** shows or hides the information associated with Nelson test excursions, for example, Alarm Rate, Alarm Counts, and Latest Alarm, in the Summary table.

e. **Choose Tests**: used to select the specific Nelson rules (1 through 8) or Range excursions to include in the Summary table.

f. **Shift Detection**: provides options for detecting shifts in the parameter over time, after outliers are removed. Shift detection includes step changes in the mean, greater or equal to one within-sigma unit, using outlier-correction and an exponentially weighted moving average (EWMA) smoothing approach for the individual observations. Options include adding shift detection statistics to the Summary table and graphs.

g. **Process Performance Graph**: produces a graph that plots P_{pk} versus the Stability Ratio for all parameters included in the Summary table. The graph is divided into four quadrants: Stable & Capable, Capable but Unstable, Incapable but Stable and Incapable and Unstable. The default boundary to classify a parameter as Capable is

$P_{pk} \geq 1.33$; while the default boundary to classify a parameter as Stable is Stability Ratio ≤ 1.5. These default boundaries can be changed.

h. **Process Performance Graph Boundaries**: this option is added to the menu shown in Figure 6.12, once the Process Performance Graph is selected. A window is launched to change the P_{pk} and Stability Ratio boundaries to create the four quadrants.

i. **Show Capability**: shows or hides the information associated with the Capabilities in the Summary table. By default, all are shown except for Spec Centered mean and Spec Scaled Dev.

7. From the red triangle at the top of the window, select **Process Performance Graph**. Highlight the parameters in the Summary table by clicking on **I** and then **G**, while holding the shift key. Right-click in the **Process Performance Graph** and select **Row Label** from the menu. The output is shown in Figure 6.13.

Figure 6.13 Process Performance Graph

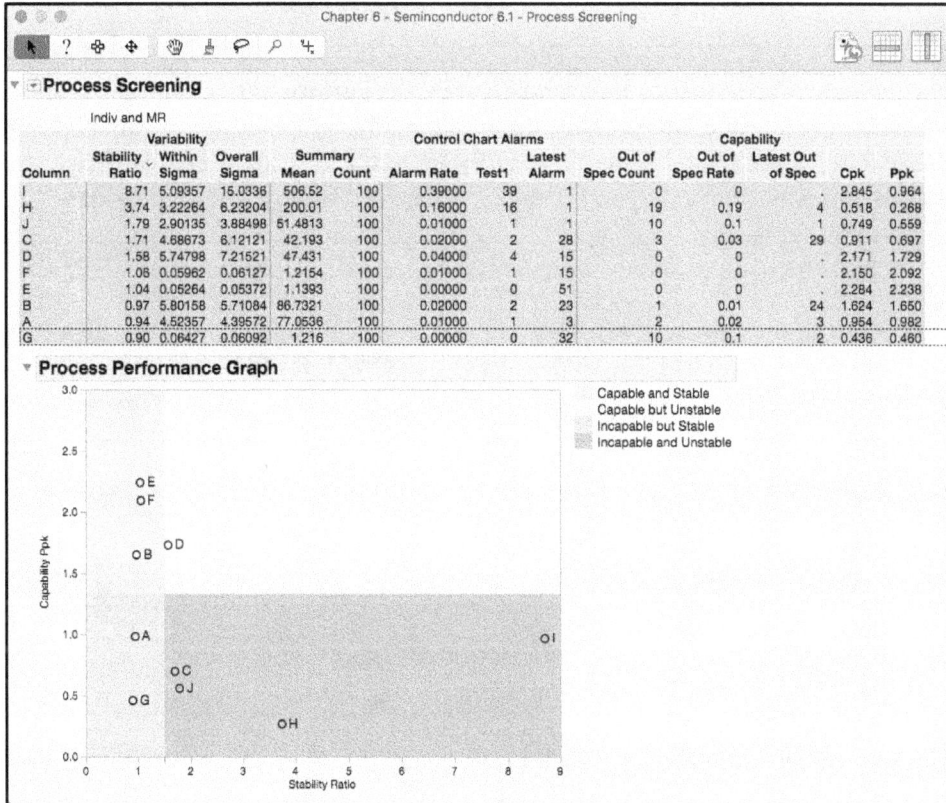

The **Summary** table in Figure 6.13 is ordered by the Stability Ratio, where the most unstable parameter (largest value) is shown first and the most stable process (smallest value) is shown last. For example, for process parameter I, SR = s^2 / s^2_{MR} = 15.0336^2/ $5.09357^2 \approx 8.71$, which would be classified as unstable. In addition, since $P_{pk} = 0.964$, Parameter I is also incapable. Therefore, Parameter I is in the most undesirable quadrant: Incapable and Unstable. Note $C_{pk} = 2.845$, is also calculated using the average moving range, and is considered the potential capability if the process is stable. Additional information for Parameter I in the **Summary** table includes the Mean = 506.52, the sample size Count = 100, the Alarm Rate = 0.39 (= 39 / 100) for Test 1, number of Test1 signals = 39, and the Latest Alarm for a Test 1 signal occurred in the last point. Also, no points were included in the Out of Spec Count, the Out of Spec Rate is zero and the Latest Out of Spec is missing.

Statistics Note 6.3: The Stability Ratio can also be calculated as a function of the capability indices. That is, $SR = \left(C_{pk} \middle/ P_{pk} \right)^2$.

For parameter I $\quad SR = \left(2.845 \middle/ 0.964 \right)^2 = 2.951^2 = 8.708$

While the individual statistics are useful, the **Process Performance Graph** in Figure 6.13 makes it much easier to view and interpret this information. This graph plots a parameter's P_{pk} versus its Stability Ratio. It is easy to see that B, E, and F are in the ideal state, D is capable but not stable, A and G are incapable but stable, and C, H, J and I are in the most undesirable state, Incapable and Unstable. We can see that for this process P_{pk} ranges from less than 0.5 to almost 2.5; while the Stability Ratio takes on values from a low of 0.9 all the way up to 8.7.

8. With all the parameters still highlighted, as is shown in Figure 6.13, click the red triangle at the top of the window, and select **Quick Graph for Selected Item**.

Figure 6.14 Quick Graph of Parameters A through

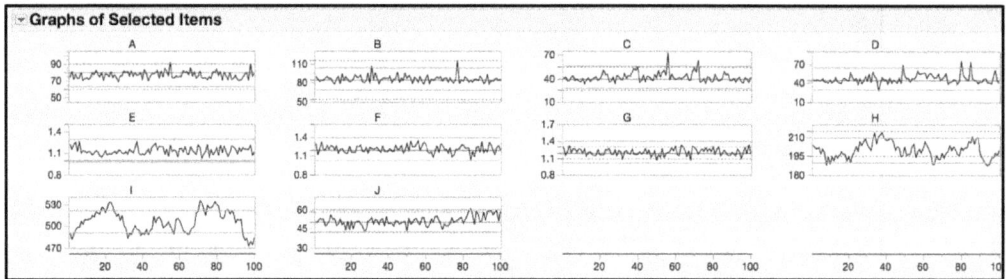

The graph in Figure 6.14 shows trend plots for the parameters, where the green lines represent the control limits for the Individuals charts. Some plots also include the specification limits, using blue dashed lines, if the scale of the Y axis allows it. From this plot, we can get a sense for the stability for each of the parameters. For example, parameters A, B, E, F, and G appear to be relatively stable about their centerlines. However, parameter H and I appear to have cycles and are nonstationary and J is trending up at the end of the series. A closer examination of the behavior can be made using the **Control Chart Builder** output.

9. From the red triangle at the top of the window, select **Control Charts for Selected Item**. Make sure all the parameters are still highlighted, as is shown in Figure 6.13. The output is in Figure 6.15.

Figure 6.15 Control Chart Builder Output for Parameter H

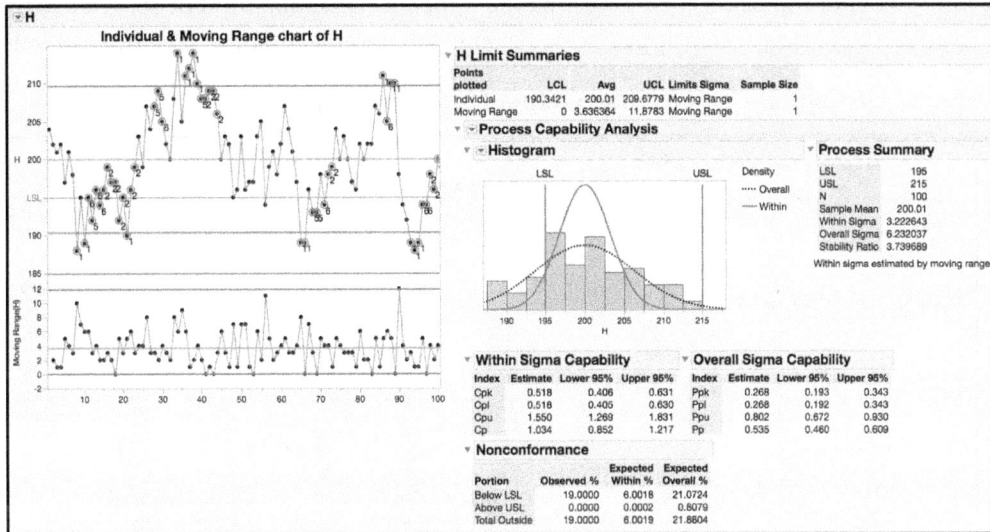

The control charts for the selected parameters will appear in the same window. The scroll bar can be used to move up and down the window. In Figure 6.15 the default output is shown for parameter H. All of the Nelson rules are automatically selected. Some or all of these can be turned off by selecting **Show Control Panel** from the red triangle next to the parameter label (**H**) and clicking on **Warnings[1]** and **Tests** and then deselecting the undesired tests. The control chart will be automatically updated. The stability ratio and process capability metrics are shown in the right-hand side of the output. Note this output is the same as the one discussed in the last section, Process Health Assessment Using the Control Chart Builder.

The control chart for Parameter H has many Nelson rule violations. Recall the alarm rate was 16%, or 16 out of 100 subgroups, from Figure 6.13. The Stability Ratio for H is 3.74, which means that the long-term variation is approximately four times as large as the short-term variation. These two results indicate excessive variability from a process operating with common cause variation, randomly about its mean. A closer examination of the charts shows a non-random trend, with cycles or drifts up and then down. This data stream seems to represent an autocorrelated nonstationary process.

The process capability for Parameter H shows that the process mean is closer to the LSL, with $P_{pl} = P_{pk} = 0.268$. Even if the process is centered between the specifications, $P_p = 0.535$, which is still very low. Nineteen out of 100 results are below the LSL and 21.88% are predicted to be outside of the specification limits, using the normal distribution approximation. This parameter was found in the incapable and unstable quadrant of the **Process Performance Graph**.

Parameter J was also in the incapable and unstable quadrant. The control chart for Parameter J is shown in Figure 6.16. It is obvious that the process mean shifted upward around observation 80. The Stability Ratio for the entire series is 1.79, which is greater than the critical value of 1.48. This means that the process is operating with special cause variation. The index $P_{pk} = 0.56$, and there were 10 out of 100 results that exceeded the upper specification limit of 58. Prior to the shift around n = 80, the process appeared to be on target and meeting the specification limits.

Figure 6.16 Control Chart Builder Output for Parameter J

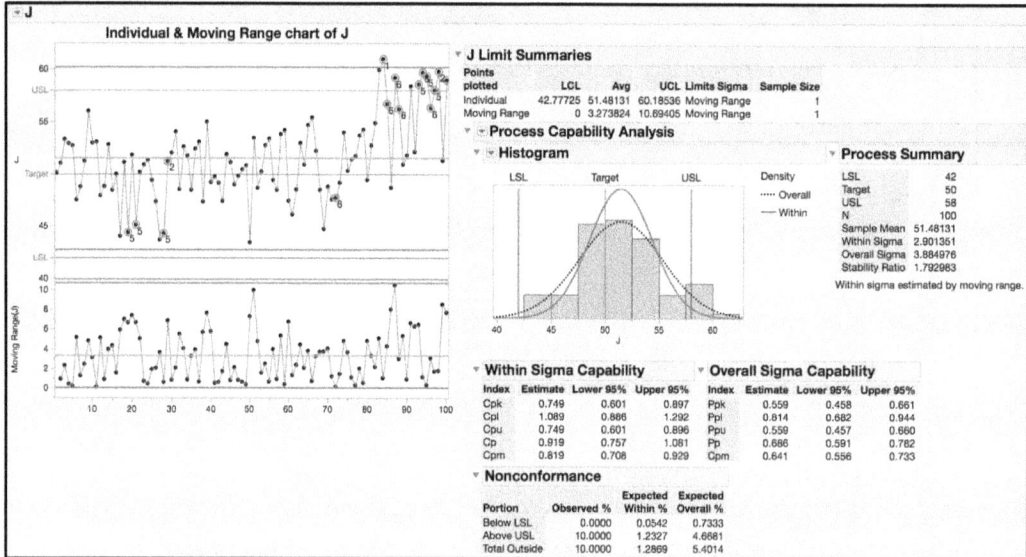

The shifts and drifts in the plots can be further explored using the tools within the **Process Screening** platform.

10. Go back to the **Process Screening** platform by clicking on it. From the red triangle at the top of the window select **Shift Detection ▶ Largest Upshift, Largest Downshift** and **Shift Graph**.

11. In the **Process Summary** table, highlight Parameter **I, H,** and **C** and then select **Quick Graph for Selected Items**, from the main menu.

12. From the red triangle at the top of the window select **Shift Detection ▶ Show Shifts in Quick Graphs**. The output is shown in Figure 6.17.

Figure 6.17 Shift Detection Output for Parameters C, H and I

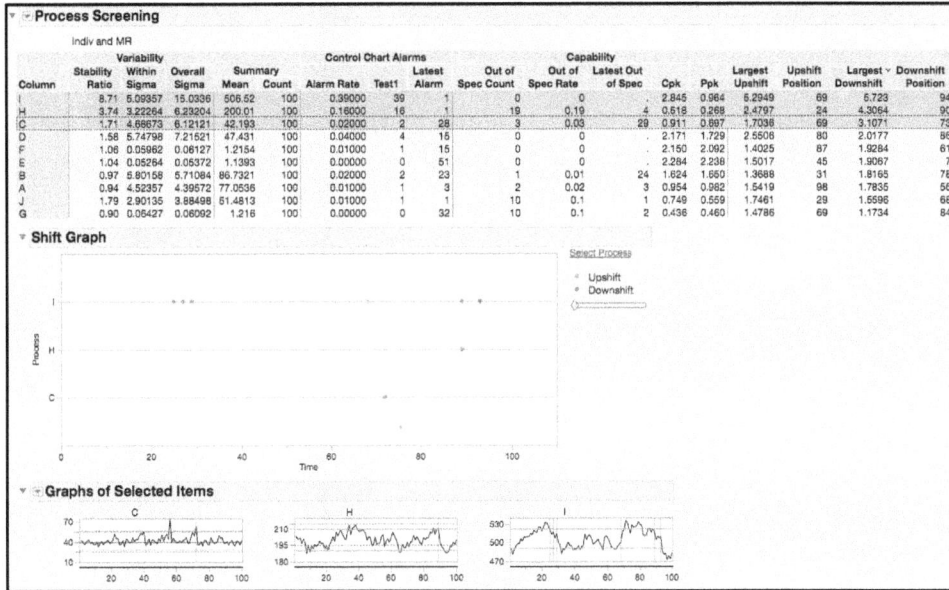

▾ ▾ **Process Screening**

Indiv and MR

Column	Variability Stability Ratio	Within Sigma	Overall Sigma	Summary Mean	Count	Control Chart Alarms Alarm Rate	Test1	Latest Alarm	Out of Spec Count	Capability Out of Spec Rate	Latest Out of Spec	Cpk	Ppk	Largest Upshift	Upshift Position	Largest Downshift	Downshift Position
I	8.71	5.09357	15.0336	506.52	100	0.39000	39	1	0	0		2.845	0.964	5.2949	69	6.723	94
H	3.74	3.22264	6.23204	200.01	100	0.16000	16	1	19	0.19	4	0.518	0.268	2.4797	24	4.3064	90
C	1.71	4.68673	6.12121	42.193	100	0.02000	2	28	3	0.03	29	0.911	0.897	1.7036	69	3.1071	73
D	1.58	5.74798	7.21521	47.431	100	0.04000	4	15	0	0	.	2.171	1.729	2.5506	80	2.0177	86
F	1.06	0.05962	0.06127	1.2154	100	0.01000	1	15	0	0	.	2.150	2.092	1.4025	87	1.9284	61
E	1.04	0.05264	0.05372	1.1393	100	0.00000	0	51	0	0	.	2.284	2.236	1.5017	45	1.9067	7
B	0.97	5.80158	5.71084	86.7321	100	0.02000	2	23	1	0.01	24	1.624	1.660	1.3688	31	1.8165	78
A	0.94	4.52357	4.39572	77.0536	100	0.01000	1	3	2	0.02	3	0.954	0.962	1.5419	98	1.7855	56
J	1.79	2.90135	3.88498	51.4813	100	0.01000	1	1	10	0.1	1	0.749	0.559	1.7461	29	1.5596	68
G	0.90	0.06427	0.06092	1.216	100	0.00000	0	32	10	0.1	2	0.436	0.460	1.4786	69	1.1734	84

▾ **Shift Graph**

Select Process
◦ Upshift
• Downshift

▾ ▾ **Graphs of Selected Items**

The shift detection statistics are added to the last four columns in the **Summary** table, shown in Figure 6.17, and the **Shift Graph** is shown below the table, with markers located at the local peaks of the shifts. For example, for Parameter I, the largest upshift occurred at subgroup 69 and is 5.29 within-sigma units, while the largest downshift occurred at subgroup 94 and is 5.72 within-sigma units. Although the graph of Parameter I shown below the **Summary** table can be used to visualize the location and size of each shift, it is easier to study them using the control chart output. From within the **Control Chart Builder** output, after removing the runs tests, position your pointer over the points and locate and click on the point for subgroup 69. Holding the Ctrl key, do the same for subgroup 94. Right-click and select **Rows ▶ Row Label**.

> **JMP Note 6.2**: In the Shift Graph, green markers indicate upshifts and red markers indicate downshifts. The markers are located at the local peaks of the shifts.

In Figure 6.18, it is easier to see the upward step change in the mean that occurred at subgroup 69. The subgroup is also highlighted in the JMP table and we can see that the results directly preceding the 69th subgroup are around 500 and at the 69th subgroup, they increase to 518 and higher. Similarly, the results directly preceding the 94th subgroup are around 510 and then drop to 487 and lower, starting at the 94th subgroup.

Figure 6.18 Largest Upshift and Downshift for Parameter I in the XmR Chart

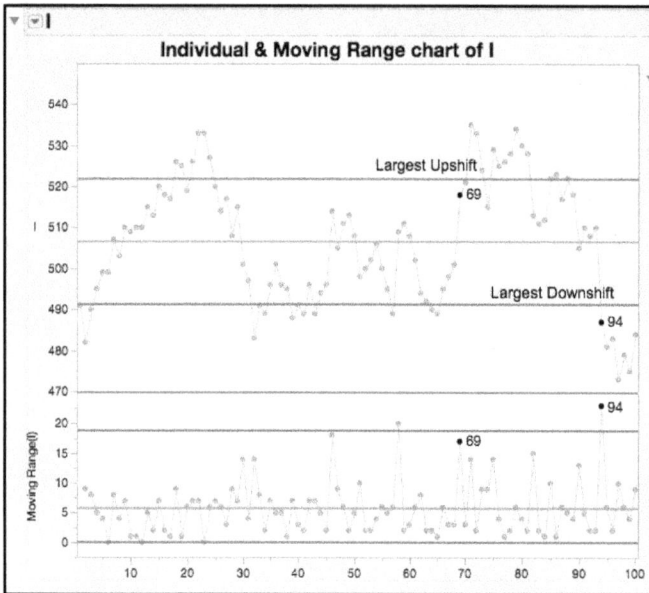

The **Shift Graph** in Figure 6.17 shows upward and downward shifts that have a size that exceeds the number of within-sigma units specified by the Shift Threshold (three by default, which you can see if you however over the diamond in the slider). For this data set, it includes shifts that exceed 3 within-sigma units for Parameters I, H and C, and they are shown in the same order as the **Summary** table. For Parameter I there are four additional downward shifts. Three of them occur between subgroups 20 and 30 and one occurs around subgroup 88 prior. In order to see the locations where a shift of 4 or greater within-sigma units occurs, right-click on the diamond in the slider bar and select **Set Value ....** In the dialog box enter **4** and click **OK**. The Shift Graph shown in Figure 6.19 is the graph that appears after we click **OK**. Now Parameter I only has one additional downshift that is at least 4 within-sigma units, occurring around subgroup 30.

> **JMP Note 6.3**: The Shift Graph does not show the positions of Largest Upshift and Largest Downshift values that appear in the Summary table if the shifts are less than the specified Shift Threshold number of within-sigma units in magnitude.

Figure 6.19 Shift Graph for Parameters C, H and I

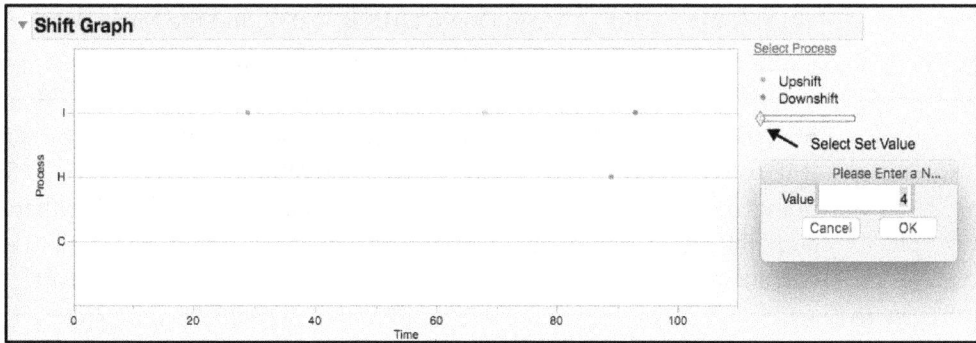

13. From the red triangle at the top of the **Process Screening** window (Figure 6.17) deselect all options associated with upshifts and downshifts. Next select **Shift Detection ▶ Drift Summaries**.

14. Right-click in the **Summary** table and select **Sort By Column...** and select **Mean Abs Drift** from the drop-down list and click **OK**.

15. Highlight the first three parameters **I, J** and **H**, in the **Summary** table and select **Shift Detection ▶ Drift Graph Selected**.

Figure 6.20 Drift Summaries for Parameters H, I and J

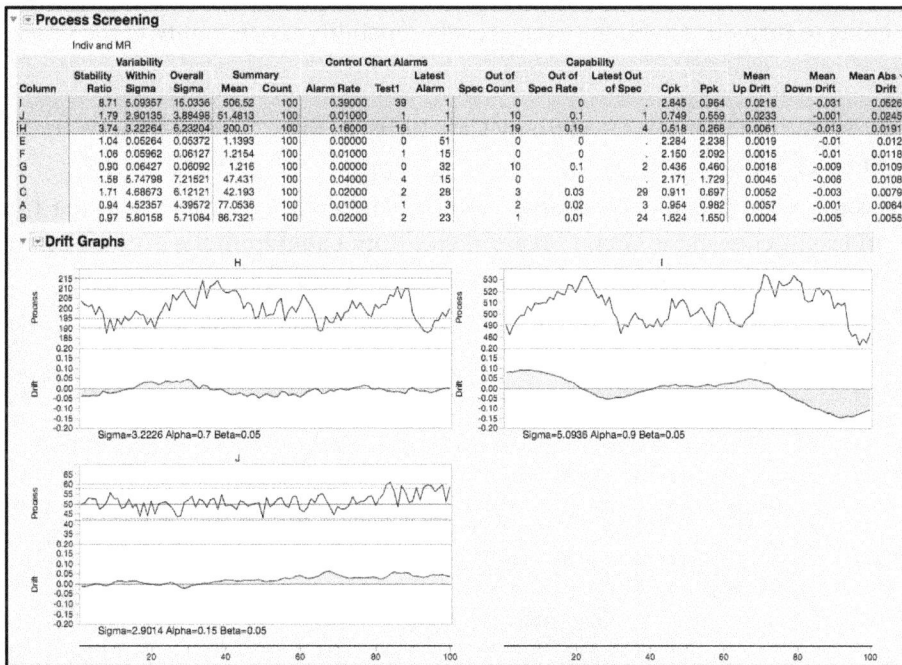

Column	Stability Ratio	Variability Within Sigma	Overall Sigma	Summary Mean	Count	Alarm Rate	Test1	Latest Alarm	Out of Spec Count	Out of Spec Rate	Latest Out of Spec	Cpk	Ppk	Mean Up Drift	Mean Down Drift	Mean Abs Drift
I	6.71	5.09367	15.0336	506.52	100	0.39000	39	1	0	0		2.845	0.964	0.0218	-0.031	0.0626
J	1.79	2.90135	3.88498	51.4813	100	0.01000	1	1	10	0.1	1	0.749	0.559	0.0233	-0.001	0.0245
H	3.74	3.22264	6.23204	200.01	100	0.16000	16	1	19	0.19	4	0.518	0.268	0.0061	-0.013	0.0191
E	1.04	0.05264	0.05372	1.1393	100	0.00000	0	51	0	0	.	2.284	2.238	0.0019	-0.01	0.012
F	1.06	0.05962	0.06127	1.2154	100	0.01000	1	15	0	0	.	2.150	2.092	0.0015	-0.01	0.0118
G	0.90	0.06427	0.06092	1.216	100	0.00000	0	32	10	0.1	2	0.436	0.460	0.0018	-0.009	0.0109
D	1.58	5.74798	7.21521	47.431	100	0.04000	4	15	0	0	.	2.171	1.729	0.0045	-0.006	0.0108
C	1.71	4.88673	6.12121	42.193	100	0.02000	2	28	3	0.03	29	0.911	0.697	0.0052	-0.003	0.0079
A	0.94	4.52357	4.39572	77.0636	100	0.01000	1	3	2	0.02	3	0.954	0.982	0.0057	-0.001	0.0064
B	0.97	5.80158	5.71084	86.7321	100	0.02000	2	23	1	0.01	24	1.624	1.650	0.0004	-0.005	0.0055

The **Drift Summaries** output is shown in Figure 6.20. Three columns are added to the end of the **Summary** table, which include the Mean Up Drift, the Mean Down Drift and the Mean Abs Drift. Parameters I, J, and H have the largest drifts. Incidentally, these also have the largest Stability

Ratios. For example, for Parameter I, the average of the positive drifts as Mean Up Drift is 0.0218, the average of the negative drifts as Mean Down Drift is -0.031, and the Mean Abs Drift is 0.0526, which is the average of the absolute value of the positive and negative drifts.

Statistics Note 6.4: Shift detection is not designed to detect slow drifts in the mean. JMP uses the classic double-exponential smoothing model of Holt. As Sall (2018) points out "the smoother is being used for characterization, rather than forecasting, and "the future is known" with off-line data". One smoother is used to estimate the current level, while the other is used to estimate the slope of the drift.

The **Drift Graphs** shown in Figure 6.20 provides a trend order plot of the data with a plot of drifts right below it. In the Drift plot, the shaded red areas indicate a downward drift, and the shaded green areas indicate an upward drift. For example, for Parameter I, the data series begins with an upward drift from subgroup 1 through 23, a downward drift through subgroup 45, an upward drift through subgroup 80, and then a downward drift through subgroup 100. In this plot, the first drift is the largest up drift and last drift is the largest down drift. Below each Drift graph, JMP displays the three parameters used in the double-smoother: the within sigma estimate (5.0936), the smoothing constant alpha (0.9) for the level smoother, and the smoothing constant beta (0.05) for the slope smoother.

We are going to turn our attention to the process capability functionality in the **Process Screening** platform.

1. From the red triangle, select **Show Capability ▶ Spec Centered Mean** and **Spec Scaled Std Dev**.
2. From the red triangle select **Goal Plot**. Highlight all of the columns in the Table Summary and right-click in the **Goal Plot** and select **Row Label** from the drop-down menu. The output is shown in Figure 6.21.

Figure 6.21 Process Capability Goal Plot for Semiconductor Data

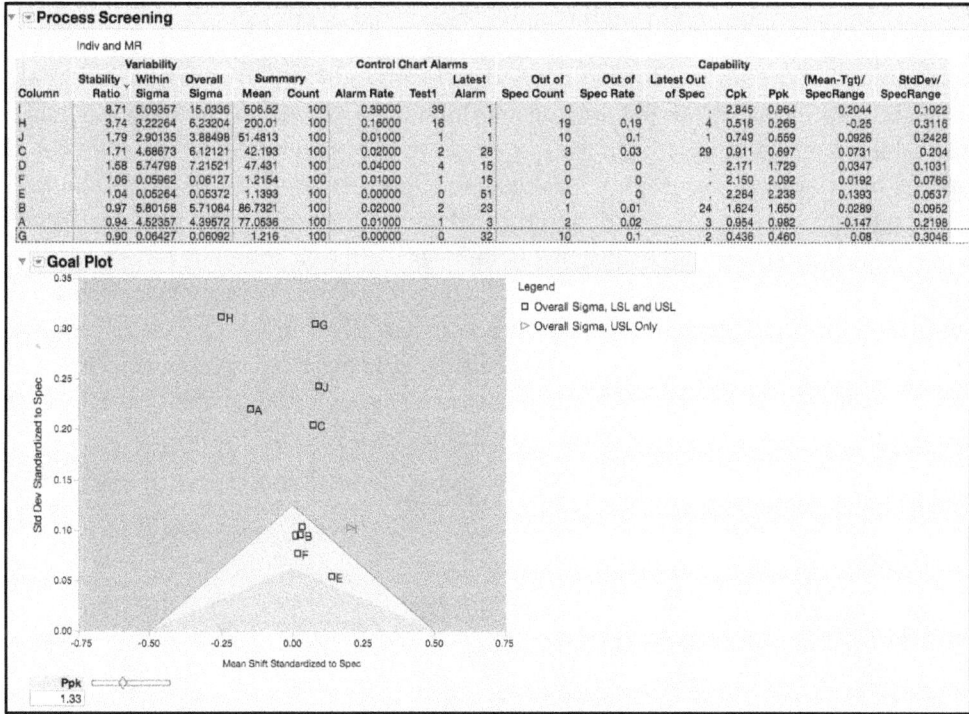

Two more process capability indices are added to the **Summary** table, as is shown in Figure 6.21. The first new column, (Mean – Tgt)/SpecRange, is the same as the Mean Shift Standardized to Spec, and the second column, StdDev/SpecRange, is the same as the Std Deviation Standardized to Spec in the **Process Capability** platform.

> **JMP Note 6.4**: The mean shift and the standard deviation standardized to the specification limits for the jth column are defined as follows:
>
> $$\text{Mean Shift Standardized to Spec} = \frac{\bar{Y}_j - T_j}{2 \times \min(T_j - LSL_j, USL_j - T_j)}$$
>
> $$\text{Std Dev Standardized to Spec} = \frac{SD(\bar{Y}_j)}{2 \times \min(T_j - LSL_j, USL_j - T_j)}$$

These two new capability statistics are used to create the **Goal Plot** shown in Figure 6.21, where the Spec Centered Mean is plotted on the X axis and the Spec Centered Standard Deviation is plotted on the Y axis. Notice that the legend in this plot indicates that all parameters have a lower and upper specification limit, indicated by Overall Sigma, LSL and USL, except for Parameter I, which only has an upper limit, indicated by Overall Sigma, USL only. The larger the absolute value of the Spec Centered Mean the more off-centered the process mean is from its target value.

In Figure 6.22, the reference line on the X axis at zero indicates a process mean equal to its target value, positive values mean that the mean is higher than the target value and negative values mean that the mean is lower than the target value. The process mean for Parameter H is the farthest from its target value, with a value of the Spec Centered Mean = (200–205)/(215–195) = -0.25 . Since a target value is not included in the JMP table, the target value was calculated as the center of the specification limits. The mean for Parameter F is the closest to its target value: Spec Centered Mean = (1.2154 – 1.2) / (1.6 – 0.8) = 0.019. In Figure 6.22, the reference line on the Y axis at 0.2 is where $P_p = P_{pk} = 1.33$, for a perfectly centered process. The process capability increases below this line and decreases above it. For example, Parameter H has the largest Spec Centered Standard Deviation: 6.232 / (215 – 195) = 0.3116. In contrast, Parameter E has the lowest Spec Centered Standard Deviation: 0.05372 / (1.5 – 0.5) = 0.0537.

Figure 6.22 Goal Plot Revised with Reference Lines

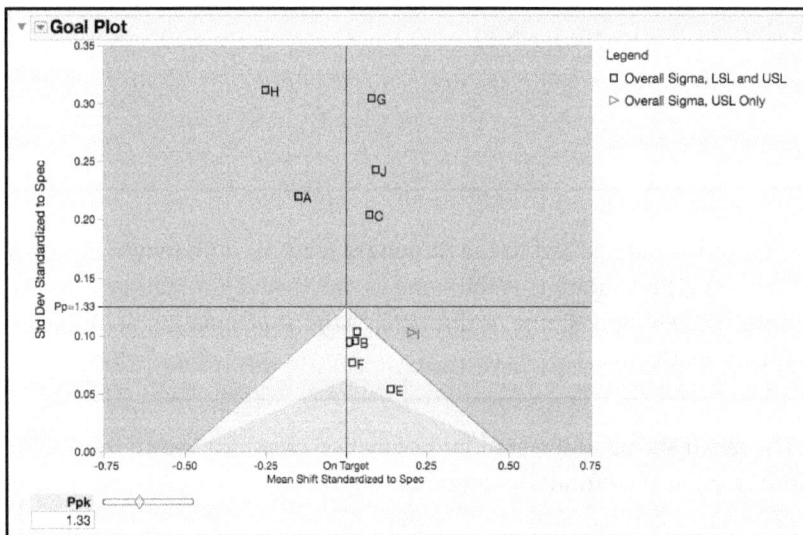

JMP Note 6.5: For parameters that only have an upper specification, and do not have a target, the Spec Centered Mean, which is the x coordinate in the Goal Plot, is defined as $\left(3\left(\dfrac{USL - \bar{Y}}{3SD(Y)} + 2\right)\right)^{-1}$.

For process parameter I, in Figures 6.21 and 6.22 this gives

$$\left(3\left(\frac{550 - 506.52}{3 \times 15.034} + 2\right)\right)^{-1} = 0.204.$$

The **Goal Plot** has three shades of color. The red shaded area indicates where $P_{pk} < 1.33$ (the default value shown under the label in the slider). The yellow shaded area in the triangle indicates where $1.33 < P_{pk} < 2.66$ (= 1.33 x 2). The greed shaded area in the triangle is where $P_{pk} > 2.66$. No parameters are in the green shaded area; however, 4 parameters are in the shaded yellow area. Parameter E has a P_{pk} = 2.238, while Parameter D has P_{pk} = 1.729. The P_{pk} for the six parameters in the red shaded area are below 1.33. For example, Parameter H and Parameter C have a P_{pk} = 0.268 and P_{pk} = 0.697, respectively. Additional information can be included in the **Goal Plot**, such as a contour for the defect rate.

3. From the red triangle next to the **Goal Plot** label (Figure 6.21) select **Defect Rate Contour**.
4. A window will launch. In the field type **0.005** (Figure 6.23) and click **OK** when finished.

Figure 6.23 Defect Rate Contour for Semiconductor Data

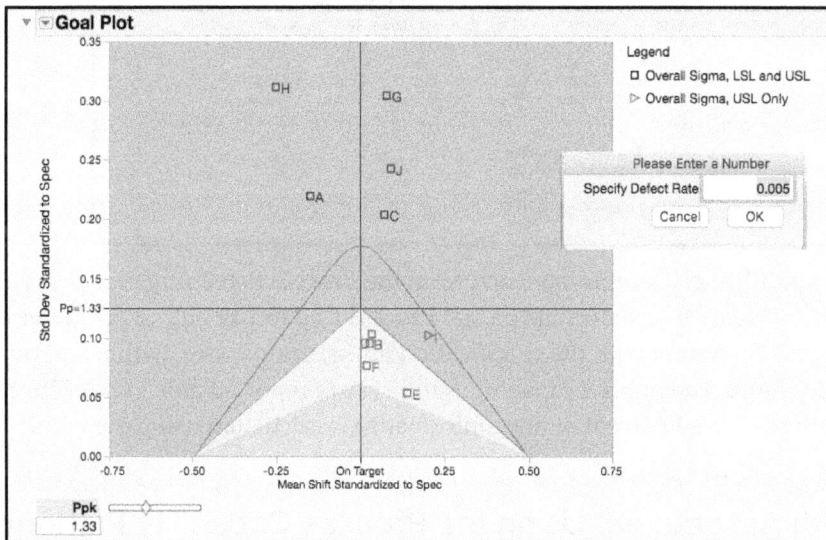

A blue line is added to the **Goal Plot** in Figure 6.23, which indicates the region where the estimated defect rate is less than the specified defect rate. The defect rate is estimated using the normal distribution. In Figure 6.23, below the blue line is where the defect rate is less than 0.005 or 0.5%; outside of the blue line is where the defect rate is greater than 0.5%. Five of the parameters (C, A, J, G, and H) might result in defect rates that exceed 0.5%.

5. In the **Summary** table, click **H** in the Column field (Figure 6.21). Alternatively, the parameter can also be selected in this table by clicking on its label in the **Goal Plot**. From the red triangle next to the **Process Screening** label select **Process Capability for Selected Items**. The partial output is shown in Figure 6.24.

Figure 6.24 Partial Process Capability Output for Parameter H

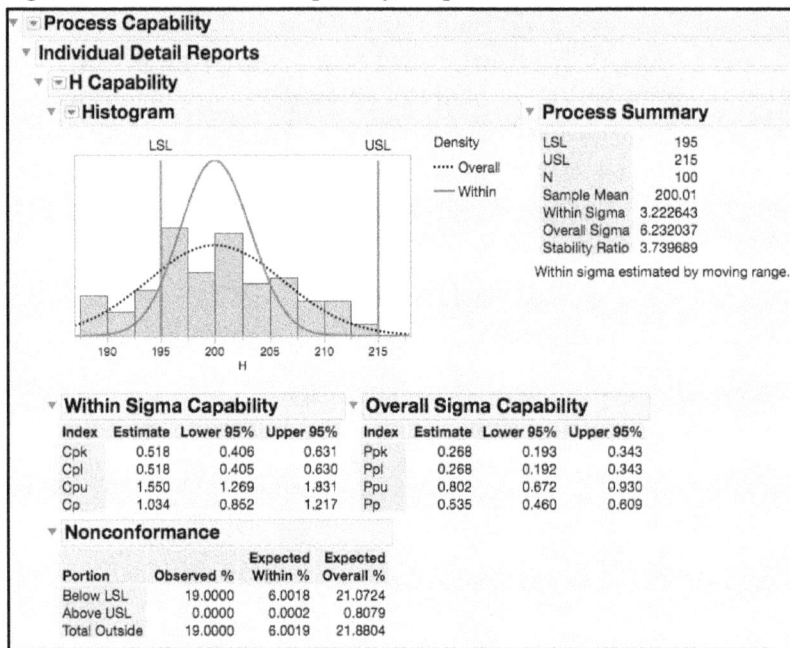

The **Process Capability** platform is launched from the **Process Screening** platform when this option is selected. Figure 6.24 shows the partial **Process Capability** output for Parameter H. This includes a **Process Summary** with the specification limits; sample size, within and overall sigma, and the Stability Ratio. The process capabilities indices are reported along with 95% confidence intervals, and the expected percent of nonconformance is calculated using the within and the overall sigma.

Process Health Assessment Using the Process Capability Platform

In this section, we will show how to carry out a process health assessment for many parameters using the **Process Capability** platform. We will use the same JMP data table, which contains 10 parameters to assess the overall health of this process. Recall, each parameter has a specification that is represents an internal release limit. The data includes one hundred measurements and represents several months of manufacturing. For each parameter, a subgroup size of n = 1 is appropriate and therefore, an XmR chart is used for monitoring purposes.

The following steps illustrate how to use the **Process Capability** platform.

1. Open Chapter 6 – Semiconductor 6.1.jmp, which has variables called *Sample ID, A, B, C, D, E, F, G, H, I,* and *J.* In this table, Sample ID is the subgroup variable, and A through J are the PHA parameters.

2. Make sure that all parameters have their **Spec Limits** included as Column Properties in the JMP table. This can be verified by double-clicking on the column label (for example, **A** and selecting **Column Properties ▶ Spec Limits (Figure 6.25)**). Alternatively, we can

click on the * symbol next to each column name in the Columns panel in the left-hand side of the JMP table. A list of the column properties will pop up. We can select **Spec Limits** to view the limits entered.

Figure 6.25 Identifying Column Properties

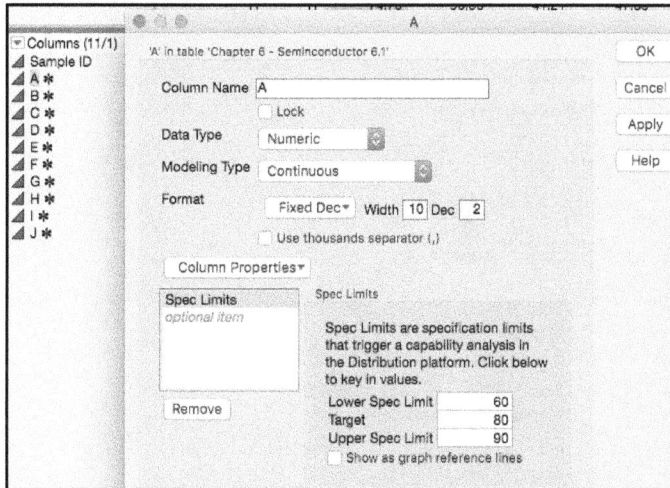

3.   Select **Analyze ▶ Quality and Process ▶ Process Capability** (Figure 6.26).

Figure 6.26 Launching the Process Capability Platform

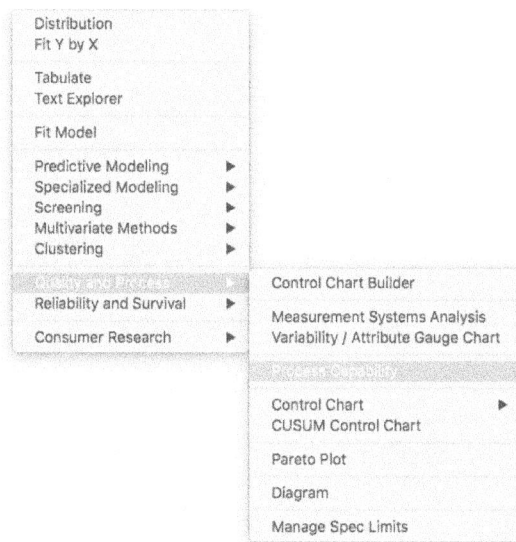

4. A launch window will appear. In the left-hand window, click **A** and hold the shift key and select **J**, so they all parameters are highlighted and then click **Y, Process**. Leave the default settings for the remaining fields (Figure 6.27).

Figure 6.27 Launch Window for Process Capability Platform

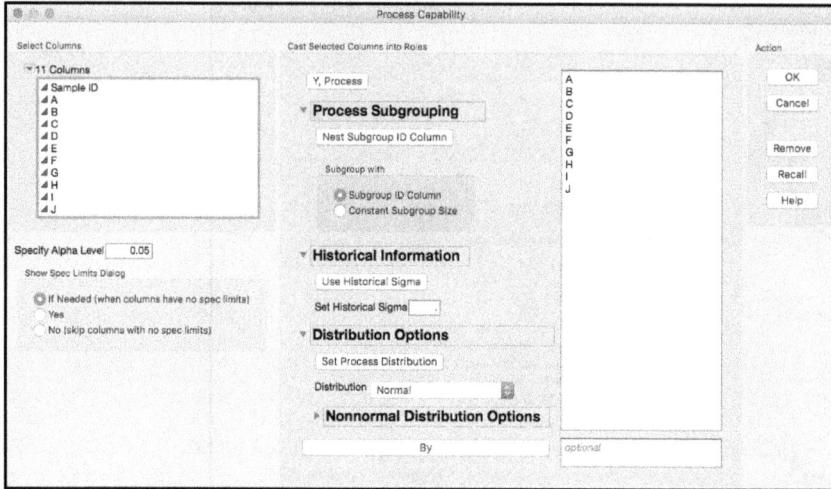

Note the dialog window has some unique features. The **Distribution Options** allows the user to identify different distributions for each parameter, such as, the normal, lognormal, gamma, or best fit. For more information see the online help.

5. Click **OK** when finished. The default output is shown in Figure 6.28.

Figure 6.28 Process Capability Default Output for Semiconductor Data

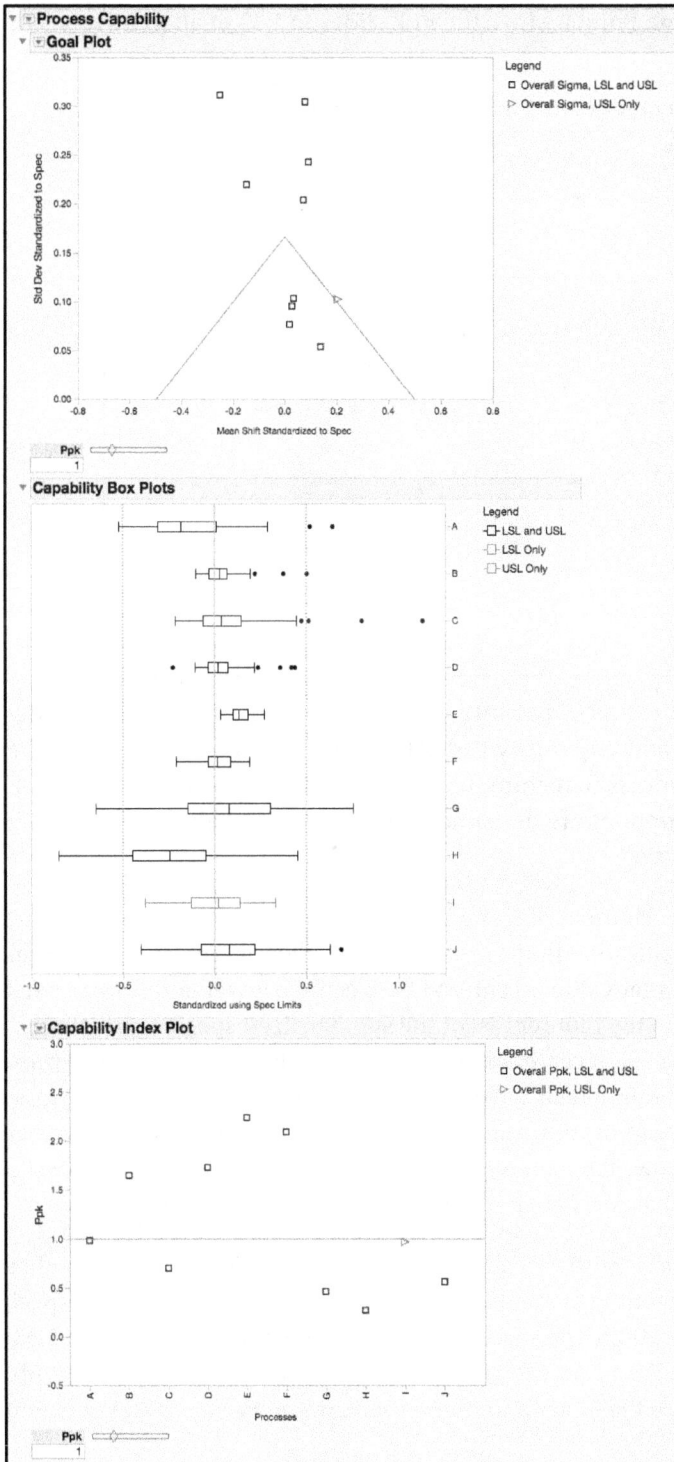

6. Click on the red triangle next to the **Capability Index Plot** and select **Shade levels** and **Label Overall Sigma Points**. Move the Ppk slider to 1.33 or alternatively, click in the box and type **1.33**.

Figure 6.29 Capability Index Plot for Semiconductor Data

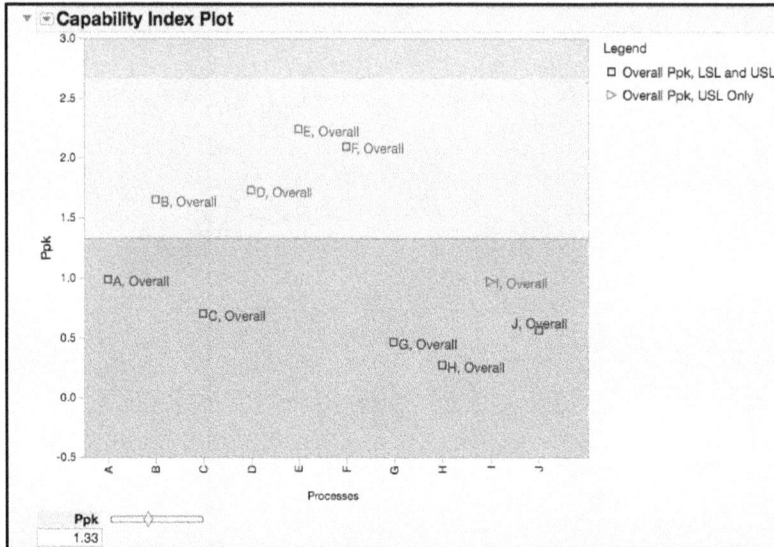

The default output for the **Process Capability** platform shown in Figure 6.28 includes the **Goal Plot**, **Capability Box Plots**, and **Capability Index Plot**. The **Goal Plot** was discussed extensively in the last section using the **Process Screening** platform (see discussion of Figure 6.21, Figure 6.22 and Figure 6.23). Since the options are the same in the **Process Capability** platform, it is not discussed further in this section.

The **Capability Box Plots** in Figure 6.28 show box plots for Parameter A through Parameter J, where the data and the specification limits are centered to their target values and scaled by the specification range. If the target value is centered between the lower and upper specification limits, then the green lines in the plot represent the standardized specification limits. For parameters where the target is not centered, only one of the dotted green lines represent the standardized specification limit, which is closer to the target value. The opposite green line represents the same distance in the opposite direction. From this plot, we can see that Parameters B, D, E, and F have the narrowest boxes, while Parameters G and H have the widest ones. We can also see that Parameter B, D, F and I are performing closest to their target values.

The **Capability Index Plot** in Figure 6.29 shows P_{pk} for each parameter. The legend on the right hand side shows different plotting symbols for those parameters with a 2-sided specification and those with a 1-sided specification. The shaded red area in the plot indicates $P_{pk} < 1.33$; the yellow shaded area is where $1.33 < P_{pk} < 2.66$; and, the green area shows the parameters with $P_{pk} > 2.66$. This plot makes it easy to see the lower and higher performing Parameters. For example,

Parameter E has the largest P_{pk}, while Parameter H has the lowest P_{pk}. There are six parameters with $P_{pk} < 1.33$ and four parameters with $P_{pk} > 1.33$.

The **Process Capability** platform has several other important options, including access to the **Process Performance Graph**.

7. Click on the red triangle next to the **Process Capability** banner at the top of the window to bring up a list of platform options (Figure 6.28). A brief description for some of the options that we will illustrate in this section is shown below. To launch the online help, click on the **?** at the top of the window and then click anywhere in the default output shown in Figure 6.28.

Figure 6.30 Process Capability Platform Options

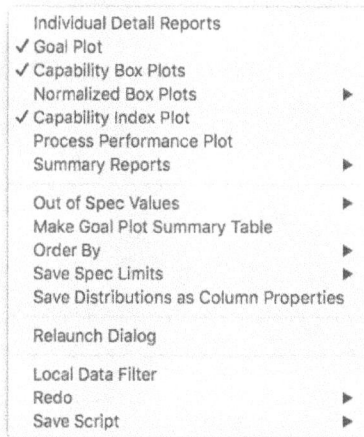

a. **Individual Detail Reports**: produces individual reports for each parameter, which includes the histogram, process summary information, process capability indices using within sigma and overall sigma and nonconformance estimates.

b. **Goal Plot**: produces a plot using the spec-normalized mean shift on the X axis and the spec-normalized standard deviation on the Y axis. The Ppk slider can be used to change the goal lines. The default is set to Ppk = 1.

c. **Capability Box Plots**: produces box plots for each parameter that are centered by their target value and scaled by specification range. The plot is helpful for comparing parameters with respect to their specification limits.

d. **Normalized Box Plots:** produces box plots for each parameter that are centered by their mean value and scaled by their standard deviation. The green lines are the normalized specification limits. The plot is helpful for comparing parameters with respect to their specification limits

e. **Capability Index Plot**: shows the Ppk for all parameters, where Ppk is plotted on the Y axis and each parameter name is on the X axis. If a non-normal distribution was used, it appears in the label for the parameter.

f. **Process Performance Plot**: produces a graph that plots P_{pk} versus the Stability Ratio for all parameters included in the Summary table. The graph is divided into four quadrants: Stable and Capable, Capable but Unstable, Incapable but Stable and Incapable and Unstable. The default boundary to classify a parameter as Capable is Ppk ≥ 1.33; while the default boundary to classify a parameter as Stable is Stability Ratio ≤ 1.5. These default boundaries can be changed.

g. **Summary Reports**: produces a table containing specifications, mean, standard deviation estimates, Stability Ratio, capability indices, and nonconformance statistics.

h. **Out of Spec Values**: this option is used to highlight and color any out-of-spec values in the JMP data table.

i. **Make Goal Plot Summary Table**: creates a JMP table for the information shown in the Goal Plot. The table includes the spec-normalized mean shift and its spec-normalized standard deviation.

8. From the red triangle at the top of the window, select **Individual Detail Reports**.

Figure 6.31 Individual Detail Report for Parameter A

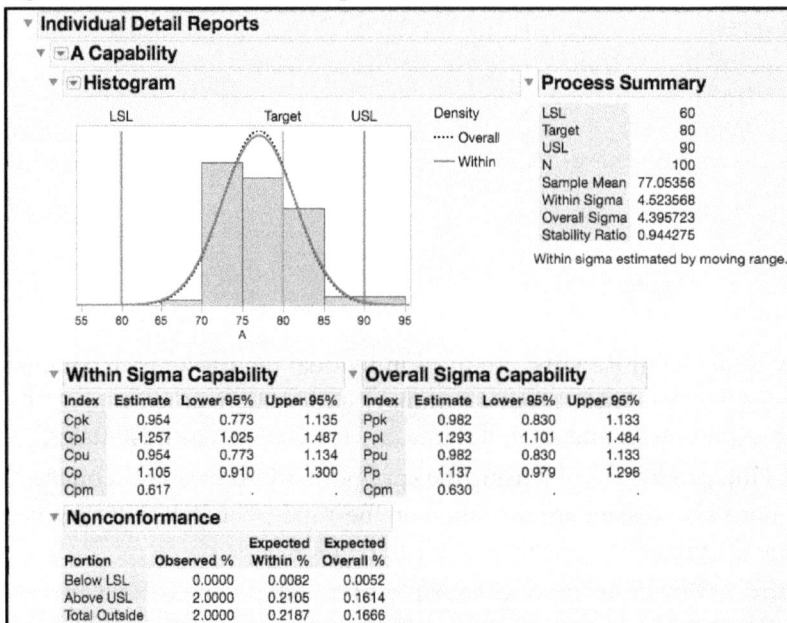

9. From the red triangle at the top of the window, select **Normalized Box Plots ▶ Overall Sigma Normalized Box Plots.** From the red triangle next to **Process Capability** select **Order By ▶ Overall Sigma Ppk Descending**. The output is shown in Figure 6.32.

Figure 6.32 Overall Sigma Normalized Box Plots

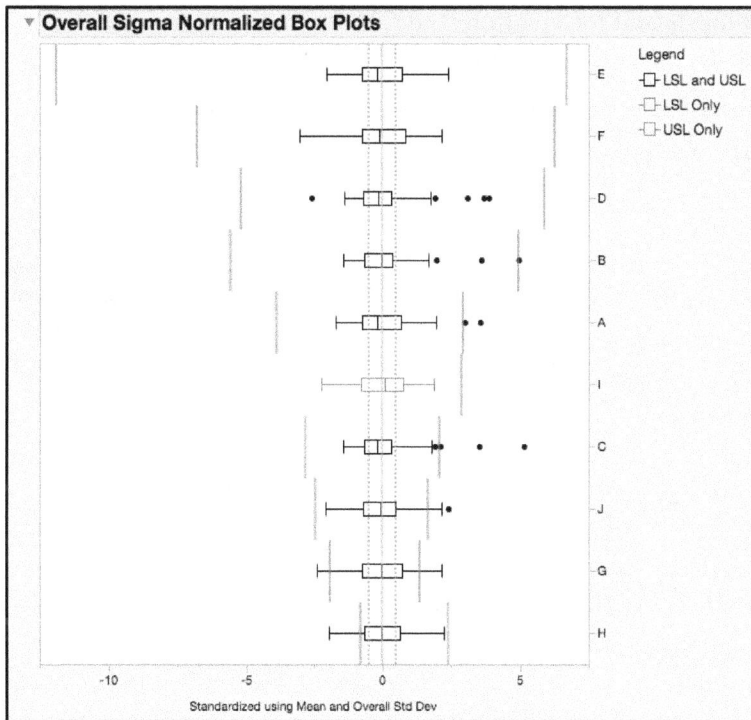

The **Individuals Details Report** for Parameter A is shown in Figure 6.31. Note the Stability Ratio and the components that comprise it are provided in the Process Summary section. Process capability indices using the overall sigma and within sigma are also presented. Note this output was discussed in the previous two sections and is not further discussed here. (See discussions related to Figure 6.4, Figure 6.6, or Figure 6.15, for example.)

The **Overall Sigma Normalized Box Plots** in Figure 6.32 shows the data and specifications normalized using the parameter's mean and standard deviation. The data and specification limits are standardized to a mean = 0 and a standard deviation = 1. The legend on the right-hand side uses different box colors that represent if the parameter has a 1- or 2-sided specification limit. For the Semiconductor data, all parameters have a 2-sided specification with the exception of Parameter I, which only has a USL. The blue lines in the plot are the standardized upper specification limits and the red lines are the standardized lower specification limits. The gray dotted lines are drawn at ± 0.5σ; which contains 50% of the distribution, using the normal distribution. This plot also makes it easy to see which parameters are performing well and which ones are too close to their specifications. For example, Parameter E, F, and D are at the top of the list, while Parameter J, Parameter G and Parameter H are at the bottom.

10. From the red triangle at the top of the window in Figure 6.28, select **Process Performance Plot.** Change the default for Ppk from 1 to 1.33. From the red triangle next to **Process Performance Plot** select **Label Points.**

Figure 6.33 Process Performance Plot in Process Capability Platform

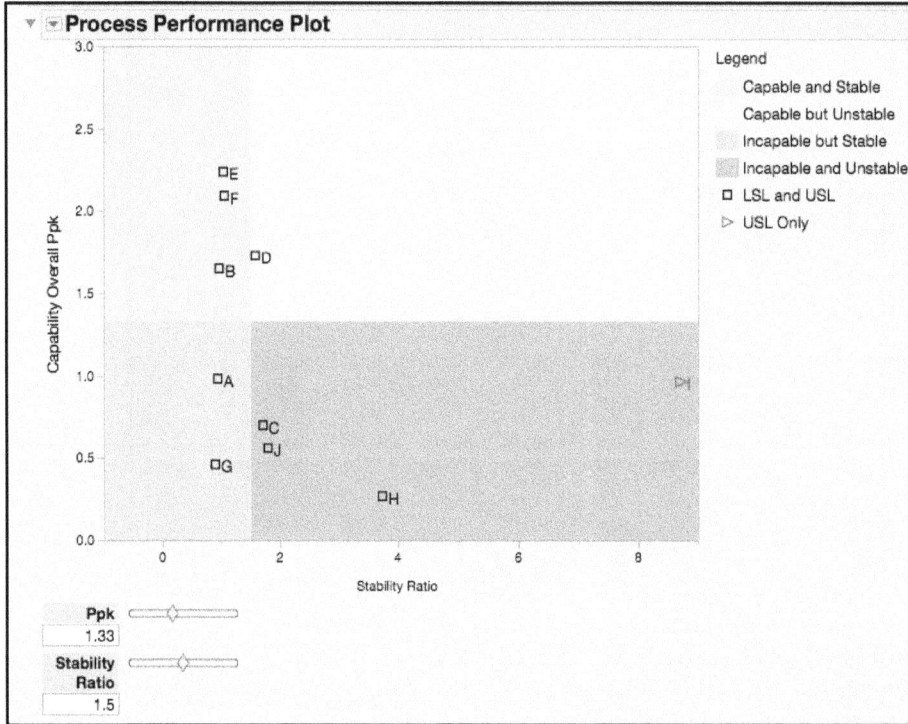

11. From the top of the window (Figure 6.28), select **Summary Reports ▶ Overall Sigma Summary Report.** Note leave it ordered by descending Ppk.

Figure 6.34 Overall Sigma Summary Report for Semiconductor Data

▾ **Overall Sigma Capability Summary Report**

Process	LSL	Target	USL	Sample Mean	Overall Sigma	Stability Ratio	Ppk	Ppl	Ppu	Pp	Cpm	Expected % Outside	Expected % Below LSL	Expected % Above USL	Observed % Outside	Observed % Below LSL	Observed % Above USL
E	0.5	1	1.5	1.1393	0.053716	1.041421	2.238	3.967	2.238	3.103	1.116	0.0000	0.0000	0.0000	0.0000	0.0000	0.0000
F	0.8	1.2	1.6	1.2154	0.061273	1.05626	2.092	2.260	2.092	2.176	2.110	0.0000	0.0000	0.0000	0.0000	0.0000	0.0000
D	10	45	90	47.43102	7.21521	1.575677	1.729	1.729	1.967	1.848	1.532	0.0000	0.0000	0.0000	0.0000	0.0000	0.0000
B	55	85	115	86.73206	5.710845	0.968965	1.650	1.852	1.650	1.751	1.676	0.0000	0.0000	0.0000	1.0000	0.0000	1.0000
A	60	80	90	77.05356	4.395723	0.944275	0.982	1.293	0.982	1.137	0.630	0.1666	0.0052	0.1614	2.0000	0.0000	2.0000
I			550	506.52	15.03362	8.711302	0.964	.	0.964	.	.	0.1913		0.1913	0.0000		0.0000
C	25	40	55	42.19304	6.12121	1.705829	0.697	0.936	0.697	0.817	0.769	2.0696	0.2487	1.8209	3.0000	0.0000	3.0000
J	42	50	58	51.48131	3.884976	1.792983	0.559	0.814	0.559	0.686	0.641	5.4014	0.7333	4.6681	10.0000	0.0000	10.0000
G	1.1	1.2	1.3	1.216	0.060919	0.898329	0.460	0.635	0.460	0.547	0.529	11.2409	2.8444	8.3965	10.0000	3.0000	7.0000
H	195		215	200.01	6.232037	3.739689	0.268	0.268	0.802	0.535		21.8804	21.0724	0.8079	19.0000	19.0000	0.0000

The **Process Performance Plot** is shown in Figure 6.33. This graph highlights the two dimensions of process health, Stability and Capability. Using the parameter's P_{pk} and Stability Ratio, the parameters are distributed in four quadrants. The ideal parameters (Capable and Stable) are shown in the green shaded region of the plot, where $P_{pk} > 1.33$ and Stability Ratio < 1.5. The Incapable and Unstable parameters are in the red-shaded region, where $P_{pk} < 1.33$ and Stability Ratio > 1.5. The legend also provided information about the parameter's specifications. This plot was discussed in great detail in the last section, using the **Process Screening** platform (see the discussion for Figure 6.13, for example).

The **Overall Sigma Capability Summary Report** is in Figure 6.34. It contains the specification limits, sample mean, overall sigma, Stability Ratio, P_{pk}, C_{pm}, and estimated and actual percent of points outside of the specification limits. The report is ordered using P_{pk}, from largest to smallest. For example, Parameter E has Ppk = 2.238, Stability Ratio = 1.04 and 0% of observed or expected values outside of the specifications. In contrast, Parameter H has P_{pk} =0.268, Stability Ratio = 3.74 and 19% of observed points outside of the specifications. Note the summary report can easily be exported to a JMP table by right clicking inside of the table and selecting **Make Into Data Table**.

The **Process Capability** platform can be used to identify out-of-spec results in the JMP table.

12. From the red triangle at the top of the window (Figure 6.28), select **Out of Spec Values ▶ Color Out of Spec Values.**

Figure 6.35 Out of Spec Values in JMP Table

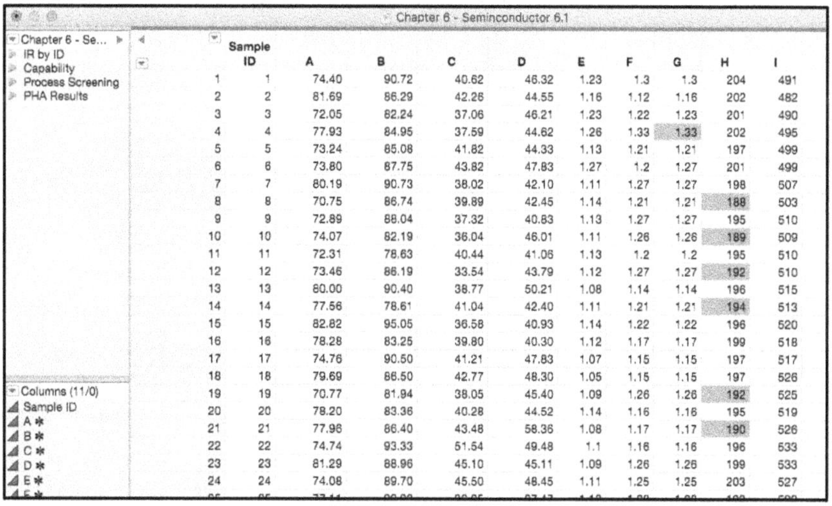

	Sample ID	A	B	C	D	E	F	G	H	I
1	1	74.40	90.72	40.62	46.32	1.23	1.3	1.3	204	491
2	2	81.69	86.29	42.26	44.55	1.16	1.12	1.16	202	482
3	3	72.05	82.24	37.06	46.21	1.23	1.22	1.23	201	490
4	4	77.93	84.95	37.59	44.62	1.26	1.33	1.33	202	495
5	5	73.24	85.08	41.82	44.33	1.13	1.21	1.21	197	499
6	6	73.80	87.75	43.82	47.83	1.27	1.2	1.27	201	499
7	7	80.19	90.73	38.02	42.10	1.11	1.27	1.27	198	507
8	8	70.75	86.74	39.89	42.45	1.14	1.21	1.21	188	503
9	9	72.89	88.04	37.32	40.83	1.13	1.27	1.27	195	510
10	10	74.07	82.19	36.04	46.01	1.11	1.26	1.26	189	509
11	11	72.31	78.63	40.44	41.06	1.13	1.2	1.2	195	510
12	12	73.46	86.19	33.54	43.79	1.12	1.27	1.27	192	510
13	13	80.00	90.40	38.77	50.21	1.08	1.14	1.14	196	515
14	14	77.56	78.61	41.04	42.40	1.11	1.21	1.21	194	513
15	15	82.82	95.05	36.58	40.93	1.14	1.22	1.22	196	520
16	16	78.28	83.25	39.80	40.30	1.12	1.17	1.17	199	518
17	17	74.76	90.50	41.21	47.83	1.07	1.15	1.15	197	517
18	18	79.69	86.50	42.77	48.30	1.05	1.15	1.15	197	526
19	19	70.77	81.94	38.05	45.40	1.09	1.26	1.26	192	525
20	20	78.20	83.36	40.28	44.52	1.14	1.16	1.16	195	519
21	21	77.96	86.40	43.48	58.36	1.08	1.17	1.17	190	526
22	22	74.74	93.33	51.54	49.48	1.1	1.16	1.16	196	533
23	23	81.29	88.96	45.10	45.11	1.09	1.26	1.26	199	533
24	24	74.08	89.70	45.50	48.45	1.11	1.25	1.25	203	527

The JMP table highlighting the out-of-specification results is shown in Figure 6.35. In this table values shaded blue are above the USL, while red shading indicates results that are below the LSL. For example, the fourth result for Parameter G of 1.33 is above the USL = 1.3, while the eighth value for Parameter H of 188 is below the LSL = 195. If the rows are also selected, then a subset table can be created for all rows with one or more out-of-spec results.

Statistical Insights

The previous sections have shown how easy it is to assess the health of our processes using the two dimensions of capability, voice of the customer, and stability, voice of the process, via the **Process Screening** or **Process Capability** platforms. In addition, the **Process Performance Graph** allows us to focus our improvements activities by helping visualize where in the four quadrants of

Stable and Capable; Capable but Unstable; Incapable but Stable; and Incapable and Unstable, our processes fall. These four quadrants are what Wheeler and Chambers (1992) describe as the four possibilities for any process.

Statistics Note 6.5: Wheeler and Chambers (1992) describe the four possibilities for any process as:

- Ideal State   —in control and 100% conforming,
- Threshold State —in control but not conforming,
- Brink of Chaos —out of control but 100% conforming,
- State of Chaos —out of control and not conforming.

A process in the Stable and Capable is in the *ideal state* where we predictably meet specifications. In the Capable but Unstable we are in a *predictability issue* because, even though our process meets specifications, it cannot do it in a predictable way. The Incapable but Stable situation presents us with a *yield issue* where the process does not meet the specifications but at least we can predict how our yields will be impacted. Finally, a process that is in the Incapable and Unstable quadrant puts us in a *double trouble* situation, where the process cannot meet specifications and is unpredictable.

Ideally, we want all our processes to be in the ideal state, and we should focus our improvement efforts in those processes that do not fall in that state. How do we decide which of the processes that have a predictability issue, yield issue, or are in the double trouble state should be tackled first? Ramírez (2018) points out that, in addition to the two dimensions of capability and stability that are displayed in the **Process Performance Graph**, we can make the size of each point proportional to a third dimension that reflects additional information about the process. This will help us decide which processes we should tackle first, based on their impact on dimensions like yield or sales, for example. A somewhat hidden feature in the **Process Screening Platform** allows us to do just that.

Figures 6.13 and 6.33 shows the **Process Performance Graph** for the 10 processes in the data set Chapter 6 – Semiconductor 6.1.jmp. We have 4 processes (C, H, I, J) in the double trouble state, two processes (A, G) that have a yield issue, one process (D) with a predictability issue, and 3 processes (B, E, F) in the ideal state. Which of the 4 processes in the double trouble state should be undertaken first? This is where additional information about our processes is helpful. For each process parameter in the data set Chapter 6 – Semiconductor 6.1.jmp, we are going to use accumulated cost of goods (ACG), in thousands of dollars, to vary the size of the points in the **Process Performance Graph.**

The **Process Screening** platform has an option that allows the size of the points to be a function of a third variable, ACG in this example. The following steps illustrate how to access and use this option in the **Process Capability** platform.

1. Open Chapter 6 – Semiconductor 6.1.jmp, which has variables called *Sample ID, A, B, C, D, E, F, G, H, I,* and *J*. In this table, Sample ID is the subgroup variable, and A through J are the PHA parameters.
2. Open Chapter 6 - Semiconductor 6.1 Limits Table with ACG.jmp, which has variables called *Process, LCL, UCL,* and *ACG* (Figure 6.36). The columns LCL and UCL are the lower and upper control limits for the individual measurements charts, corresponding to each process defined by the Process column. The column ACG is the accumulated costs of goods, in thousands of dollars, for each of the process steps defined by column Process.

Figure 6.36 Semiconductor Data Limits Table with Accumulated Cost of Goods

Chapter 6 - Seminconductor 6.1 Limits Table with ACG

	Process	LCL	UCL	ACG (Thousands)
1	A	63.482854932	90.624261738	2
2	B	69.327319597	104.13680041	10
3	C	28.1328585	56.253214836	5
4	D	30.187077732	64.674965612	2
5	E	0.9813904751	1.2972095249	3
6	F	1.0365432932	1.3942567068	3
7	G	1.0231785053	1.4088214947	8
8	H	190.3420699	209.6779301	7
9	I	491.23929938	521.80070062	0.5
10	J	42.777253133	60.185359127	1

3. Alternatively, to create the Semiconductor 6.1 Limits Table with ACG shown in Figure 6.36, we can save the lower and upper control limits for the individual measurements charts by clicking on the red triangle next to **Process Screening** and selecting **Save Details Table** (Figure 6.37).

Figure 6.37 Saving Control Limits

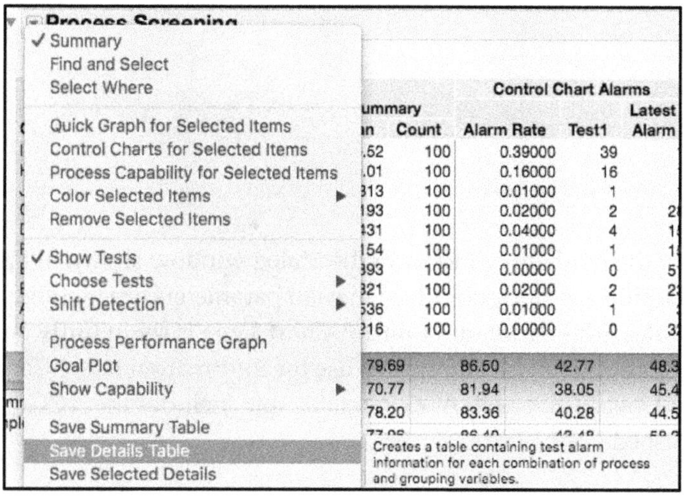

4. The resulting table is displayed in Figure 6.38. From this table we can subset the LCL and UCL for each process parameters to create the dataset Chapter 6 - Semiconductor 6.1 Limits Table with ACG.jmp in Figure 6.36.

Figure 6.38 Details Table with Control Limits for Semiconductor Data

	Column	Subgroup	Individual	Moving Range	LCL	UCL	MR Limit	Subgroup Size	Alarms	Drift
1	A	1	74.401	•	63.482854932	90.624261738	16.67335779	1		0.0064597864
2	A	2	81.6876667	7.2866667	63.482854932	90.624261738	16.67335779	1		0.0079918843
3	A	3	72.051	9.6366667	63.482854932	90.624261738	16.67335779	1		0.0088337993
4	A	4	77.9345	5.8835	63.482854932	90.624261738	16.67335779	1		0.0085971727
5	A	5	73.235	4.6995	63.482854932	90.624261738	16.67335779	1		0.0086455822
6	A	6	73.796	0.561	63.482854932	90.624261738	16.67335779	1		0.0075076385
7	A	7	80.1896667	6.3936667	63.482854932	90.624261738	16.67335779	1		0.0082480016
8	A	8	70.7543333	9.4353334	63.482854932	90.624261738	16.67335779	1		0.0081039708
9	A	9	72.8855	2.1311667	63.482854932	90.624261738	16.67335779	1		0.0058972234
10	A	10	74.0696667	1.1841667	63.482854932	90.624261738	16.67335779	1		0.0045615801
11	A	11	72.3146667	1.755	63.482854932	90.624261738	16.67335779	1		0.003022697
12	A	12	73.4585	1.1438333	63.482854932	90.624261738	16.67335779	1		0.0012690145
13	A	13	80.0006667	6.5421667	63.482854932	90.624261738	16.67335779	1		0.0015929514
14	A	14	77.556	2.4446667	63.482854932	90.624261738	16.67335779	1		0.0030029407
15	A	15	82.8233333	5.2673333	63.482854932	90.624261738	16.67335779	1		0.0051470062
16	A	16	78.2816667	4.5416666	63.482854932	90.624261738	16.67335779	1		0.0074461794
17	A	17	74.7643333	3.5173334	63.482854932	90.624261738	16.67335779	1		0.0074735913
18	A	18	79.6856667	4.9213334	63.482854932	90.624261738	16.67335779	1		0.0078445308
19	A	19	70.7713333	8.9143334	63.482854932	90.624261738	16.67335779	1		0.0070681279

5. Now launch the **Process Screening** platform from select **Analyze ▶ Screening ▶ Process Screening,** as shown in Figure 6.39.

Figure 6.39 Launching the Process Screening Platform

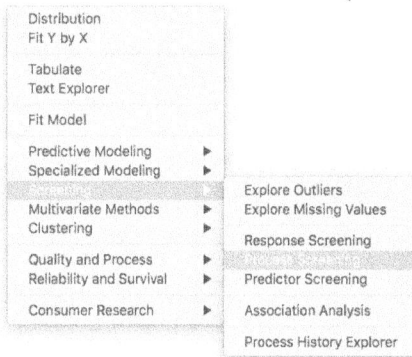

6. As we did before, in the left-hand window of the dialog window shown in Figure 6.40, click **A** and hold the shift key and select **J**, so they all parameters are highlighted and then click **Process Variables**. Make sure that **Control Chart Type** is set to **Indiv and MR**. Click **Use Limits Table**. This option will allow us to use the information in the Chapter 6 - Semiconductor 6.1 Limits Table with ACG.jmp table that includes the **ACG** variable. Click **OK** when finished.

Figure 6.40 Launch Window for Process Screening

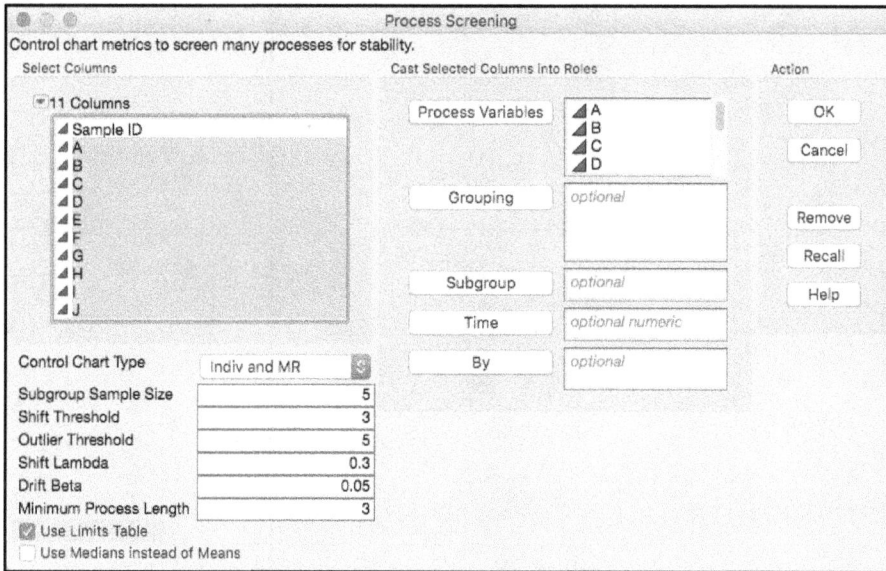

7. A launch window appears (Figure 6.41) to select the limits table. Select Chapter 6 - Semiconductor 6.1 Limits Table with ACG.jmp. Click **OK** when finished.

Figure 6.41 Choose Limits Table Dialog Window

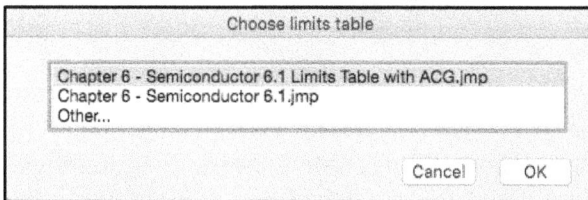

By default, the following window will populate the Process Variable, LCL and UCL. We now select **ACG** and click the **Importance** field at the bottom of the window, as shown in Figure 6.42. Click **OK** when finished.

Figure 6.42 Limits Specifications Table Dialog Window

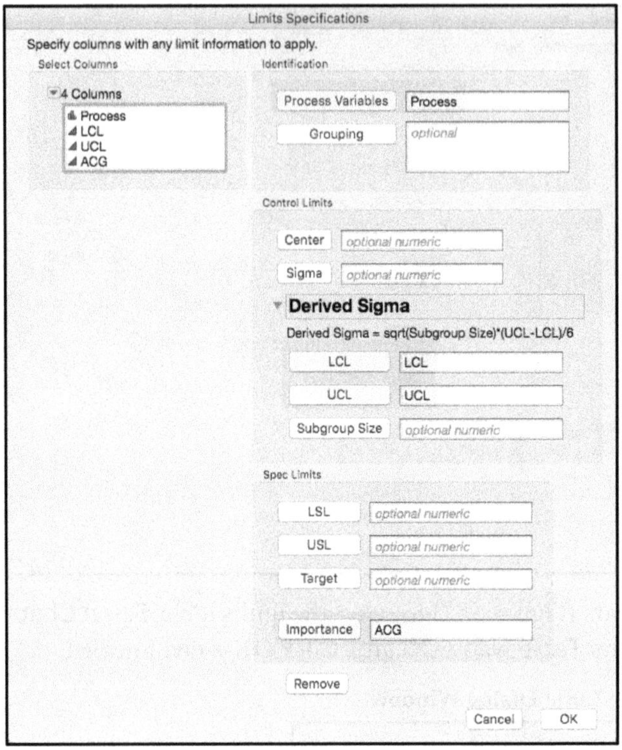

The **Process Screening** report appears sorted by the **Importance** variable ACG. From the red triangle at the top of the window, select **Process Performance Graph**. Highlight the parameters in the **Summary** table by clicking on B and then I, while holding the shift key. Right-click in the Process Performance Graph and select **Row Label** from the menu. Figure 6.43 shows the **Process Performance Graph**, which is similar to Figure 6.13, but with the size of the points proportional to the values in the **Importance** column ACG. The large the circle, the more ACG associated with that process.

Figure 6.43 Process Performance Graph With Points Sized by ACG

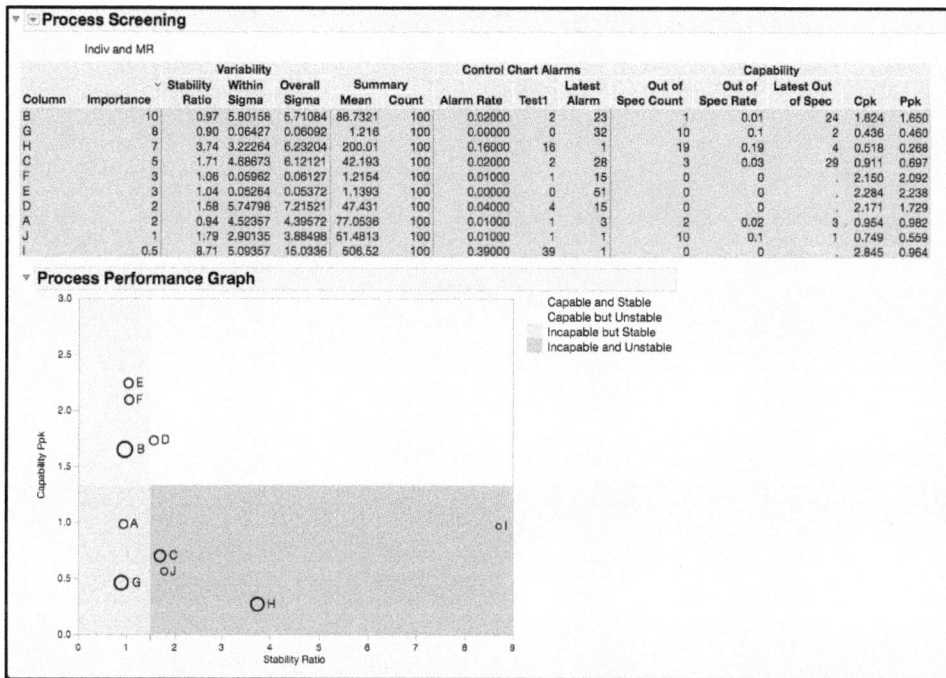

The Process Screening **Summary** table, shown in Figure 6.43, includes the Importance value for each process parameter. As was noted in Step 9, the table is sorted by the Importance value, instead of the Stability Ratio. The parameters in Double Trouble (Incapable and Unstable) are the ones that should be scrutinized first. These parameters can be further prioritized using the ACG variable, in parenthesis, resulting in the following ordering: H ($7K) & C ($5K) and J ($1K) & I ($0.5K). Not only are parameters H and C incapable and unstable, but they also have larger accumulated cost of goods. In contrast, Parameters I and J, which are also in Double Trouble, have the lowest ACG. If it is not possible to provide resources to improve the performance of these four parameters simultaneously, then efforts might focus on H and C.

The Incapable and Stable quadrant should be considered next, since it might result in a yield issue. Using the ACG, Parameter G ($8K) has more Importance than Parameter A ($2K), and has the second highest ACG among all parameters. Once again, if resources are limited then Parameter G should move to the top of the list.

Parameter B ($10K), which has the highest importance, or ACG, is in the ideal quadrant. This is the ideal situation and efforts should be placed on maintaining its current performance. Since E ($3K) and F ($3K) are also in the ideal quadrant, focus should also be placed on maintenance. The last parameter, D ($2K), which is capable but unstable, and has a lower ranked Importance, should be closely monitored to ensure that it doesn't slip to a lower performance quadrant.

Chapter 7: Cumulative Sum and Exponentially Weighted Moving Average Control Charts

Overview

This chapter illustrates how to generate CUSUM and EWMA control charts using examples from Chapter 9, Cumulative Sum and Exponentially Weighted Moving Average Control Charts, of *Introduction to Statistical Quality Control* (ISQC), and includes discussions, tips and statistical insights on some of the fundamental concepts of statistical process control.

These control chart techniques are readily available for data measured on a continuous scale, where smaller mean shift detection is desired. They can also be adopted for attribute data, such as, those modeled with a binomial or Poisson distribution. These chart types may be more often used during phase 2 monitoring, when special cause variation has been reduced.

Two JMP platforms are highlighted in this chapter, the **CUSUM Control Chart** and **Control Chart** platforms.

CUSUM and EWMA Control Chart Review

The Shewhart control charts presented in Chapter 3, such as, the Individual Measurement and Moving Range (XmR) or XBar and R charts, use three sigma control limits to detect shifts in the process mean. These charts are known to detect larger shifts in the process mean, on the order of at least 2σ or more. Runs tests, like Western Electric Rules or Nelson rules, can be added to the chart to increase their sensitivity to detect smaller shifts in the process mean. However, if smaller mean shift detection is required, then this approach is not optimal due to the significant increase in false signals (see Britt, Ramírez, and Mistretta (2016)).

For small shift detection, the cumulative sum (CUSUM) control chart or exponentially weighted moving average (EWMA) control chart should be used. These charts are designed to detect smaller shifts (1.5σ or less) in the process mean quite efficiently. They can be used with individual points, where the rational subgroup is of size 1, or with subgroup means. Unlike Shewhart charts, these charts use charting statistics that build upon previous values, which contribute to their ability to detect shifts more quickly.

The CUSUM chart is designed to rapidly detect small shifts in the mean of a process. For example, in ISQC Chapter 9, Montgomery illustrates the approach using a 1σ shift for a process with a mean that shifts from $\mu = 10$ to $\mu = 11$. The data being monitored, x_i, are transformed into a charting statistic by summing up the deviations of each point from a specified target value, μ_0:

$$C_i = \sum_{j=1}^{i}(x_j - \mu_0) \qquad (1)$$

and C_i is plotted against the sample number i. If the process mean is close to the given target μ_0, then this sum will bounce around zero, and when a shift happens in one direction, the sum will start to become positive and rapidly increase, or become negative and rapidly decrease, depending on the direction of the shift. ISQC equation 9.1 is similar, but uses means instead.

While the basic CUSUM algorithm is shown in equation (1), there are many ways that this approach is further described and advanced in Chapter 9 of ISQC. For example, a tabular form of the CUSUM provides an easier way to identify shifts that are specifically above the target value (μ_0), or specifically below the target value (μ_0), by creating two charting statistics, C_i^+ and C_i^-, as shown in ISQC equations (9.2) and (9.3). These two charting statistics can be overlaid in one plot with "control limits", similar to an Individual Measurement chart, which makes it easier to detect changes. A reference value (or slack value) K can be included in the charting statistic and used with a decision interval H, to further refine the performance of the chart to detect the shift size of interest. The data can also be normalized so that the charting statistics are standardized so that they are not scale-dependent and the same values of K and H can be used for many parameters. Finally, the CUSUM can be adjusted for short runs, use with the binomial or Poisson distribution or to monitor the process variability (σ) instead of the process mean (μ).

The EWMA chart is also designed to detect smaller shifts in the mean of a process. As the name suggests, the charting statistic is a weighted average of the individual points, x_i. The data are averaged using weights that depend on the "age" of the sample mean. The weight of the most recent x_t is the largest and each subsequent weight decreases geometrically with the age of the sample. The charting statistic z_i has the following form (ISQC equation (9.22):

$$z_i = \lambda x_i + (1-\lambda)z_{i-1} \qquad (2a)$$

If we substitute the value of z_{i-1} recursively in equation (2a) we obtain ISQC equation 9.23.

$$z_i = \lambda \sum_{j=0}^{i-1}(1-\lambda)^j x_{i-j} + (1-\lambda)^i z_0 \qquad (2b)$$

In equation (2b) z_0 is a starting value, such as a target value or the average of preliminary data. The statistic z_i is plotted against the sample number i. In equation (2a), $0 < \lambda < 1$ and all of the weights, $\lambda(1-\lambda)^i$, in the weighted average decrease geometrically. For example, the weights in ISQC Figure 9.6 are calculated as: $\lambda_0 = 0.2$, $\lambda_1 = 0.2 \times (1-0.2)^1 = 0.16$, $\lambda_2 = 0.2 \times (1-0.2)^2 = 0.128$, $\lambda_3 = 0.2 \times (1-0.2)^3 = 0.1024$, $\lambda_4 = 0.2 \times (1-0.2)^4 = 0.0819$, $\lambda_5 = 0.2 \times (1-0.2)^5 = 0.0066$, $\lambda_6 = 0.2 \times (1-0.2)^6 = 0.0052$, $\lambda_7 = 0.2 \times (1-0.2)^7 = 0.0042$, $\lambda_8 = 0.2 \times (1-0.2)^8 = 0.0034$, $\lambda_9 = 0.2 \times (1-0.2)^9 = 0.0027$, and so on.

The control limits for the EWMA charting statistic are not static; they increase for the first several periods and then reach a steady state when the number of observations gets larger. The control limits, shown in ISQC equations (9.25) and (9.26), depend on the choice of the weighting factor λ and a factor L, which multiplies the standard deviation, σ. The ability of the EWMA to detect a shift of a certain size depends on the value of these two factors and tables are provided that show the average run length for different combinations of λ and L.

Just like the CUSUM, the EWMA can be adapted for short runs, data that follows a binomial or Poisson distribution or to monitor variability. In addition, the EWMA is robust to departures from normality (see ISQC Section 9.2.3), since it is a weighted average of many points. Some people also prefer it to a CUSUM because of its natural extension to time series analysis; since the EWMA provides a forecast of where the process mean will be at the next time period, assuming an IMA (1, 1) model.

Finally, a simple unweighted moving average of the data can be used to detect smaller shifts in the mean. This control chart is referred to as a Moving Average control chart. The charting statistic M_i is a simple average of the observations in the number of periods included in the span w, as is shown in ISQC equation 9.37:

$$M_i = \frac{x_i + x_{i-1} + x_{i-2} + \cdots + x_{i-w+1}}{w} \qquad (3)$$

where the charting statistic is plotted against the sample number. For example, if $w = 5$ then at time period i, the most recent five observations are averaged and the oldest observation is removed. Similar to the EWMA, the control limits reach a steady state as more data is added to the chart. However, the limits are wider in the beginning and then tighten later on. The Moving Average chart is more effective than the Shewhart chart in detecting smaller shifts; however, it is not as sensitive as the CUSUM or EWMA.

JMP Small Shift Detection Control Chart Platforms

The JMP **Control Chart** menu provides several platforms to create small shift detection control charts, such as the CUSUM, EWMA or UWMA. These platforms were introduced in Chapter 2. In this chapter, we focus on the use of these platforms for small shift detection applications. We also use a new generation of quality platforms within JMP, the **CUSUM Control Chart**, which makes the generation and interpretation of CUSUM charts easier. Table 7.1 provides a summary of the features we find most useful for the three chart types presented in Chapter 9 of ISQC. Note the Moving Average control chart in ISQC is referred to as UWMA in JMP.

Table 7.1 Comparison of JMP Features for Small Shift Control Charts

JMP Feature	CUSUM Control Chart	EWMA	UWMA
Charting Parameters	Target, Sigma, Head Start, H and K, sides, and data units.	Weight, Sigma, Mean, KSigma, and Alpha.	Moving average span, Sigma, Mean, KSigma, and Alpha.
Saving Limits	Export specified charting parameters and summaries.	Export and import previously specified charting parameters.	Export and import previously specified charting parameters.
ARL Information	Provided as an option in charting output. An ARL profiler is also available.	NA	NA
Interactivity	Relaunch or Redo analysis, Column Switcher.	Relaunch or Redo analysis, Column Switcher.	Relaunch or Redo analysis, Column Switcher.

Examples from ISQC Chapter 9

The examples presented here from Chapter 9 of ISQC are shown in Table 7.2. The examples will be reproduced using JMP, as are shown in ISQC. For some examples, additional output not provided in ISQC is shown to illustrate JMP functionality or elaborate on important points considered by the authors.

Table 7.2 Summary of Examples from ISQC Chapter 9

ISQC Example Number	JMP Table Name	JMP Platform	Key Points
9.1 Random Normal Data - CUSUM	Chapter 7 – ISQC Table 9.1	CUSUM CUSUM Control Chart	Demonstrate upper and lower tabular CUSUMs, reference value k, and FIR to improve detection. Create a CUSUM for variability. The ARL is shown using another data set.
9.2 Random Normal Data - EWMA	Chapter 7 – ISQC Table 9.1	EWMA	Demonstrate the EWMA chart and how to enter the values in the launch window.
9.3 Random Normal Data - UWMA	Chapter 7 – ISQC Table 9.1	UWMA	Demonstrate the UWMA charting platform.

ISQC Example 9.1 Random Normal Data - CUSUM

In this example, we will show how to construct a CUSUM chart in JMP. The data in Table 9.1 of ISQC consists of thirty randomly generated results. The first twenty results were generated using a normal distribution with $\mu = 10$ and $\sigma = 1$ and the next ten results were generated using a random normal distribution with $\mu = 11$ and $\sigma = 1$. A subgroup size of n = 1 is assumed. The data were constructed to evaluate the ability of a CUSUM chart to detect a 1σ shift in the mean in the last 10 observations. Since we know what the true mean and sigma are, the limits for the first 20 observations become LCL=10-3×1=7; CL=10; UCL=10+3×1=13. We need to input these as a columns property for the variable Xi.

The following steps illustrate how to construct a CUSUM control chart using the **Control Chart** platform. But first we will reproduce the Individual and Moving Range chart in ISQC Figure 9.1.

1. Open the JMP data table Chapter 7 – ISQC Table 9.1.jmp, which has columns called *Sample Number, Xi,* and *True Mean.* In this table, Sample Number is the subgroup variable, Xi is the response of interest, and True Mean is the mean used for the data simulation.
2. Select **Rows ▶ Color or Mark by Columns** … and highlight **True Mean**, and for **Markers** select **Standard**. Click **OK**.
3. Right click on **Xi** and select **Control Properties ▶ Control Limits**. In the **Control Limits** section of the dialogue window, select **Individual Measurement** then type in **10** for **Avg, 7** for **LCL** and **13** for **UCL**. Click **OK**.
4. Select **Analyze ▶ Quality and Process ▶ Control Chart Builder**.
5. Drag-and-drop **Sample Number** to the **Subgroup** zone, and **Xi** to the **Y** zone as shown in Figure 7.1. By default, the **Control Chart Builder** platform generates an IR chart. Click **Done**.

Figure 7.1 Control Chart Builder Dialog Window

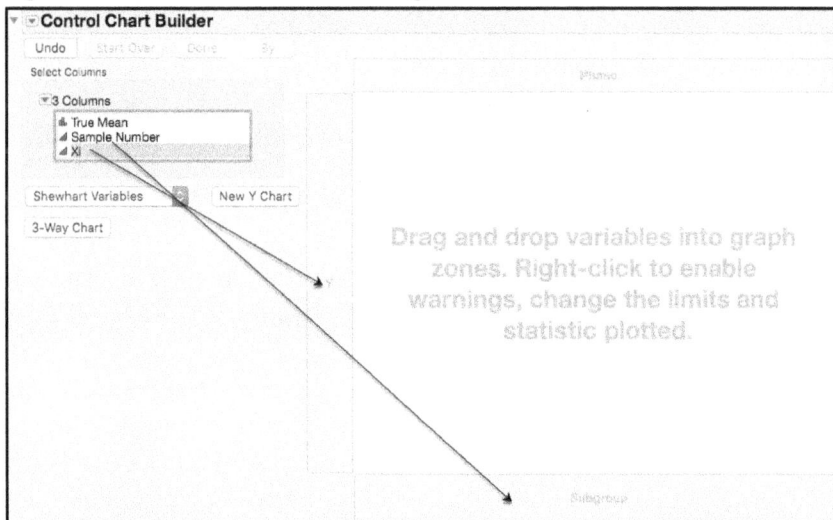

6. Right-click on the graph area to reveal the contextual menu. Select **Warnings** ▶ **Tests** ▶ **All Tests** as shown in Figure 7.2a. The chart is shown in Figure 7.2b.

Figure 7.2a Selecting All Tests from Individual Chart for Xi

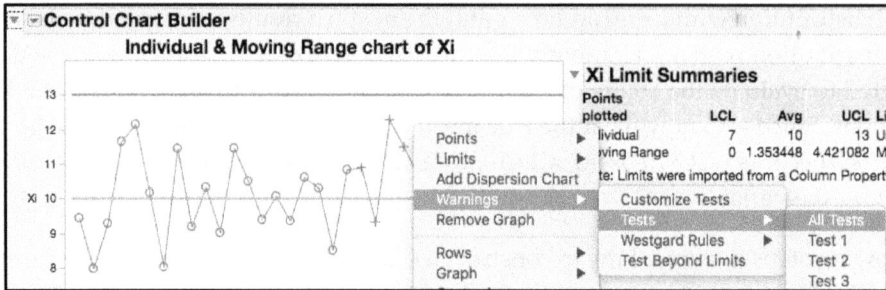

Figure 7.2b Individual and Moving Range Chart for Xi with All Tests

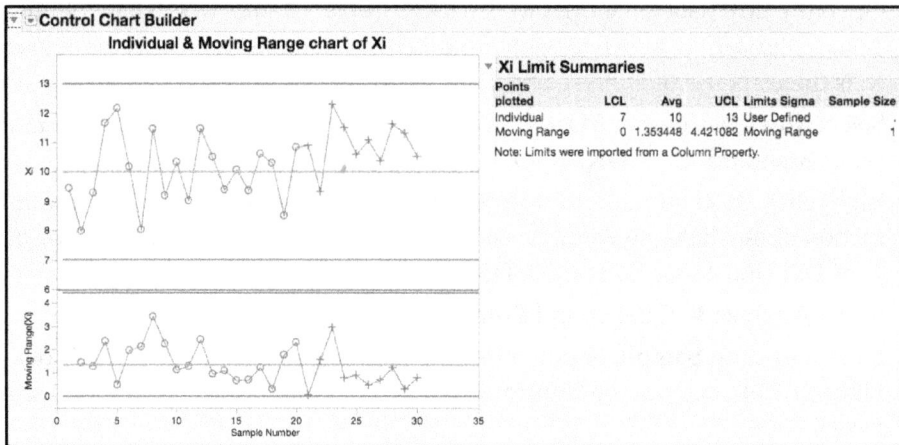

The chart in Figure 7.2b matches the chart in ISQC Figure 9.1. Since there are no signals from the runs tests, the 1σ mean shift in the last 10 observations is undetected using this approach. Next we show how a CUSUM chart can be generated to detect a 1σ mean shift. The CUSUM statistic will be created manually using the **Formula Editor**, to display the calculations involved, and then the **Control Chart** platform will be used to create the CUSUM chart.

1. Click on Chapter 7 – ISQC Table 9.1.jmp to make it active. From the main menu, select **Rows** ▶ **Clear Row States**. Make sure the Control Limits for **Xi** are cleared out too.

2. Double click on at the top of the table to the right of the **True Mean** column to add a new column. Double click on the new column to bring up the Column dialog box. Enter **Xi – 10** for the **Column Name** then click **OK**.

3. Right click on the column name in the JMP data table and select **Formula**. This will launch the formula editor window. Select **Xi** from the list of column names. Then click on the minus sign at the top of the window and type in **10** (Figure 7.3). Click **OK** when done.

Figure 7.3 Formula Editor to Create New Variable, Xi - 10

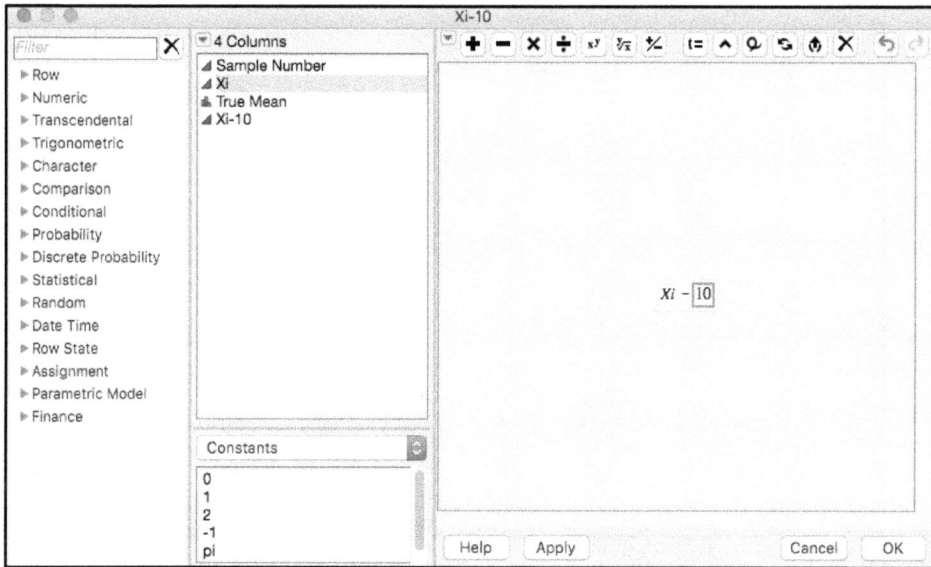

4.   Double click in the table again to add another new column. Label the new column **Ci**. Right click on the label and select **Formula**. From the formula editor window, on the left hand side, click on **Statistical ▶ Col Cumulative Sum** and select **Xi-10** to enter it in the field (Figure 7.4).

Figure 7.4 Formula Editor to Create New Variable, Ci

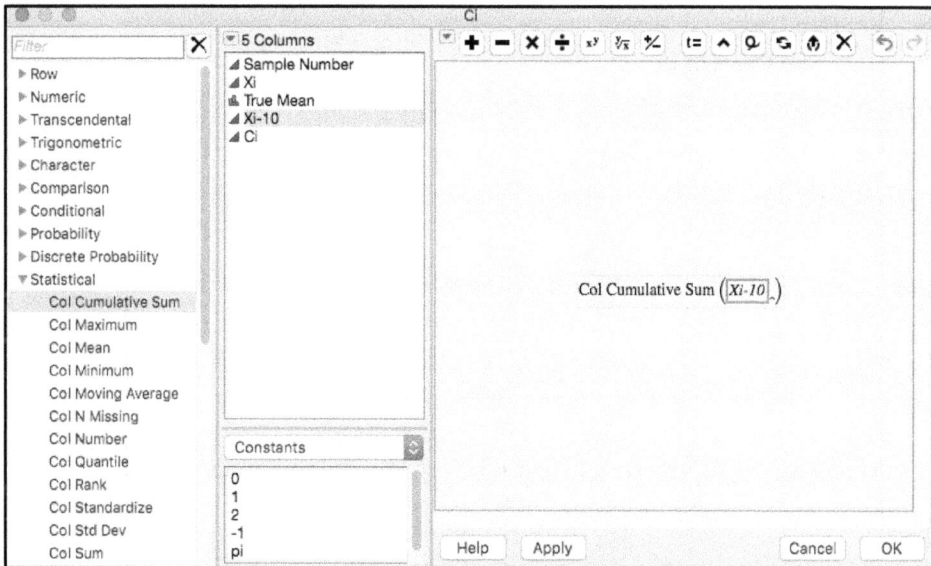

5.   Click **OK** when done. The resulting table is shown in Figure 7.5.

Figure 7.5 Manually Created Cumulative Sum Charting Statistic

	True Mean	Sample Number	Xi	Xi-10	Ci
1	10	1	9.45	-0.55	-0.55
2	10	2	7.99	-2.01	-2.56
3	10	3	9.29	-0.71	-3.27
4	10	4	11.66	1.66	-1.61
5	10	5	12.16	2.16	0.55
6	10	6	10.18	0.18	0.73
7	10	7	8.04	-1.96	-1.23
8	10	8	11.46	1.46	0.23
9	10	9	9.2	-0.8	-0.57
10	10	10	10.34	0.34	-0.23
11	10	11	9.03	-0.97	-1.2
12	10	12	11.47	1.47	0.27
13	10	13	10.51	0.51	0.78
14	10	14	9.4	-0.6	0.18
15	10	15	10.08	0.08	0.26
16	10	16	9.37	-0.63	-0.37
17	10	17	10.62	0.62	0.25
18	10	18	10.31	0.31	0.56

6. To differentiate the shift in means, add a plotting symbol using **True Mean** before creating the chart, as we did in Step 2 in the last example. Select **Analyze ▶ Quality and Process ▶ Control Chart ▶ Run Chart**. Select **Ci** as the **Process** (response) variable. Then select **Sample Number** and click **Sample Label** to identify the subgroup variable. Click **OK** when finished. The plot is shown in Figure 7.6.

Figure 7.6 Plot of Cumulative Sum Statistic, Ci

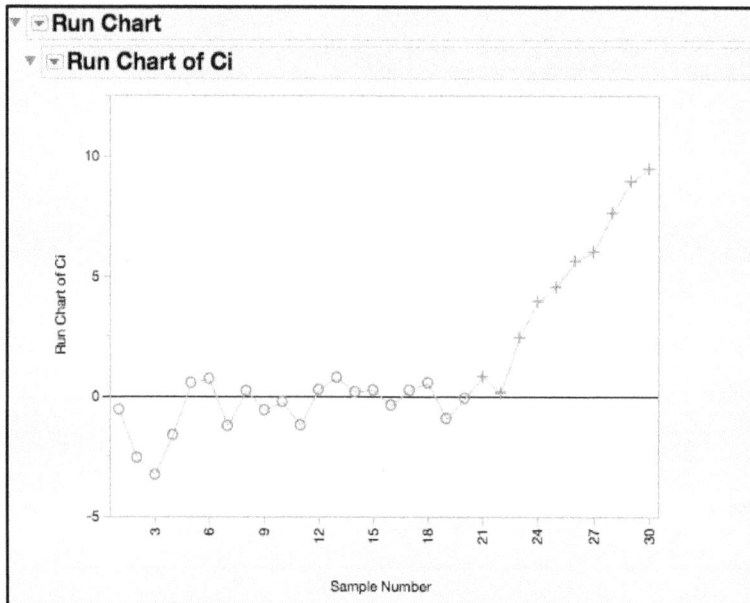

The results Figure 7.5, match the ones shown in ISQC Table 9.1, and the CUSUM chart in Figure 7.6 matches ISQC Figure 9.2. In this example, one form of the CUSUM statistic (ISQC equation (9.1)), known as a 2-sided CUSUM, was generated 'by hand' using the raw data. First, the given target mean of 10 was subtracted from each individual result, in the column $X_i - 10$. A positive result indicates that the result is above the target and a negative result indicates the result is below the given target. In the last column, the cumulative sums of the differences are calculated. For example, in the 3rd row $C_i = -3.27 = (-0.55) + (-2.01) + (-0.71)$.

The cumulative sum is plotted and evaluated against zero. By the third observation following the shift, $N = 23$, the cumulative sum starts growing quickly throughout the remaining runs. This is a very visual display of the 1σ mean shift, which occurred from run 20 to run 21. This form of the CUSUM charting statistic can also be generated in JMP using the new **CUSUM Control Chart** platform, or **CUSUM** platform in the **Control Chart** menu. We first illustrate how to generate a CUSUM chart using the **Control Chart ▶ CUSUM** platform.

1. Select **Analyze ▶ Quality and Process ▶ Control Chart ▶ CUSUM**.

    Figure 7.7 JMP Menu Selections for a CUSUM Chart

    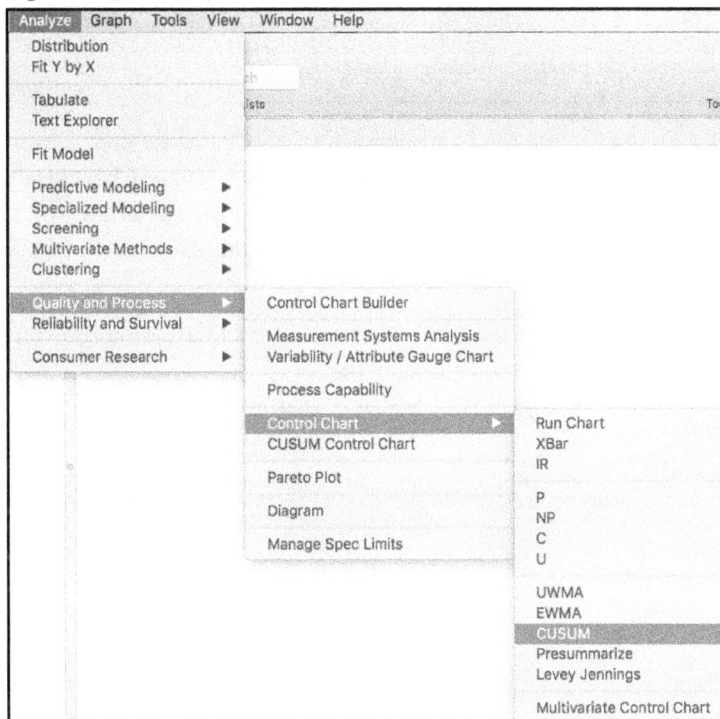

2. A launch window will appear. Select **Xi** as the **Process** (response) variable. Then select **Sample Number** and click **Sample Label** to identify the subgroup variable. Make sure **Two Sided** and **Data Units** are checked on the left-hand side of the window and there is a

missing value for **K**. Click on **Specify Stats** and enter **10** in the file for Target (Figure 7.8). Click **OK** when finished.

Figure 7.8 Launch Window for a 2-Sided CUSUM Chart

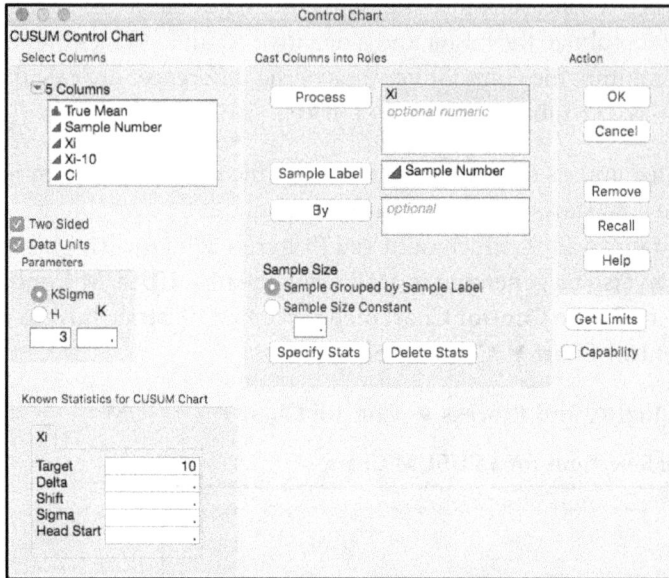

3. From the drop down next to the red triangle for **CUSUM of Xi** deselect **Show V Mask**. To export the charting statistic, from the red triangle next to **Control Chart** select **Save Summaries**. This will create a JMP data table with the Cumulative Sums of Xi. Note the original data is not included in this table. The summary table and the CUSUM chart are shown in Figure 7.9.

Figure 7.9 CUSUM Chart for Xi in Control Chart Platform and JMP Output

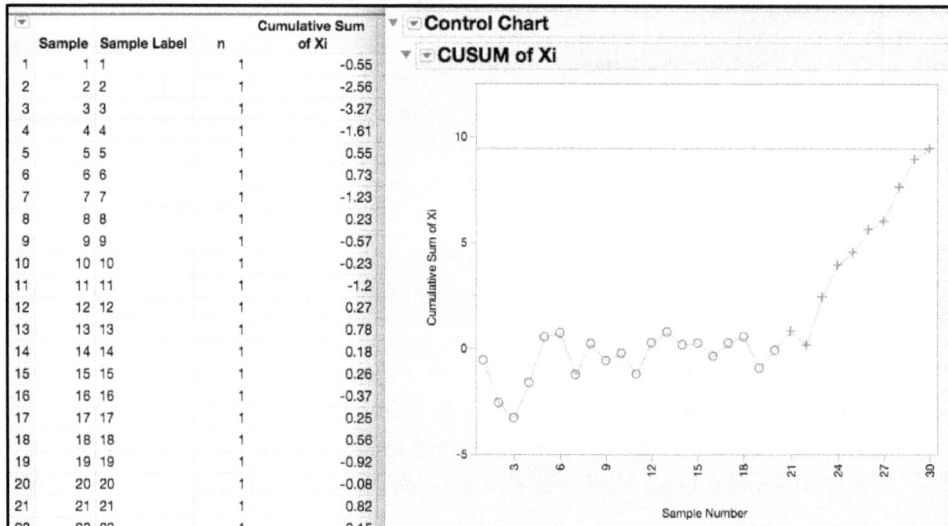

The Tabular CUSUM is another form of the charting statistic, which consists of two one-sided CUSUMs C⁺ and C⁻. One statistic, C⁺, charts deviations above the target value while the statistic C⁻ charts deviations below the target value. The charting statistic for the two one-sided CUSUM is determined by the maximum of 0 and the current CUSUM value. In addition, a reference value K can be used to adjust the values. These equations are shown in ISQC equations (9.2) and (9.3). In the steps below, the launch window for the **Control Chart ▶ CUSUM** platform is explained in more detail.

1. Open Chapter 7 – ISQC Table 9.2.jmp. This is the same data as was used previously; which has variables called *Sample Number, Xi,* and *True Mean.* In this table, Sample Number is the subgroup variable and Xi is the response of interest.
2. Select **Analyze ▶ Quality and Process ▶ Control Chart ▶ CUSUM.**
3. A launch window will appear, which has many options for creating CUSUM charts (Figure 7.10).

Figure 7.10 Features of the CUSUM Dialog Window

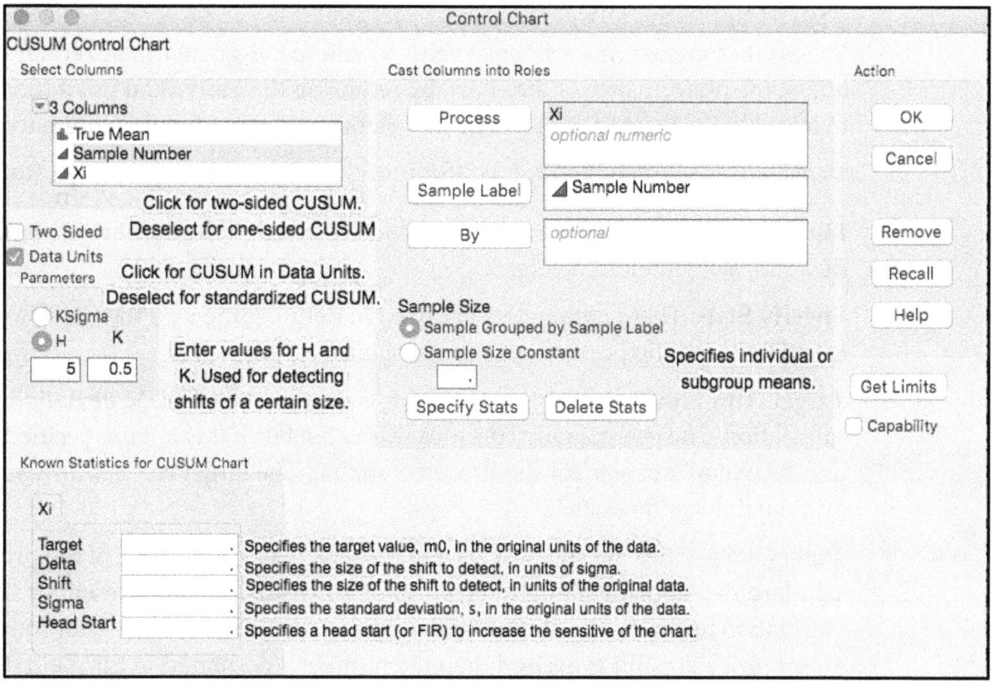

For the reader's benefit, the definitions for some of the fields are repeated here.

○ **Two Sided:** When checked, requests a two-sided CUSUM with a V-mask to detect shifts. If unchecked, a one- sided CUSM is produced with H to detect shifts.

○ **Data Units**: If checked, the CUSUM charting statistic will be computed in the original units of the data. If unchecked, then the CUSUM charting statistic is standardized by normalizing the process variable using *Sigma.*

○　**K Sigma:** Specifies control limits in multiples of the standard deviation or standard error, for the Two-Sided V-mask chart. Enter the K Sigma value in the box that appears below H. Control limits are specified at k sample standard deviations (for individual points) or standard errors (for means) above and below the expected value, which shows as the shift.

> **JMP Note 7.1:** When you specify K Sigma, the lead distance of the V-mask is computed using the type I error, alpha, corresponding to K standard deviations in a normal curve. For example, for K = 3, alpha = 0.0027.

○　**H:** For a two-sided CUSUM, H is the vertical distance between the origin for the V-mask and the upper or lower arm of the V-mask. For a one-sided chart, H is the decision interval. When *Data Units* is selected then H = hσ (for individual points) or $H = h\sigma / \sqrt{n}$ (for means). If *Data Units* is not selected then H = h, which is stated in the number of standard deviations or standard errors.

○　**K:** Specifies the use of a reference value k, where k is greater than zero. It is used to adjust the performance of the chart, by requiring the individual point (or mean) to be at least k units beyond the target value before accumulating the differences. When *Data Units* is selected then K = kσ (for individual points) or $K = k\sigma / \sqrt{n}$ (for means). If *Data Units* is not selected then K = k, which is stated in the number of standard deviations or standard errors.

○　**Specify Stats:** These parameters are used in defining the way the CUSUM statistic is calculated, and the performance of the chart.

○　**Target:** This is the target value, μ_0, which is used in the CUSUM deviations from target calculation. The user usually specifies this value but if it is left unspecified, JMP will use the overall average for the Process variable. The target value is always in the original units of the data.

○　**Delta (δ):** Specifies the smallest shift to be detected as a multiple of the process standard deviation or standard error. In other words, Delta represents a shift in the population mean, or as a shift in the sampling distribution of the subgroup mean. For example, if a 2σ shift is desired then the number 2 is entered in this field. For a two-sided chart or positive one-sided chart, enter a positive value. For a negative one-sided chart, enter a negative value. If left blank, the default is 1. You can choose either Delta or Shift.

○　**Shift (Δ):** This is the minimum value that you want to detect on either side of the target value. The shift value is entered in the same units of the data. For example, if a 2σ shift is desired and σ = 2 then the number 4 is entered in this field. If left blank the default value is 1σ. You can choose either Shift or Delta.

- ○ **Sigma**: This is an estimate of the known standard deviation for the Process parameter, σ_0. If left unspecified then JMP estimates it from all of the data. It is always in the original units of the data. If Data Units is unchecked, then the data are normalized using the estimate for sigma, producing a normalized CUSUM statistic. The value is also used to define the V-mask. If Data Units, Two Sided and KSigma are all checked, then sigma is used to define the V-mask limits for the two-sided CUSUM.

- ○ **Head Start**: Specifies an initial value for the cumulative sum, S_0, for a one-sided CUSUM chart. Note S_0 is usually zero. Enter the Head Start value as a multiple of the standard deviation (individual) or standard error (mean), if Data Units is not selected. Otherwise, enter it in the units of the data.

The CUSUM platform has many inputs that interact together. Unlike a Shewhart control chart, the user must specify the control limits for a tabular (or one-sided) CUSUM chart. This is done using H (decision limit) and K (reference value). There are endless ways to combine these two parameters in order to increase or decrease the sensitivity of the chart to detect mean shifts, as is measured by the Average Run Length (ARL). This topic is further discussed in the Statistical Insights section of this chapter. We will continue recreating ISQC Table 9.2 using a tabular CUSUM.

4. Select **Xi** as the **Process** (response) variable. Then select **Sample Number** and click **Sample Label** to identify the subgroup variable. Make sure **Two Sided** is unchecked and **Data Units** is checked on the left-hand side of the window. For **H** enter **5** and for **K** enter **0.5**. Click on **Specify Stats** and enter **10** in the file for Target and **1** for **Sigma**. Click **OK** when finished.

5. Click on the **CUSUM of Xi** red triangle menu and select **Show Parameters**. Click on the red triangle menu at the top of the window next to **Control Chart** and select **Save Summaries**. This creates a data table with the upper CUSUM values (Figure 7.11).

Figure 7.11 Tabular Upper One-Sided CUSUM Chart

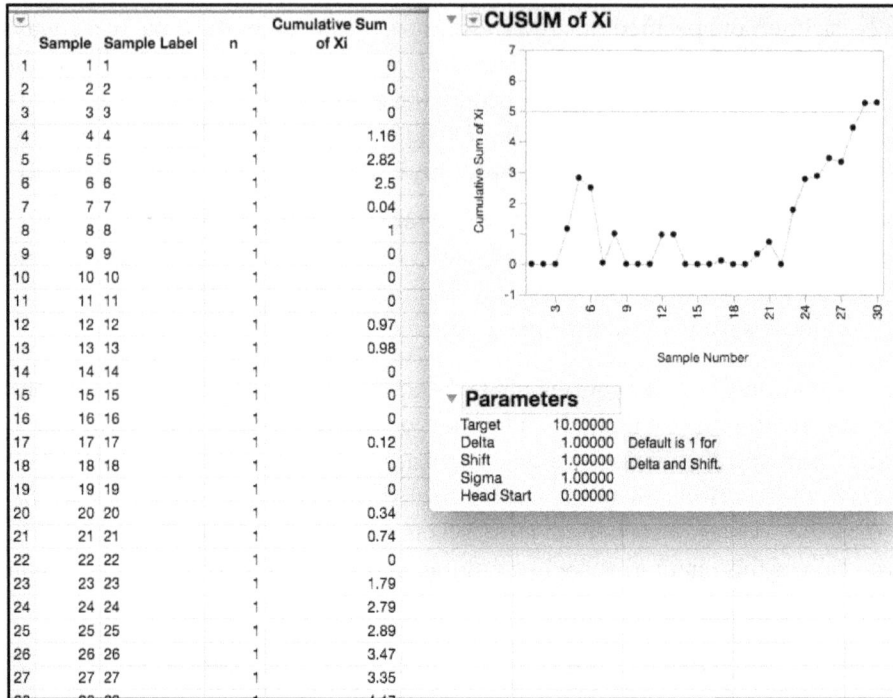

Sample	Sample Label	n	Cumulative Sum of Xi
1	1 1	1	0
2	2 2	1	0
3	3 3	1	0
4	4 4	1	1.16
5	5 5	1	2.82
6	6 6	1	2.5
7	7 7	1	0.04
8	8 8	1	1
9	9 9	1	0
10	10 10	1	0
11	11 11	1	0
12	12 12	1	0.97
13	13 13	1	0.98
14	14 14	1	0
15	15 15	1	0
16	16 16	1	0
17	17 17	1	0.12
18	18 18	1	0
19	19 19	1	0
20	20 20	1	0.34
21	21 21	1	0.74
22	22 22	1	0
23	23 23	1	1.79
24	24 24	1	2.79
25	25 25	1	2.89
26	26 26	1	3.47
27	27 27	1	3.35

CUSUM of Xi

(Chart: Cumulative Sum of Xi vs. Sample Number)

Parameters

Target	10.00000	
Delta	1.00000	Default is 1 for
Shift	1.00000	Delta and Shift.
Sigma	1.00000	
Head Start	0.00000	

6. Highlight the JMP data table, Chapter 7 – ISQC Table 9.2.jmp, by clicking on it and re-launch the CUSUM platform. Select **Xi** as the **Process** (response) variable. Then select **Sample Number** and click **Sample Label** to identify the subgroup variable. Make sure **Two Sided** is unchecked and **Data Units** is checked on the left-hand side of the window. For **H** enter **5** and for **K** enter **0.5**. Click on **Specify Stats** and enter **10** in the file for Target, **-1** for **Delta** and **1** for **Sigma**. Click **OK** when finished.

7. Click on the red triangle next to **CUSUM of Xi** and select **Show Parameters**. Click on the red triangle at the top of the window next to **Control Chart** and select **Save Summaries**. This creates a data table with the lower CUSUM values (Figure 7.12).

Figure 7.12 Tabular Lower One-Sided CUSUM Chart

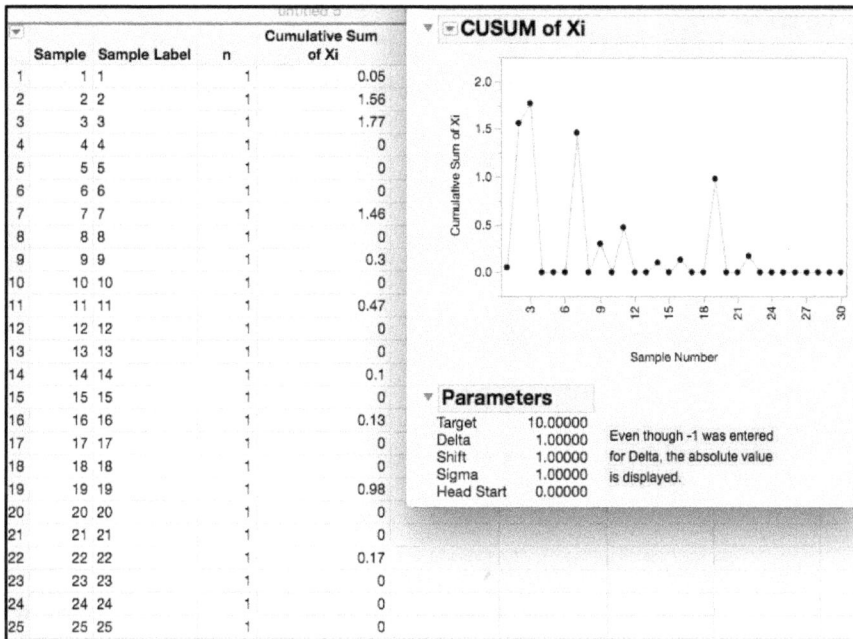

In order to plot both the upper and lower CUSUM statistics in one chart, we need to merge the two data sets. First, we will change the CUSUM values in the lower one-sided CUSUM table from Step 7 to negative values. This is accomplished by double clicking in the column header row to add a new column. Right click on the Column 5 label and select **Formula**. In the formula window, select **Cumulative Sum of Xi** and click on the multiplication symbol at the top. Double click in the open field and type in **-1**. Click **OK** when done. Highlight the new column and copy it, using CNTRL-c (⌘C on a Mac), for example.

Add a new column to the first untitled table from Step 5 by double clicking in the column area of the table and paste in the negative CUSUM values. Change the names of the charting statistics to **Upper Cumulative Sum of Xi** and **Lower Cumulative Sum of Xi**. Save the table as **Chapter 7 – ISQC Tabular CUSUM for 9.1.jmp**.

8. Select **Graph ▶ Graph Builder**. Select **Upper Cumulative Sum of Xi and Lower Cumulative Sum of Xi** and drag-and-drop both to the **Y** zone. Select **Sample** and drag-and-drop to the **X** zone, as shown in Figure 7.13a.

Figure 7.13a Graph Builder Dialog for ISQC Example 9.1

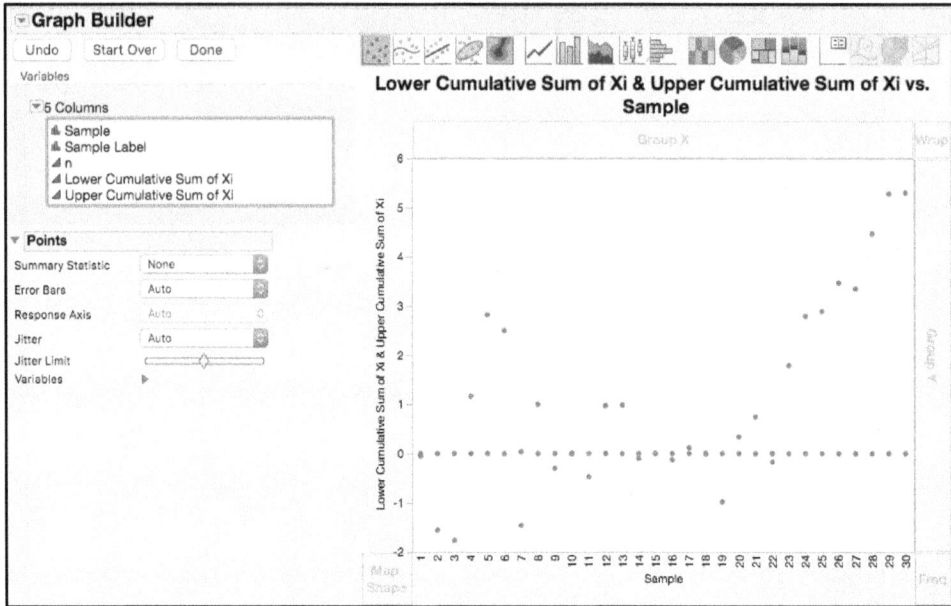

9. Right click on the graph area to reveal the contextual menu. Select **Add ▶ Lines** as shown in Figure 7.13b.

Figure 7.13b Adding Lines to Two-Sided CUSUM for ISQC Example 9.1

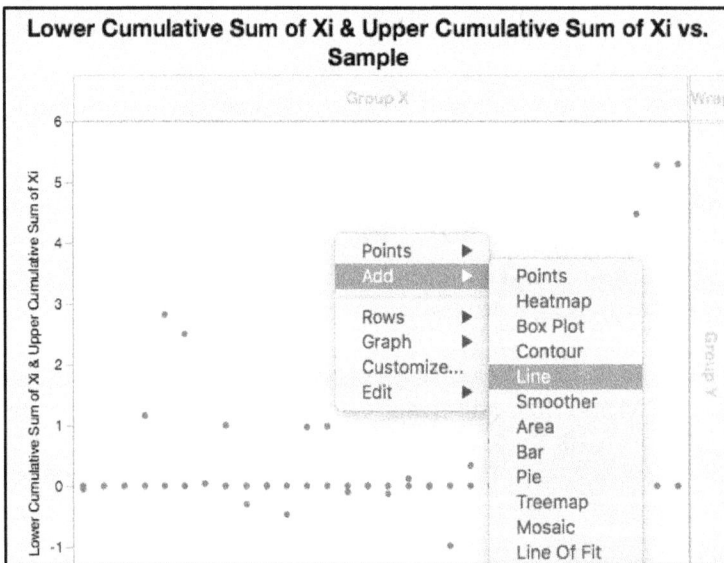

10. Double click on the left axis to bring up the **Y Axis Settings** dialog window. Re-scale the axis by typing **-6** for **Minimum:** and **6** for **Maximum:**. Add a reference line by typing **5** into **Value:** and **Upper CUSUM** into **Label:** click **Add**. Repeat using **-5** and **Lower**

CUSUM. Add an unlabeled reference line at 0. Click **Done**. Figure 7.13c shows two-sided CUSUM chart.

Figure 7.13c Two One-sided CUSUM Chart for ISQC Example 9.1

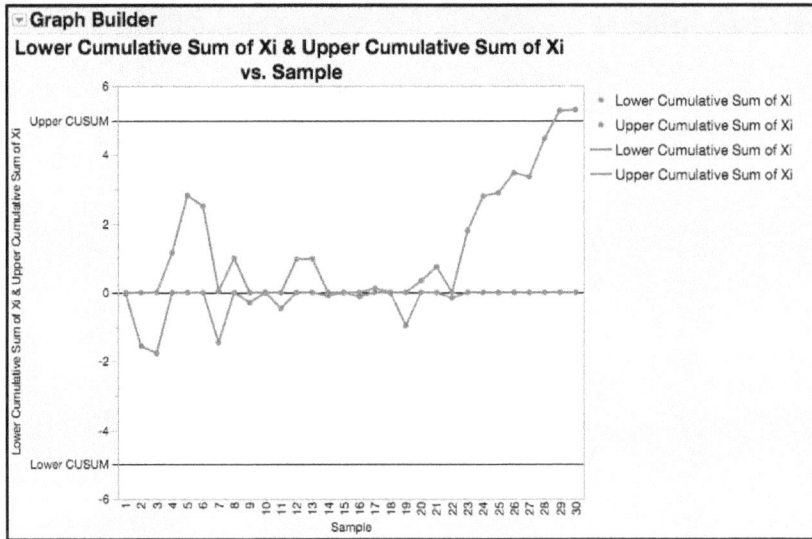

The chart in Figure 7.13c corresponds to ISQC Figure 9.3(b). The chart finally produces a signal at the 28th sample, which is 7 subgroups after the shift occurred, at sample 21. While one major benefit of this chart is is to show the shifts away from the target value, there is still a need to visually assess the raw data in order to interpret the CUSUM signal. If the CUSUM continues to increase, as it does in Figure 7.13c, a sustained mean shift is most likely the root cause. However, a spike in the CUSUM statistic could be indicative of one or two large outliers, followed by a return to the target value.

> **JMP Note 7.2**: To generate a tabular CUSUM chart in the CUSUM platform, deselect Two Sided. Also, deselecting Data Units will standardize the CUSUM charting statistic.

It should be noted that the two-sided CUSUM charts can be standardized by subtracting the target value from the xi variable and dividing by the standard deviation. This way, everything is in the units of sigma and the same values for k and h can be used for many different parameters. This is accomplished in JMP by deselecting the *Data Units* option in the CUSUM launch window. For the data in ISQC Example 9.1, since $\sigma = 1$, the tabular standardized CUSUM charting statistics would have the same exact values for each sample as those presented above and will not be reproduced here.

The **Quality and Process ▶ Control Chart ▶ CUSUM** platform does not produce both charts simultaneously, and some manual manipulation is required to view them in the same plot. The new **CUSUM Control Chart** platform is a new generation of quality platforms within JMP that makes the creation and interpretation of CUSUM charts easier. This new platform generates the tabular, two one-sided CUSUM in one plot, so no extra manipulation is required.

The following steps illustrate how to construct the CUSUM control chart in Figure 7.13c using the **CUSUM Control Chart** platform.

1. Open the JMP data table Chapter 7 – ISQC Table 9.1.jmp, which has columns called *Sample Number, Xi,* and *True Mean.* In this table, Sample Number is the subgroup variable, Xi is the response of interest, and True Mean is the mean used for the data simulation.

2. Select **Analyze ▶ Quality and Process ▶ CUSUM Control Chart**. The launch window is shown in Figure 7.14. This launch window requires minimal input from the user.

Figure 7.14 CUSUM Control Chart Launch Window

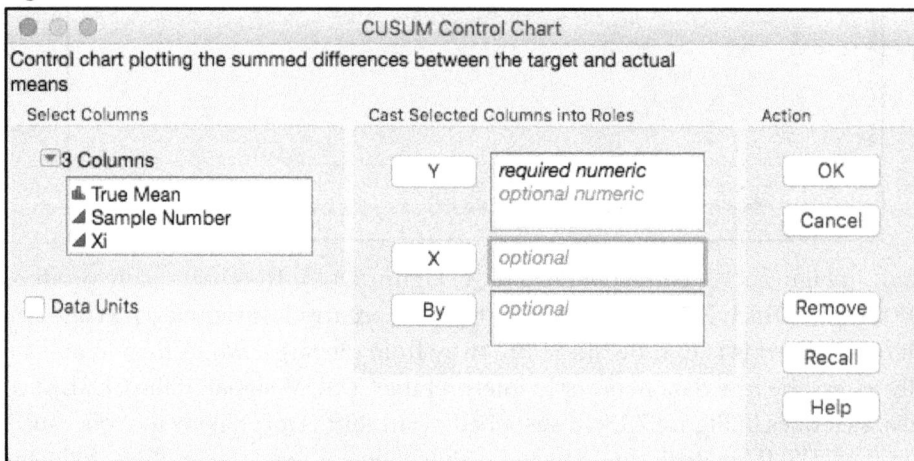

3. Select **Xi** and click **Y**. Then select **Sample Number** and click **X** and leave **Data Units** unchecked, as shown in Figure 7.15.

Figure 7.15 CUSUM Control Chart Launch Window for ISQC Example 9.1

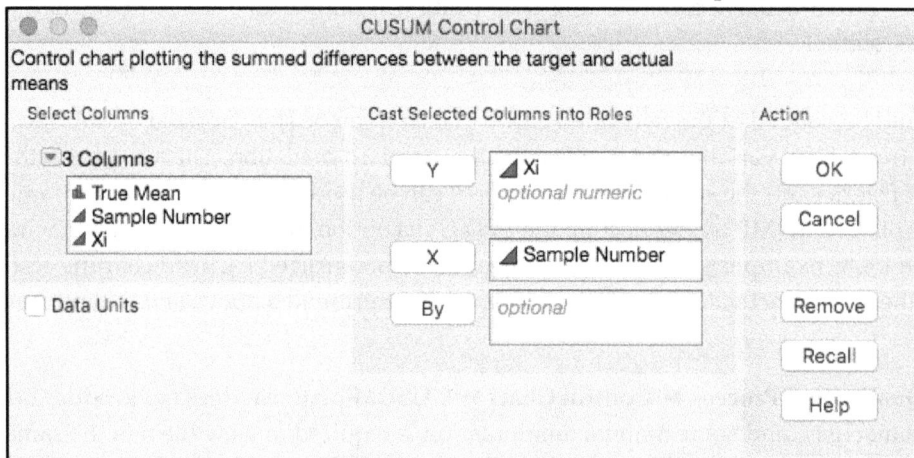

4. Click **OK**. The default output is shown in Figure 7.16.

Figure 7.16 Default CUSUM Control Chart **Platform Output for ISQC Example 9.1**

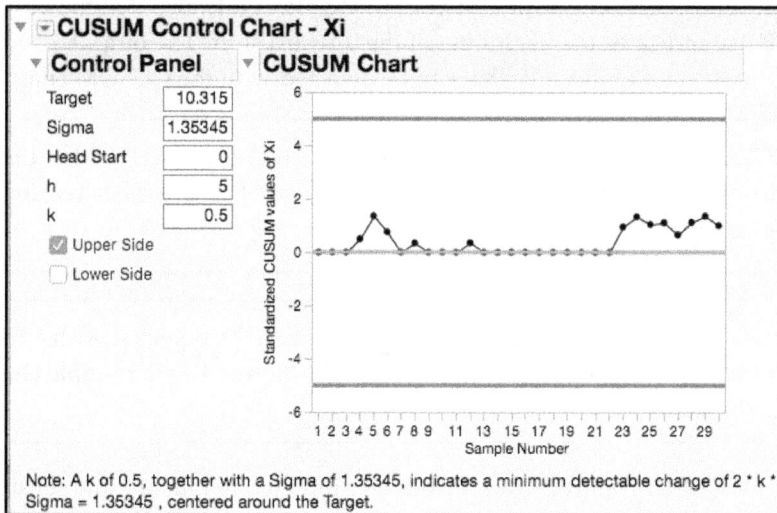

By default, the **CUSUM Control Chart** platform displays the Upper Side **CUSUM Chart**, and uses the mean and standard deviation of the data as the **Target** and **Sigma** values. The decision limit, **h**, is set to 5, and the reference value, **k**, to 0.5. The lower case h and k indicate that the CUSUM chart is in standardized units. These defaults values generate a chart that is helpful as a starting point. Since for this example we know the values for **Target** and **Sigma**, we can input those directly in the platform.

5. Type **10** in **Target** and **5** in **Sigma**, as shown in Figure 7.17. Click also on **Lower Side** to generate the Lower Side CUSUM.

Figure 7.17 Two-Sided CUSUM Control Chart **Platform Output for Example 9.1**

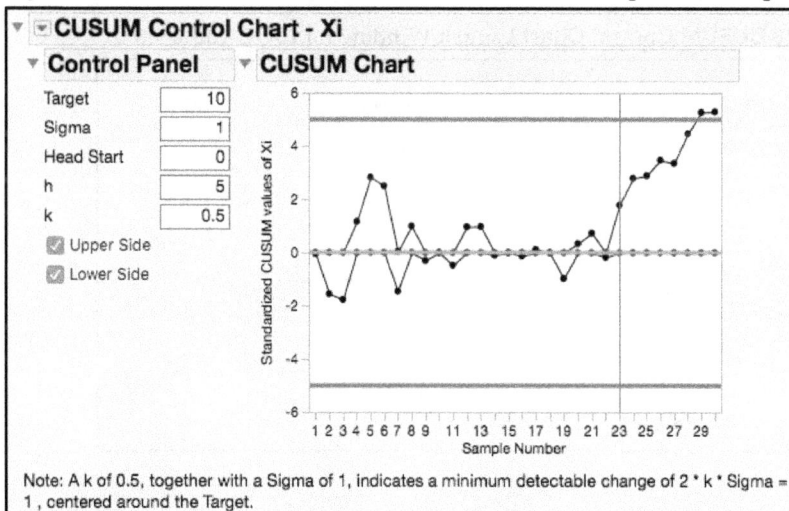

The chart in Figure 7.17 corresponds to ISQC Figure 9.3(b), and Figure 7.13c. However, in this case, we did not have to manipulate the y-axis to get the right range from -6 to 6 or add reference lines at the decision values of ±5 and 0, or add lines to the plot. Note also that the plot has a line at Sample Number 23 that indicates the beginning of the upward shift. The output also has a footnote that explains the size of the shift that can be detected with the CUSUM chart.

> ✦ **JMP Note 7.3**: The reference line in the x-axis of the two-sided CUSUM of the **CUSUM Control Chart** platform indicates the first point where an increasing, or decreasing, trend started.

Statistics Note 7.1: The size of the change that can be detected with a CUSUM chart is a function of the reference value, k, and the Sigma. The detectable change is given by $2 \times \text{Sigma} \times k$.

The parameters of the CUSUM chart can be further manipulated to increase the sensitivity to detect shifts faster at the start up of the chart by specifying a head-start value. This is referred to as the fast initial response or FIR (see ISQC Section 9.1.6). This procedure starts the initial CUSUM charting statistics at the FIR value specified by the user. The following steps illustrate how to use this feature in the **CUSUM Control Chart** platform.

1. Open Chapter 7 – ISQC Table 9.6.jmp. It has a variable *Period*, which is the subgroup variable and *Xi*, the response of interest.
2. From the main menu, select **Analyze ▶ Quality and Process ▶ CUSUM Control Chart**.
3. A launch window will appear. Select **Xi** as the **Y** (response) variable. Then select **Period** and click **X** to identify the subgroup variable. Make sure **Data Units** is checked (Figure 7.18).

Figure 7.18 CUSUM Control Chart Launch Window for ISQC Table 9.6

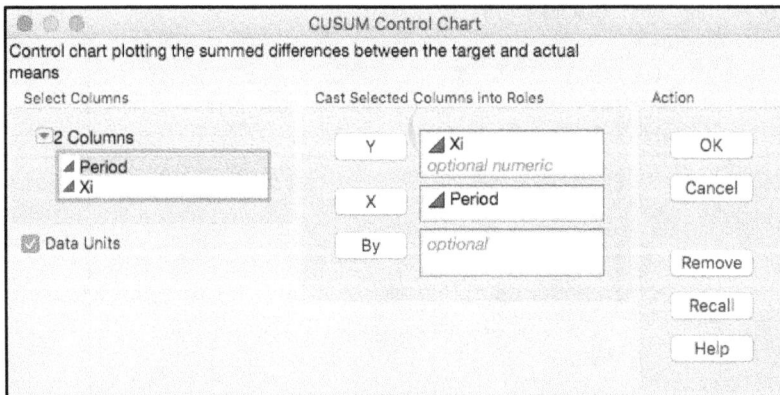

4. Click **OK**. By default, the **CUSUM Control Chart** platform displays the Upper Side CUSUM Chart, and uses the mean and standard deviation of the data as the **Target** and **Sigma** values, as shown in Figure 7.19. Note the message, "Using Data Units," since we clicked on **Data Units** on the dialog window.

Figure 7.19 Default Upper One-Sided CUSUM for ISQC Table 9.6

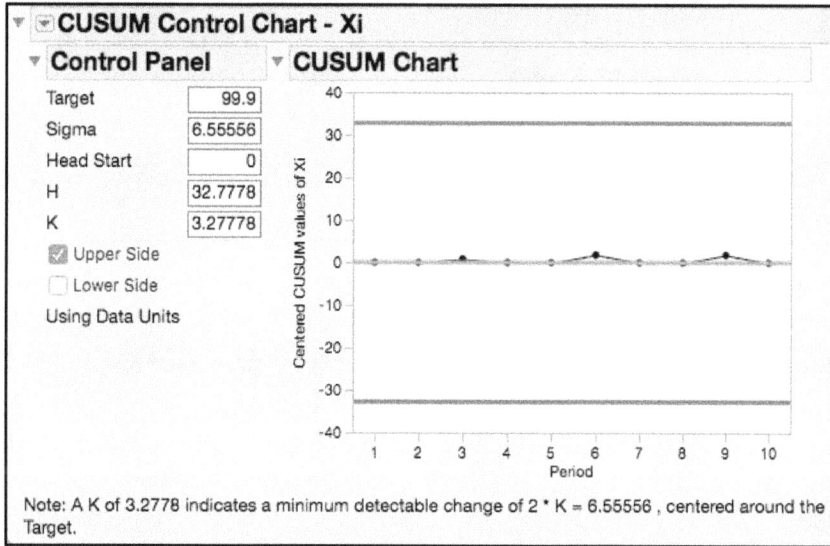

Note: A K of 3.2778 indicates a minimum detectable change of 2 * K = 6.55556 , centered around the Target.

5. To create the two-sided headstart CUSUM enter **100** for **Target, 1** for **Sigma, 6** for **Head Start, 12** for **H, 3** for **K**, and click on **Lower Side**. Figure 7.20 displays the two-sided CUSUM chart.

Figure 7.20 Headstart CUSUM for ISQC Table 9.6

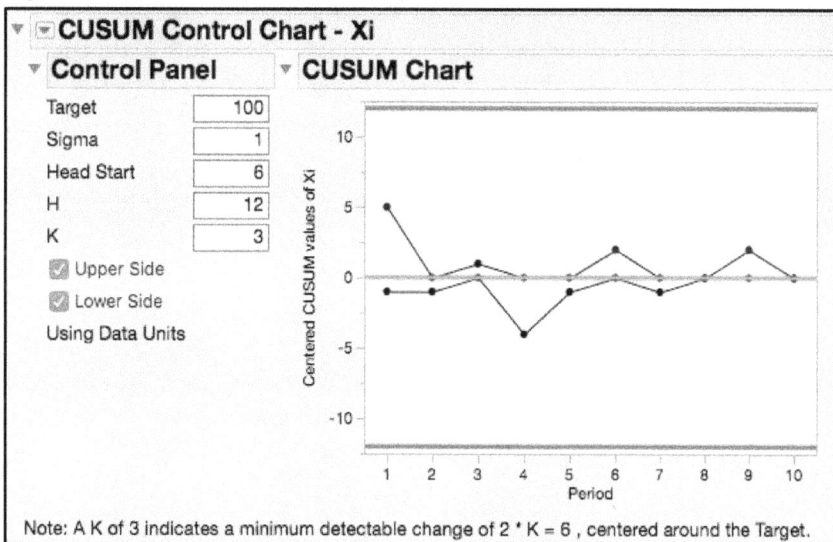

Note: A K of 3 indicates a minimum detectable change of 2 * K = 6 , centered around the Target.

6. Click on the red triangle next to **CUSUM Control Chart - Xi** and select **Save Summaries**.

Figure 7.21 Headstart CUSUM Summaries for ISQC Table 9.6

Sample Label	n	Sample Mean Xi	Upper Cumulative Sum of Xi	Positive Runs	Lower Cumulative Sum of Xi	Negative Runs
1	1	102	5	1	-1	1
2	1	97	0	0	-1	2
3	1	104	1	1	0	0
4	1	93	0	0	-4	1
5	1	100	0	0	-1	2
6	1	105	2	1	0	0
7	1	96	0	0	-1	1
8	1	98	0	0	0	0
9	1	105	2	1	0	0
10	1	99	0	0	0	0

The **Summaries** dataset is shown in Figure 7.21. The columns **Upper Cumulative Sum of Xi, Positive Runs,** match the C_i^+ and N^+ in section (a) of ISQC Table 9.6, while the **Lower Cumulative Sum of Xi** and **Negative Runs** match the C_i^- and N^- in section (b). The headstart benefits the upper CUSUM charting statistic more than the lower one, with an initial value of 5. However, since the raw data does not contain a shift, i.e., it appears to be randomly distributed about the target value of 100, the CUSUM charting statistics quickly converge towards zero. The headstart approach works well when the mean is further away from the specified target, as is shown in the next example.

1. Open Chapter 7 – ISQC Table 9.7.jmp. The JMP data table has a variable *Period*, which is the subgroup variable and *Xi*, the response of interest.
2. Select **Analyze ▶ Quality and Process ▶ CUSUM Control Chart**.
3. In the launch window that appears, select **Xi** as the **Y** (response) variable. Then select **Period** and click **X** to identify the subgroup variable. Make sure **Data Units** is checked.
4. Click **OK**. The default Upper Side CUSUM is created with the default values of the mean and standard deviation of the data as the **Target** and **Sigma**. We need to update these values to generate the headstart CUSUM.
5. Enter **100** for **Target, 1** for **Sigma, 6** for **Head Start, 12** for **H, 3** for **K**, and click on **Lower Side.** The two-sided CUSUM is shown in Figure 7.22.

Figure 7.22 Headstart CUSUM for ISQC Table 9.7

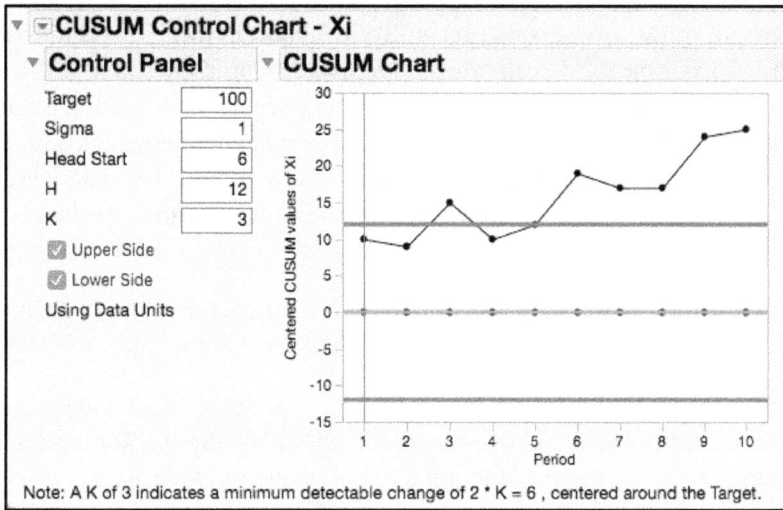

Note: A K of 3 indicates a minimum detectable change of 2 * K = 6 , centered around the Target.

6. Click on the red triangle next to **CUSUM Control Chart - Xi** and select **Save Summaries**.

Figure 7.23 Headstart CUSUM Summaries for ISQC Table 9.6

Chapter 7 - ISQC Table 9.7 - Xi Summaries							
Sample Label	n	Sample Mean Xi	Shift Start	Upper Cumulative Sum of Xi	Positive Runs	Lower Cumulative Sum of Xi	Negative Runs
1	1	107	*	10	1	0	0
2	1	102		9	2	0	0
3	1	109		15	3	0	0
4	1	98		10	4	0	0
5	1	105		12	5	0	0
6	1	110		19	6	0	0
7	1	101		17	7	0	0
8	1	103		17	8	0	0
9	1	110		24	9	0	0
10	1	104		25	10	0	0

In the **Summaries** dataset, shown in Figure 7.23, the columns **Upper Cumulative Sum of Xi, Positive Runs,** match the C_i^+ and N^+ in section (a) of ISQC Table 9.7, while the **Lower Cumulative Sum of Xi** and **Negative Runs** match the C_i^- and N^- in section (b). The mean of this data set is 105, which is considerably higher than the target value of 100. The first upper CUSUM statistic is 10 using the FIR approach and without it, it would have been 4. The upper CUSUM without the FIR crosses H = 12 at the 6th subgroup, compared to the 3rd subgroup with the FIR. As an exercise the reader can re-create this CUSUM without the FIR and study its impact on this data set.

In ISQC Chapter 9, Montgomery also discusses the standardized two-sided CUSUM statistics, a CUSUM statistic for monitoring process variability, the V-mask procedure for tabular CUSUM and a self-starting CUSUM for phase I monitoring or for short run scenarios. Some of these topics will be discussed in the Statistical Insights section in this chapter.

ISQC Example 9.2 Random Normal Data - EWMA

In this example, we will show how to construct an EWMA chart in JMP. This is the same data set that was used in ISQC Example 9.1. Recall, the data reproduced in ISQC Table 9.10 consists of thirty randomly generated results. The first twenty results were generated using a normal distribution with $\mu = 10$ and $\sigma = 1$ and the next ten results were generated using a random normal distribution with $\mu = 11$ and $\sigma = 1$. A subgroup size of $n = 1$ is assumed. The data were constructed to evaluate the ability of an EWMA chart to detect a 1σ shift in the mean in the last 10 observations.

The following steps illustrate how to construct an EWMA control chart using the **Control Chart** platform.

1. Open Chapter 7 - ISQC Table 9.10.jmp, which has variables called *Subgroup Number, Xi,* and *True Mean.* In this table, Sample Number is the subgroup variable and Xi is the response of interest. Note trend plot of the data is shown in Figure 7.2.

2. Select **Analyze ▶ Quality and Process ▶ Control Chart ▶ EWMA.**

3. A launch window will appear. To create the EWMA, select **Xi** as the **Process** (response) variable. Then select **Subgroup Number** and click **Sample Label** to identify the subgroup variable. For **Weight** enter **0.1**, leave **KSigma** selected and enter **2.7** below **Alpha**. Click on **Specify Stats** and enter **1** in the field for **Sigma** and **10** for **Mean** (Figure 7.24).

Figure 7.24 Launch Window for EWMA Example

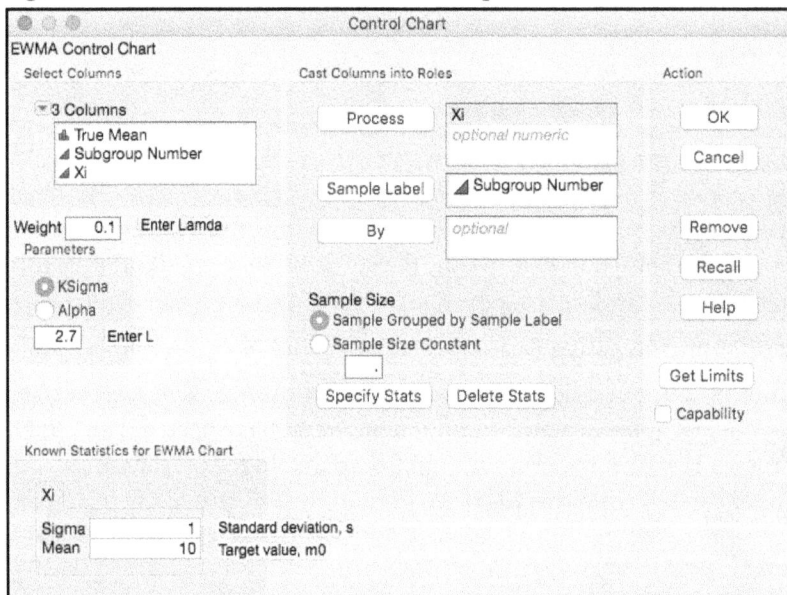

4. Click **OK** when finished. The control chart will appear. Click on the red triangle next to **EWMA of Xi** and select **Test Beyond Limits** from the drop down menu. The chart is shown in Figure 7.25.

Figure 7.25 EWMA Chart for ISQC Example 9.2

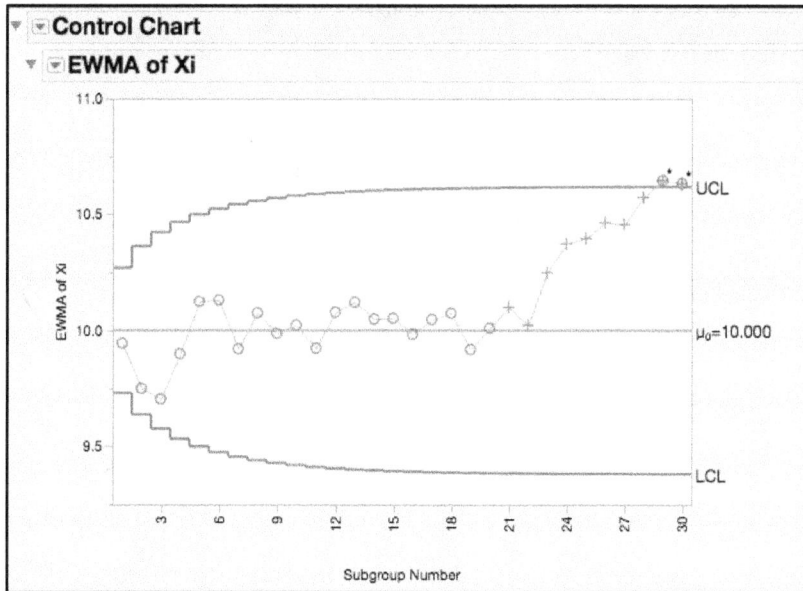

5. Click on the red triangle next to **EWMA of Xi** and select **Save Summaries** from the drop down menu. Then select **Save Limits ▶ In New Table**. The output is shown in Figure 7.26.

Figure 7.26 EWMA Charting Statistics and Parameters for ISQC Example 9.2

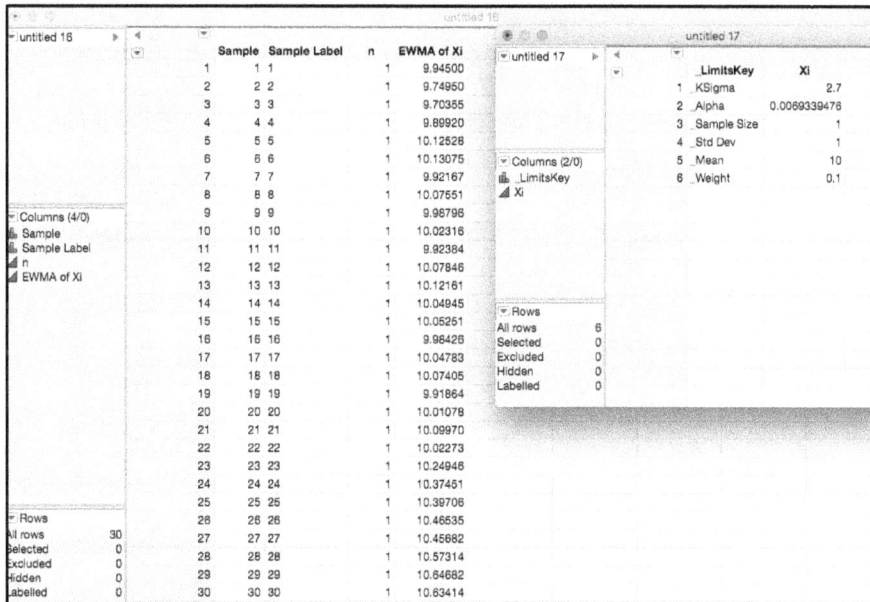

6. Add a new column to the outputted table shown in Figure 7.26 by double clicking in the column area and call it **Xi**. Copy the raw data from Chapter 7 - ISQC Table 7.9.jmp and paste it into the new table.

7. From the main menu select **Graph ▶ Graph Builder**. Select **EWMA of Xi** and **Xi** and drag then onto the **Y** zone, then select **Sample** and drag it onto the **X** zone. Add lines to the plot and a reference line to the y-axis at 10.621. Click **Done** when finished. The graph is shown in Figure 7.27.

Figure 7.27 Overlay of EWMA Charting Statistic and Raw Data for ISQC Example 9.2

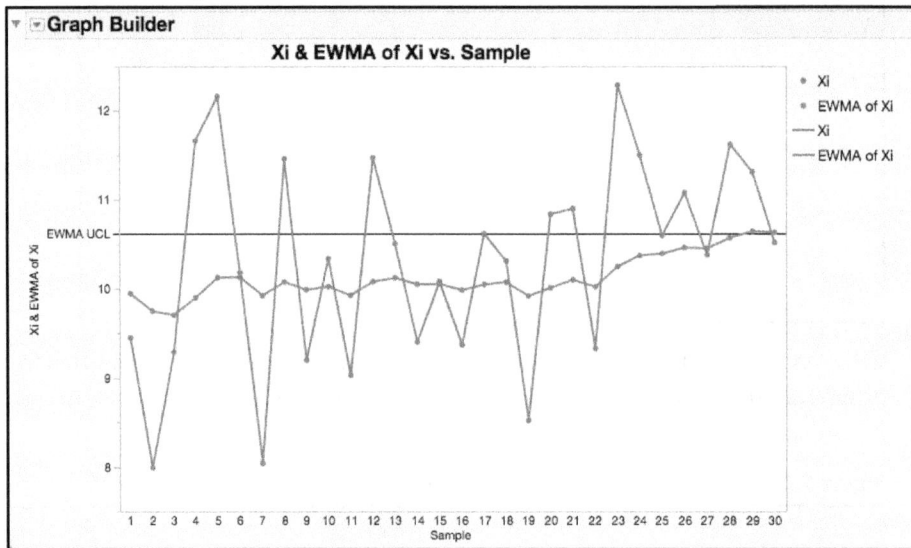

Figure 9.7 in ISQC is reproduced in Figure 7.25. There is a signal at subgroup number 29 and one at subgroup number 30, which occurred eight points after the shift in the mean at subgroup 21. Interestingly, the mean shift was detected at the same subgroup number for EWMA and the CUSUM chart shown in Figure 7.13c. The EWMA charting statistic, which is the exponentially weighted moving average, is overlaid with the original data in Figure 7.27. The charting statistic is a smoothed representation of the raw data and more readily shows the mean shift.

Statistics Note 7.2: CUSUM charts and EWMA charts are equally effective for detecting smaller shifts in the process mean. The choice is usually a user preference for one over the other.

ISQC Example 9.3 Random Normal Data - UWMA

In this this example, we will show how to construct a Uniformly Weighted Moving Average (UWMA), or Moving Average, chart in JMP. This is the same data set that was used in ISQC Example 9.1 and ISQC Example 9.2. Recall, the data reproduced in ISQC Table 9.14 consists of thirty randomly generated results. The first twenty results were generated using a normal distribution with μ = 10 and σ = 1 and the next ten results were generated using a random normal distribution with μ = 11 and σ = 1. A subgroup size of n = 1 is assumed. The data were constructed to evaluate the ability of an UWMA chart to detect a 1σ shift in the mean in the last 10 observations.

The following steps illustrate how to construct a UWMA control chart using the **Control Chart** platform.

1. Open Chapter 7 – ISQC Table 9.14.jmp, which has variables called *Observation Number*, *Xi*, and *True Mean*. In this table, Observation Number is the subgroup variable and Xi is the response of interest.
2. Select **Analyze ▶ Quality and Process ▶ Control Chart ▶ UWMA**.
3. A Launch window will appear. To create the UWMA, select **Xi** as the **Process** (response) variable. Then select **Observation Number** and click **Sample Label** to identify the subgroup variable. For **Moving Average Span** enter **5**, leave **KSigma** selected and the default value of **3** below **Alpha**. Click on **Specify Stats** and enter **1** in the field for **Sigma** and **10** for **Mean** (Figure 7.28).

Figure 7.28 Launch Window for UWMA Chart

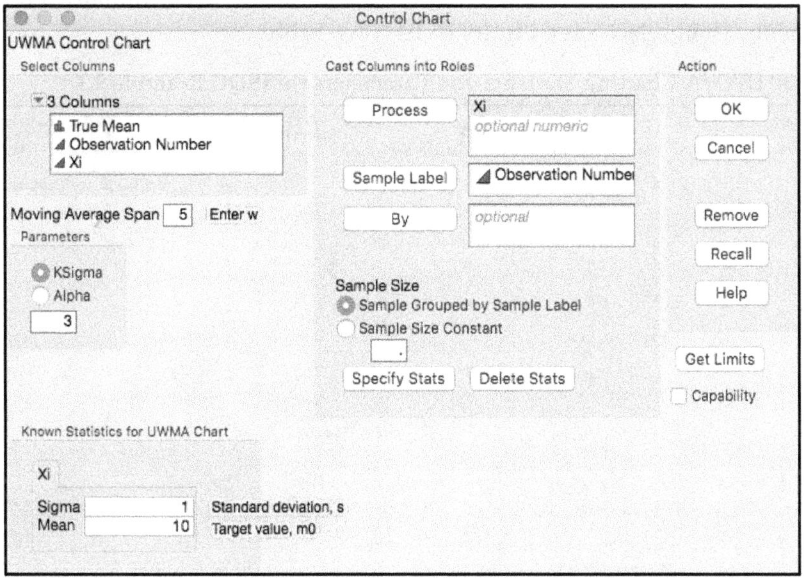

4. Click **OK** when finished. The control chart will appear. Click on the red triangle next to **UWMA of Xi** and select **Test Beyond Limits** from the drop down menu. The chart is shown in Figure 7.29.

Figure 7.29 UWMA Chart for ISQC Example 9.3

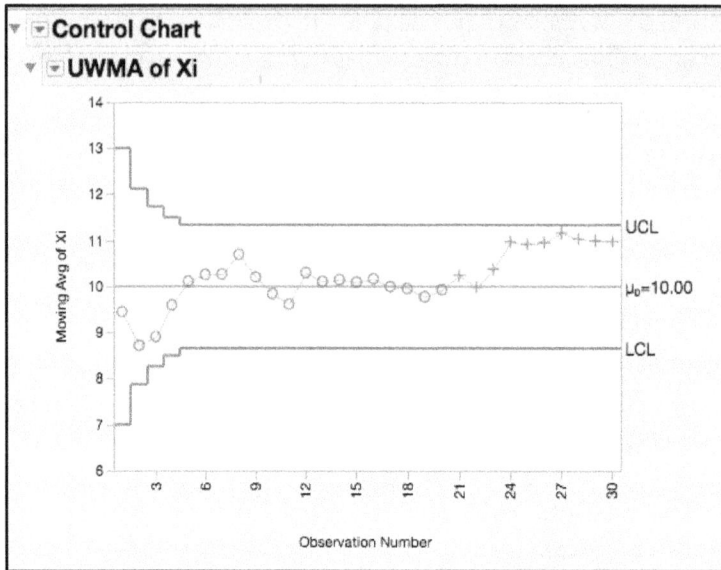

5. Click on the red triangle next to **UWMA of Xi** and select **Save Summaries** from the drop down menu. Then select **Save Limits ▶ In New Table**.

Figure 7.30 UWMA Charting Statistics and Parameters for ISQC Example 9.3

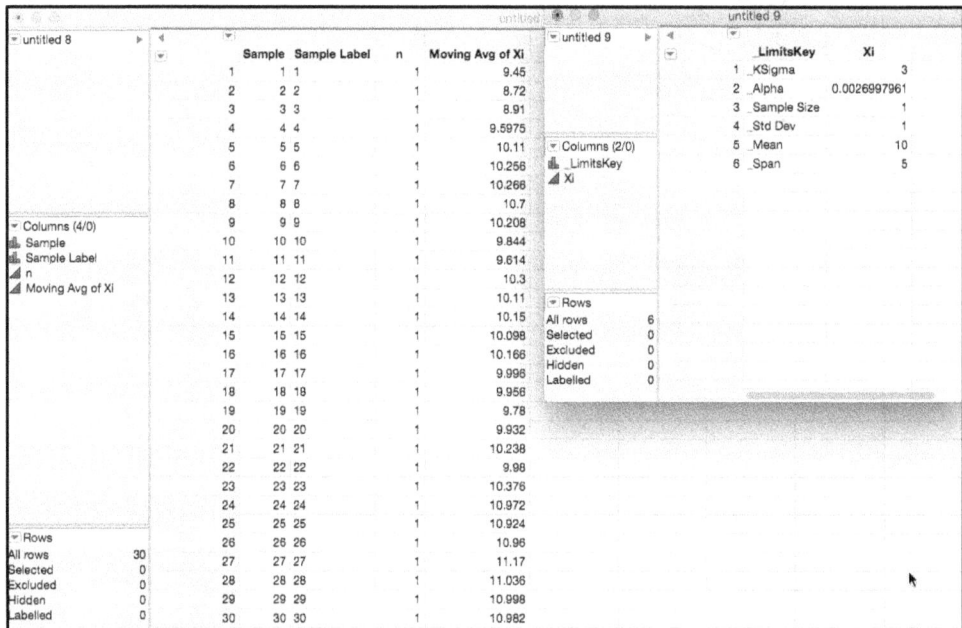

6. Add a new column to the outputted table shown in Figure 7.30 by double clicking in the column area and call it **Xi**. Copy the raw data from Chapter 7 - ISQC Table 7.9.jmp and paste it into the new table.

7. From the main menu select **Graph ▶ Graph Builder**. Select **Moving Avg of Xi** and **Xi** and drag them onto the **Y** zone, then select **Sample** and drag it onto the **X** zone. Add lines to the plot and a reference line to the y-axis at 11.38. Click **Done** when finished.

Figure 7.31 Overlay of UWMA Charting Statistic and Raw Data for ISQC Example 9.3

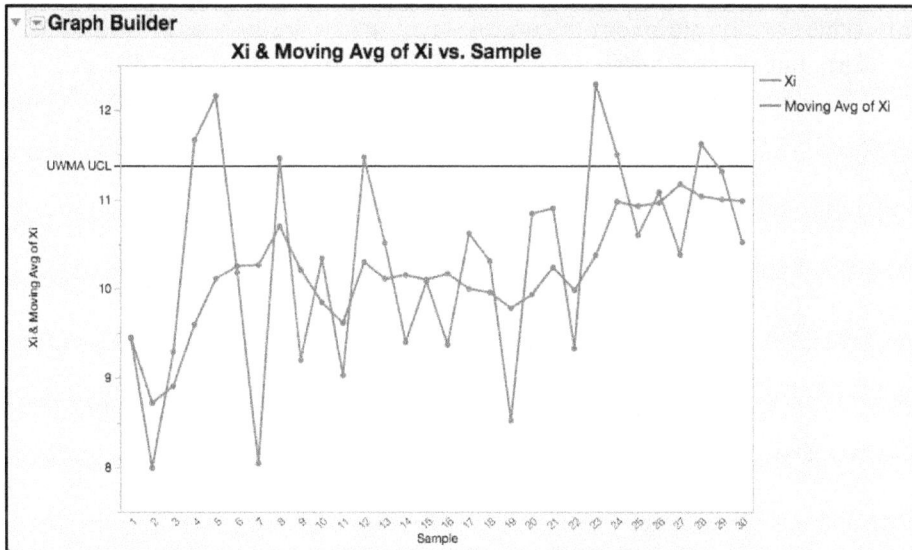

Figure 9.8 in ISQC is reproduced in Figure 7.29. Unlike the CUSUM and EWMA, there are no signals in the chart. The UWMA charting statistic, which is the moving average for five consecutive points, is overlaid with the original data in Figure 7.31. The charting statistic is a smoothed representation of the raw data and more readily shows the mean shift. However, it is not as effectively shown as it is in the EWMA.

Statistics Note 7.3: The CUSUM chart can be adapted for other data types, including parameters that can be modeled using a binomial or Poisson distribution. They can also be adopted for short run scenarios or to detect shifts in the variance.

Statistical Insights

In this section, we elaborate upon some of the examples provided in ISQC Chapter 9. The examples highlighted in this section include several important concepts we have encountered over our many years of applying SPC successfully to a variety of industries. For most of these examples, additional output not provided in ISQC is included to illustrate JMP functionality or further elaborate on important points considered by the authors.

Selecting H (h) and K (k) for CUSUM Charts

CUSUM charts are known for their ability to detect small shifts in the mean, on the order of 2σ or less. However, they are also quite effective at detecting larger shift sizes too. The ability of the chart to detect different shift sizes is controlled through the use of the reference value K (k) and the decision limit H (h).

The performance of a CUSUM chart, or any type of control chart, can be evaluated by means of the Average Run Length (ARL). The average run length describes how long it takes, on average, for the control chart to indicate an out-of-control signal. When the process is in-control, ARL(0) shows the false alarm rate of the selected parameters. For example, an ARL(0) = 370 indicates that a false signal will be generated, on average, every 370 observations. For a particular shift size of interest Delta, ARL(Delta) is used to evaluate the ability of the chart to detect the specified shift size. For example, ARL(1σ) = 10.4 implies that it will take, on average, 10.4 observations for the chart to signal a 1σ shift in the mean.

h and k relationships

ISQC Section 9.1.3 discusses the design of CUSUM charts in terms of ARL and the values of the decision interval H= hσ, and the reference value K=kσ. The values of h and k are key in the design of CUSUM charts and the evaluation of the ARL. The parameters h and k are related to the value of the mean shift, δ, and the probabilities of a false alarm, α, and the probability of not detecting a shift, β:

$$k = \frac{\delta\sigma}{2} \qquad\qquad (4)$$

$$h = dk \qquad\qquad (5)$$

where the lead distance d is given in ISQC equation 9.20,

$$d = \frac{2}{\delta^2}\ln(\frac{1-\beta}{\alpha}) \qquad\qquad (6)$$

The value of k=1/2 is a common value used, which implies a detection of a $\delta\sigma$=1, or one sigma shift. ISQC Table 9.3 provides the ARL performance of the two-sided tabular CUSUM chart as a function of different values of h for k=1/2, and Table 9.4 provides the values of h and k that give ARL(0) ~ 370.

The **CUSUM** platform can be used to evaluate the performance of the CUSUM chart. For this we generate a table of random normal values with mean 0 and standard deviation 1, and use that data to evaluate the performance of a CUSUM chart in terms of h and k. Note that the CUSUM chart launch window calls for the values of decision interval H= hσ, and the reference value K=kσ. By setting the value of σ=1 the parameters H and K are the same as h and k.

The following steps illustrate how to generate a random sample of 50 observations from a Normal(0,1), and how to specify the parameters in the **CUSUM** platform to evaluate the performance of the chart in terms of ARL.

1. Create a new data table by selecting **File ▶ New ▶ New Data Table**. Then select **Cols ▶ New Columns**, or double click of the column area of the data table to generate a new column. Call this column **y**. Add 50 rows to the table by selecting **Rows ▶ Add Rows**, or by double clicking on the rows area of the table, and enter **50** in the dialogue window.

2. Add a **Formula** to the column y and select **Random > Random Normal** from the Formula editor. Click **OK**. This will generate 50 random samples from a Normal(0,1).

3. Select **Analyze ▶ Quality and Process ▶ Control Chart ▶ CUSUM**.

4. Once the CUSUM launch window appears input the values of H=4, K=0.5, Target=0, Delta=1, Sigma=1 and leave Two Sided selected as shown in Figure 7.32. Note that we set the **Sample Size Constant** and equal to 1. Click **OK** to generate the CUSUM chart.

Figure 7.32 CUSUM Chart Dialog for Evaluating h=4 and k=0.5.

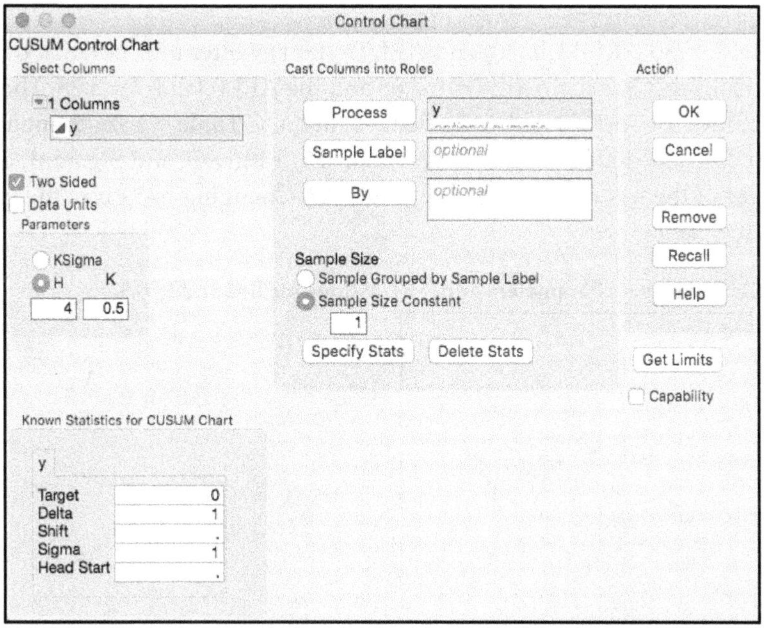

5. In the CUSUM window, click on the red triangle next to **CUSUM of y** and select **Show ARL** from the drop down menu as shown in Figure 7.33. Click on the red arrow again and select **Show Parameters**. The output is shown in Figure 7.34.

Figure 7.33 CUSUM Chart for Generated Normal(0,1) Data

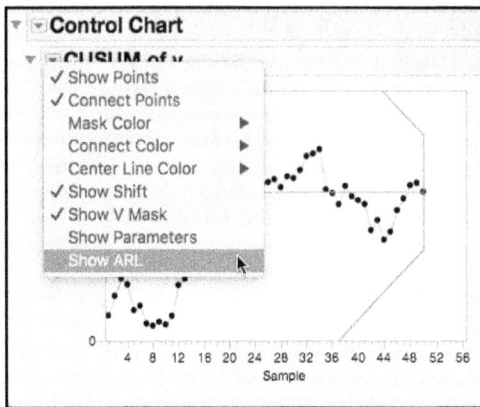

The CUSUM parameters and ARL(Delta=0) and ARL(Delta=1) values are shown at the bottom of the CUSUM chart in Figure 7.34. The ARL(0)= 167.68 and the ARL(Delta=1)= 8.38. These values corresponds to the values for ARL(0) and ARL(Delta=1) in ISQC Table 9.3. By changing the value of **Delta** in the CUSUM chart dialog and keeping h=4, k=0.5, Target=0, and Sigma=1, one can generate all the values of the h=4 column of ISQC Table 9.3. Changing h=5 gives the value of the h=5 column in the ISQC table.

Figure 7.34 CUSUM Chart Parameters and ARL Values for h=4 and k=0.5

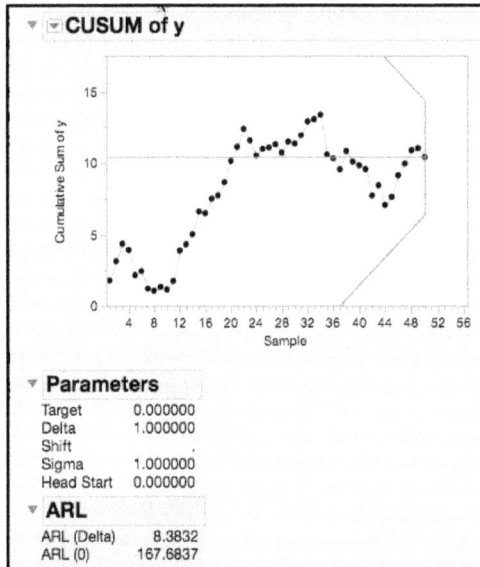

Figure 7.35 shows the ARL values for h=1.99 and k=1.25 given in ISQC Table 9.4. Here we keep Target=0, Delta=1 and Sigma=1. As expected, the ARL(0) is close to the stated 370, ARL(0)= 373. Note that with these parameters ARL(Delta=1)=18. This means that, on average, we expect to detect a shift of 1σ units in about 18 observations.

Figure 7.35 CUSUM Chart Parameters and ARL Values for h=1.99 and k=1.25

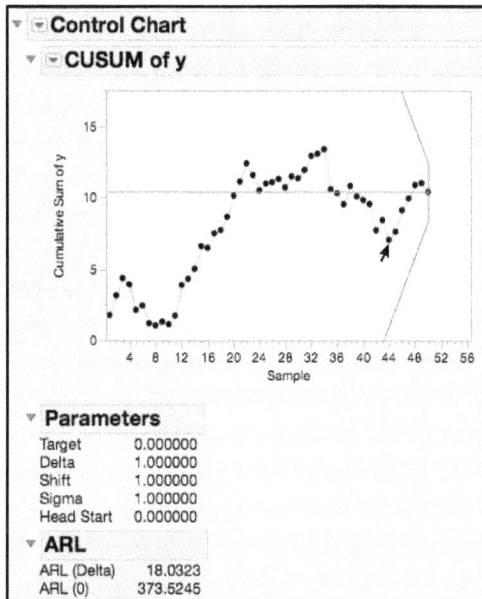

We wrote a JMP script to make it easier to calculate the ARL for different combinations of h and k. There is no need to use a data set; you can just pass the parameters for the CUSUM chart.

1. Open CUSUM ARL.jsl and click on **Run Script** at the top of the window. A dialog window will appear.
2. Enter **h = 4** and **k = 0.5** as shown in Figure 7.36. Click **OK**. The output is in Figure 7.37.

Figure 7.36 CUSUM ARL Script Dialog Window

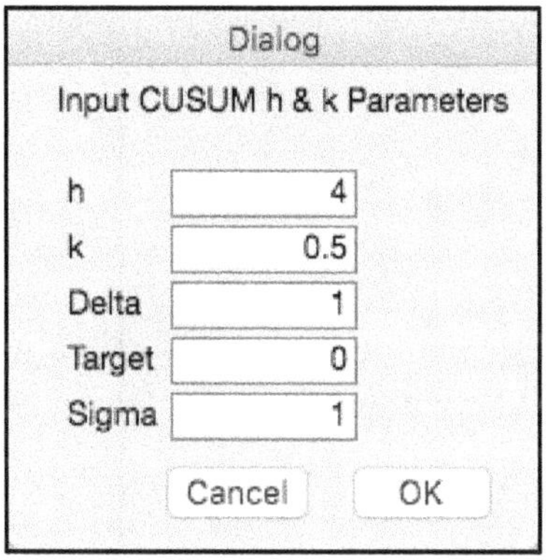

Figure 7.37 ARL Values for h=4 and k=0.5

The script output is shown in Figure 7.37. The output shows the parameters of the CUSUM chart and the ARL(0) and ARL(δ) values. Note that the ARL(0) values include both the one- and two-sided CUSUM. The Two-Sided values corresponds to the values for ARL(0) and ARL(Delta=1) in ISQC Table 9.3.

Statistics Note 7.4: The CUSUM is a very flexible control chart. Although, it is known for its ability to detect small mean shifts, h and k can be chosen to detect larger shifts effectively. For example, with h = 2.5, k=1.5, and Delta=2.5, the ARL(Delta) is 3.25 and ARL(0) = 5.429.

The new **CUSUM Control Chart** platform can also generate the ARL values for a given combination of h and k.

1. For the 50 random samples from a Normal(0,1) table that we created before, select **Analyze ▶ Quality and Process ▶ CUSUM Control Chart**.

2. In the launch window that appears, select **y** as the **Y** (response) variable, as shown in Figure 7.38.

Figure 7.38 CUSUM Control Chart Dialog for Generated Normal(0,1)

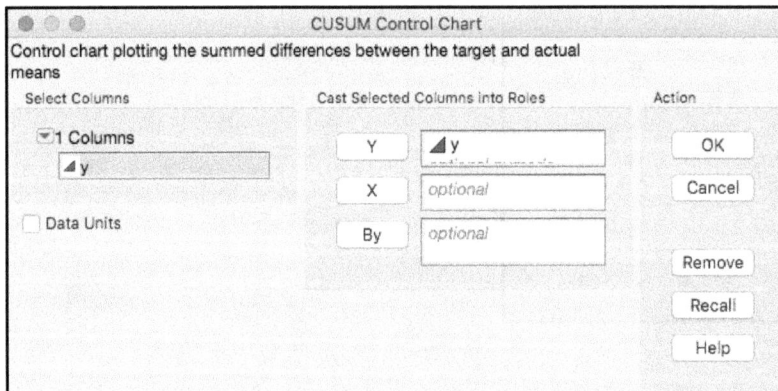

3. Click **OK**. The default Upper Side CUSUM is created with the default values of the mean and standard deviation of the data as the **Target** and **Sigma**.

4. Enter **0** for **Target, 1** for **Sigma, 4** for **h**, **0.5** for **k**, and click on **Lower Side.** The two-sided CUSUM is shown in Figure 7.39.

Figure 7.39 CUSUM Chart for Generated Normal(0,1)

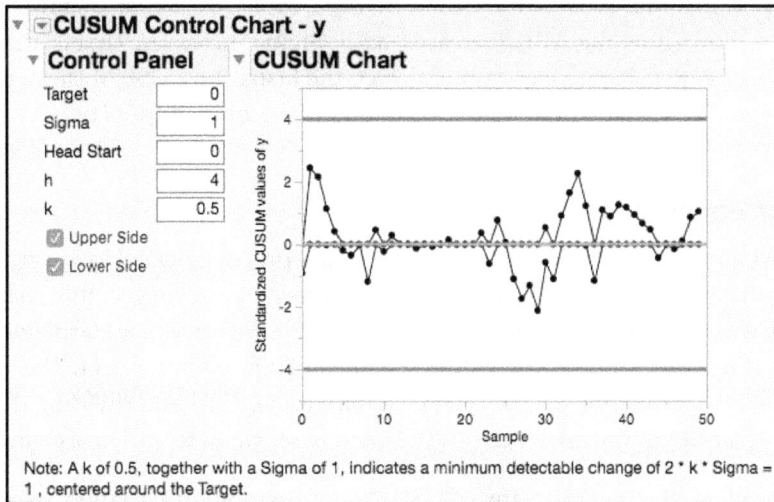

5. From the red triangle next to **CUSUM Control Chart – y** select **Show ARL**. The output is shown in Figure 7.40.

Figure 7.40 CUSUM Chart for Generated Normal(0,1) with ARL

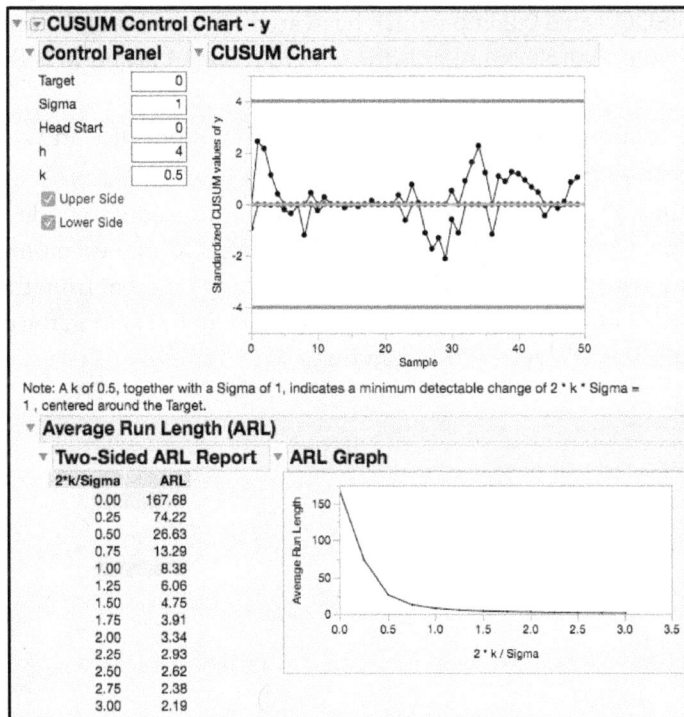

The **Average Run Length (ARL)** report in Figure 7.40 shows two columns corresponding to Delta=2k/Sigma, and ARL. The ARL(0)= 167.68 and the ARL(Delta=2k/Sigma=1)= 8.38, which again corresponds to the values for ARL(0) and ARL(Delta=1) in ISQC Table 9.3 and in Figure 7.37. In fact, the report includes most of the values in ISQC Table 9.3, except Delta=4. The **ARL Graph** in the report shows how the ARL decreases, as 2k/Sigma increases. That is, the higher the value of Delta=2k/Sigma, the change we want to detect, the faster the CUSUM chart, on average, will detect the change. The **CUSUM** platform, on the other had, only outputs two ARL values and no ARL graph.

CUSUM Chart for Variability

CUSUM charts are known for detecting smaller shifts in the process mean. However, they can be altered to detect shifts in the process variability. One approach to accomplish this, described in ISQC Section 9.1.8, was originally developed by Hawkins (1981). A new standardized quantity v_i is created from the standardized response values (y_i), using ISQC equation 9.11. The standardized quantity v_i is created by subtracting 0.822 from the square root of $|y_i|$ and dividing this difference by 0.349. This new quantity is approximately N(0, 1) and is sensitive to variance changes.

The following steps illustrate how to create a CUSUM chart to monitor variation, using the data presented in ISQC Table 9.2. Recall the first 20 observations are N(10, 1) and the last 10 observations are N(11, 1). We will use the same CUSUM parameters (H and K) used in Example 9.1.

1. Open Chapter 7 – ISQC Table 9.2.jmp, which has variables called *Sample Number, Xi*, and *True Mean*. In this table, Sample Number is the subgroup variable and Xi is the response of interest.
2. Double click in the column header to add one new column. Right click and select **Column Info** and name the column **Yi** and click **OK**.
3. Right click on Column **Yi** and select **Formula....** In the formula editor window, first click on the division symbol at the top of the window. Then select **Xi** in the Columns window to populate the numerator. Highlight **Xi** and select the minus symbol from the top of the window and then click in the open field and type **10**. Click in the field in the denominator and type in **1** (Figure 7.41). Click on **OK** when done.

Figure 7.41 Formula Editor with Formula to Normalize Xi

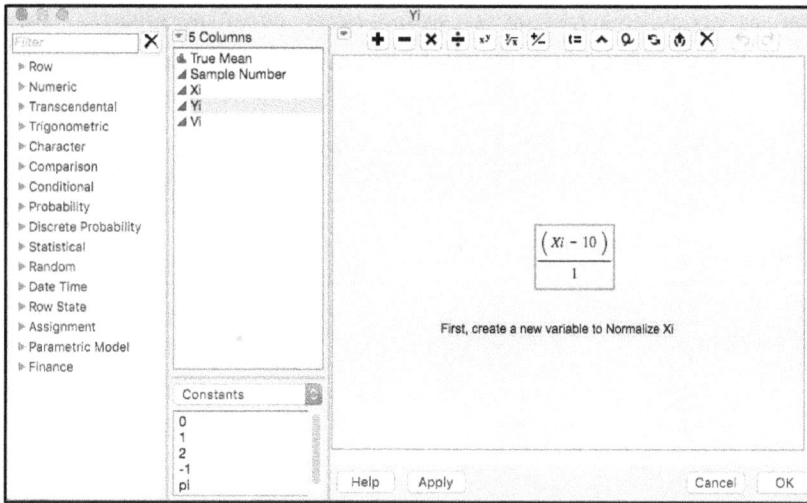

$$\frac{(Xi - 10)}{1}$$

First, create a new variable to Normalize Xi

4. Double click in the header to add another new column. Right click and select **Column Info** and name the column **vi** and click **OK.**

5. Right click on Column **vi** and select **Formula....** In the formula editor window, first click on the division symbol at the top of the window. With the numerator highlighted, click on the square root symbol at the top of the window. Next, from the far left window, select **Numeric ▶ Abs** and then select **Yi** in the Columns window to populate the numerator. Highlight the outer box in the numerator and select the minus symbol from the top of the window and then click in the open field and type **0.822**. Click in the field in the denominator and type in **0.349** (Figure 7.42). Click on **OK** when done.

Figure 7.42 Formula Editor to Normalize Yi

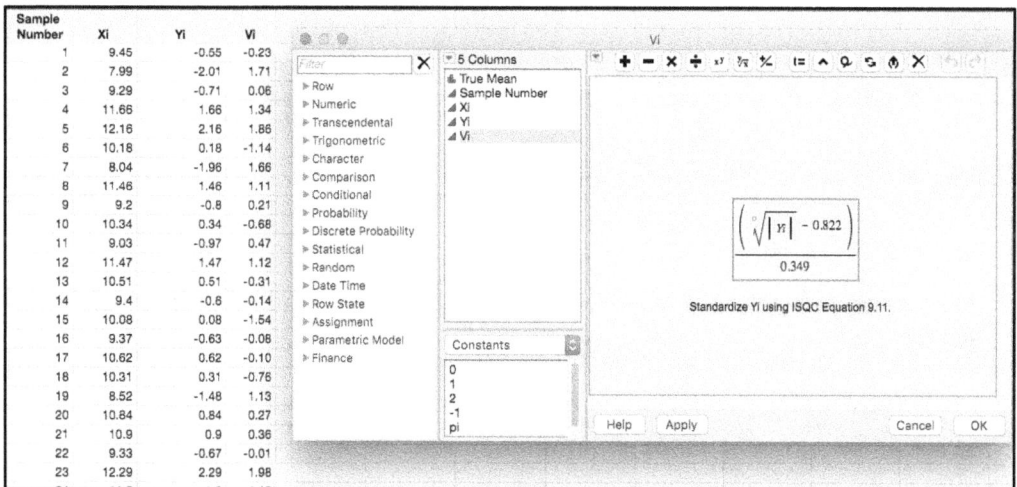

Sample Number	Xi	Yi	Vi
1	9.45	-0.55	-0.23
2	7.99	-2.01	1.71
3	9.29	-0.71	0.06
4	11.66	1.66	1.34
5	12.16	2.16	1.86
6	10.18	0.18	-1.14
7	8.04	-1.96	1.66
8	11.46	1.46	1.11
9	9.2	-0.8	0.21
10	10.34	0.34	-0.68
11	9.03	-0.97	0.47
12	11.47	1.47	1.12
13	10.51	0.51	-0.31
14	9.4	-0.6	-0.14
15	10.08	0.08	-1.54
16	9.37	-0.63	-0.08
17	10.62	0.62	-0.10
18	10.31	0.31	-0.76
19	8.52	-1.48	1.13
20	10.84	0.84	0.27
21	10.9	0.9	0.36
22	9.33	-0.67	-0.01
23	12.29	2.29	1.98

$$\left(\frac{\sqrt{|Yi|} - 0.822}{0.349}\right)$$

Standardize Yi using ISQC Equation 9.11.

6. Select **Analyze ▶ Quality and Process ▶ CUSUM Control Chart.**

7. Select **vi** as the **Y** (response) variable. Then select **Sample Number** and click **X** to identify the subgroup variable. Make sure **Data Units** is checked. Click **OK**.

8. The default Upper Side CUSUM is created with the default values of the mean and standard deviation of the data as the **Target** and **Sigma**. We need to update these values to generate the variability CUSUM.

9. Enter **0** for **Target, 1** for **Sigma, 5** for **H, 0.5** for **K,** and click on **Lower Side**. The output is shown in Figure 7.43.

Figure 7.43 CUSUM for Variability Using ISQC Example 9.1

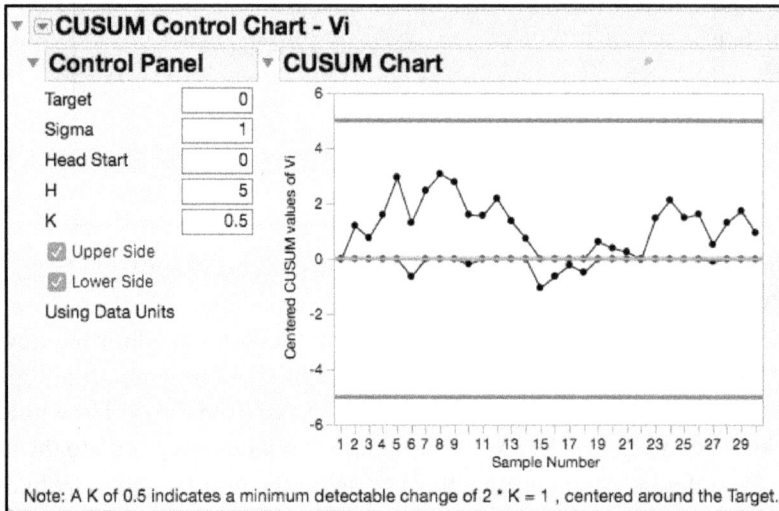

Note: A K of 0.5 indicates a minimum detectable change of 2 * K = 1 , centered around the Target.

The variability CUSUM chart shown in Figure 7.43 does not show any signals. This is not surprising since there was no shift in the standard deviation, just the mean.

Other approaches are available to detect shifts in the process variation. For instance, Ramírez (1989) and Box and Ramírez (1991) suggest an approach based on the cumulative sums of the square deviations from target. The cumulative sums S_k are given by

$$S_k = \sum_{i=1}^{k}(X_i - \mu)^2 - k \tag{7}$$

where the reference value k is chosen to detect a change in variation from a given baseline value σ_0 to a new value σ_1. The reference value k provides maximum discrimination between σ_0 and σ_1 and is given by

$$k = \frac{\log\left(\sigma_1^2 \middle/ \sigma_0^2\right)}{\dfrac{1}{\sigma_0^2} - \dfrac{1}{\sigma_1^2}} \tag{8}$$

The decision interval h for the Ramírez CUSUM for variance is chosen to give desired ARL values. For a given values of σ_0 and σ_1 and reference value k, the nomogram in Ramírez and Juan (1989) can be used select the appropriate value for h.

Let us examine a dataset from ISQC Chapter 6, show in Table 6.4. The data consists of Piston Ring Diameter and it was revised from its original source to center around a target = 4 mm with standard deviation of 0.01 mm. The data are plotted in Figure 7.44. There are a total of 125 observations with an increase in variation starting around N = 70. Both approaches, the transformed v_i and the Ramírez CUSUM for variance, are applied to this data set.

For the Ramírez CUSUM the reference value k is set to detect a change in standard deviation

from 0.01 mm to 0.02 mm, or $k = \dfrac{\log\left(0.02^2 \big/ 0.01^2\right)}{1\big/0.01^2 - 1\big/0.02^2} = 0.000185.$

The CUSUM statistic becomes $(\textbf{\textit{Diameter}} - 4)^2 - 0.000185.$

Figure 7.44 Modified Piston Ring Data for Variability CUSUM

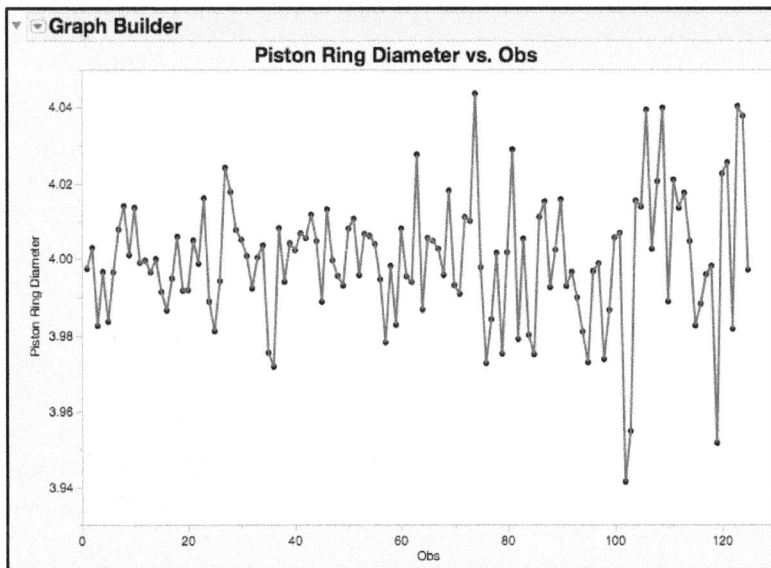

The following steps show how to produce CUSUM charts for variability for the Piston Ring data.

1. Open Chapter 7 – ISQC Piston Ring Data Set.jmp, which has variables called *Obs* and *Piston Ring Diameter*. Obs is the subgroup variable and Piston Ring Diameter is the parameter of interest. The normalized parameter, v, is also included in this JMP table. Note Piston Diameter was normalized using ISQC equation 9.11.
2. Select **Analyze ▶ Quality and Process ▶ CUSUM Control Chart**.

3. Select **v** as the **Y** (response) variable. Then select **Obs** and click **X** to identify the subgroup variable. Make sure **Data Units** is checked.

4. Click **OK**. The default Upper Side CUSUM is created with the default values of the mean and standard deviation of the data as the **Target** and **Sigma**. We need to update these values to generate the variability CUSUM.

5. Enter **0** for **Target, 1** for **Sigma**, **5** for **H**, **0.5** for **K**, and click on **Lower Side.**

6. From the red triangle next to **CUSUM Control Chart – v** select **Show ARL** (Figure 7.45).

Figure 7.45 Variability CUSUM for Piston Ring Data Using Transformed v

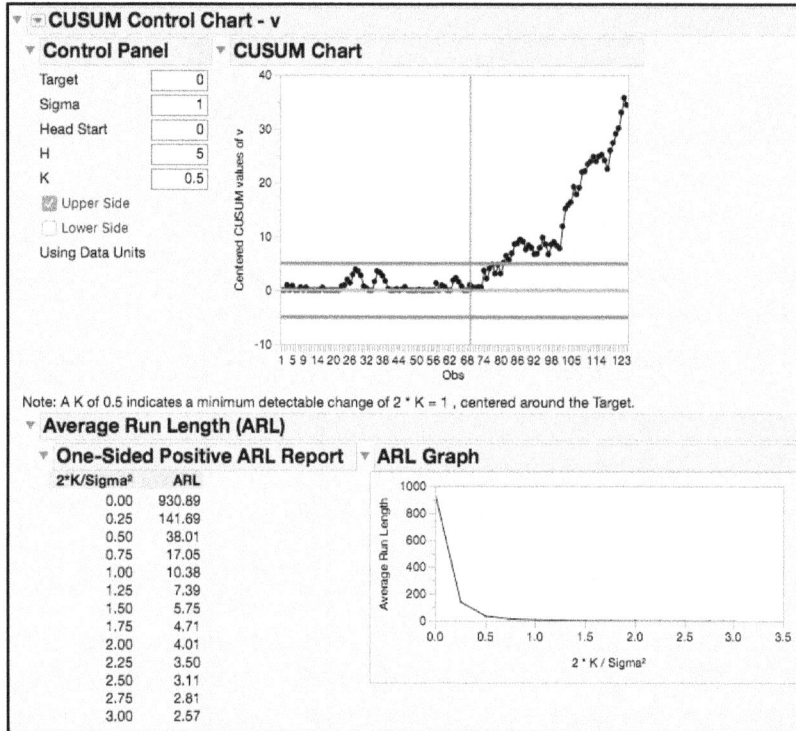

7. In order to create the Ramírez CUSUM for variance, click on the **CUSUM Variance** script in the left-hand side of the JMP table Chapter 7 – ISQC Piston Ring Data Set.jmp. The output is shown in Figure 7.46.

Figure 7.46 Ramírez CUSUM for Variance

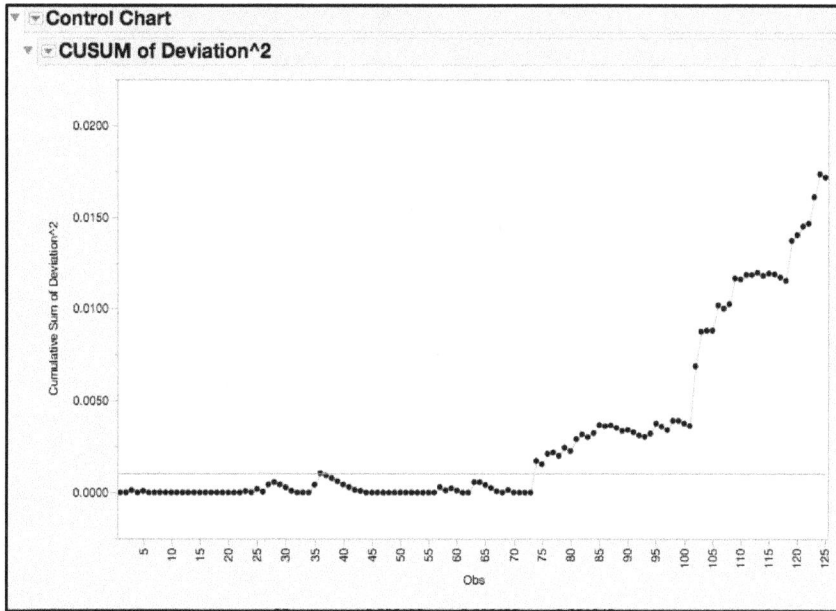

8. An overlay plot of the two CUSUM charts is generated by selecting the **CUSUM Variance & CUSUM v** (overlay script).

Figure 7.47 Overlay of CUSUM Variance Approaches for Piston Ring Data

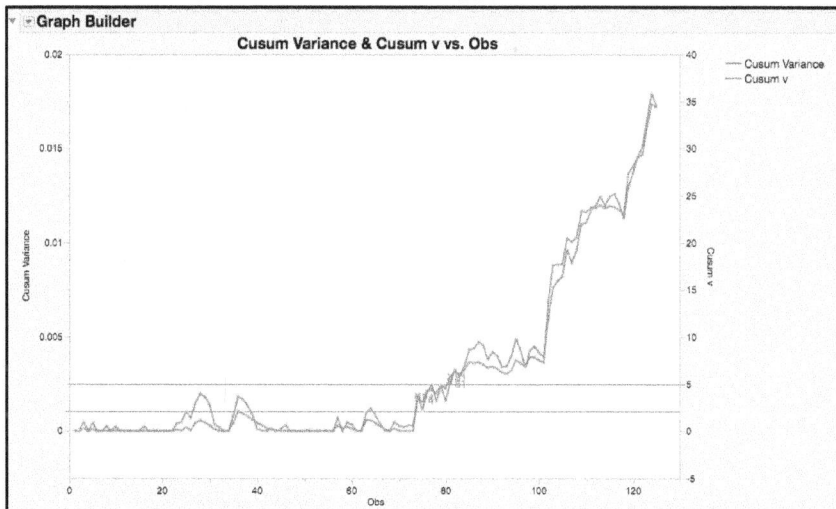

The two CUSUM charts overlaid in in Figure 7.47 produce signals at observations 74 (Ramírez) and 81 (Hawkins). The ARL for v_i indicates a signal in about 10 observations from the actual shift in the variance (the ARL(1)=10.38 in Figure 7.45), while the Ramírez's CUSUM for variance detected a shift after only 4 observations.

Chapter 8: Other Univariate Statistical Process Monitoring and Control Techniques

Overview

This chapter illustrates several univariate statistical process monitoring and control techniques, which are covered in Chapter 10, Other Univariate Statistical Process-Monitoring and Control Techniques, of *Introduction to Statistical Quality Control* (ISQC). The two techniques covered in this chapter include SPC for autocorrelated process data and Cuscore charts for monitoring changes in time series parameters.

The techniques presented in this chapter are discussed for measurements using a continuous scale. The concepts will be illustrated using several platforms in JMP, including, **Time Series** and the **Control Chart Builder**, and examples based on data from the Semiconductor Industry.

Special Topics Review

This chapter illustrates monitoring techniques for autocorrelated data. These techniques are discussed in ISQC Chapter 10, Other Univariate Statistical Process-Monitoring and Control Techniques. As Montgomery points out, an underlying assumption for the use of control charts is that, when the process is in control, the data can be represented by a normal distribution with a constant mean μ and standard deviation σ. This is shown algebraically in ISQC equation (10.9), or equation (1) below.

$$x_t = \mu + \varepsilon_t, \qquad\qquad t = 1, 2, \cdots \qquad\qquad (1)$$

where ε_t is normally and independently distributed, with mean 0 and standard deviation σ.

In practical terms, the model in equation (1) means that, when the process is in control, the plotted data on a Shewhart chart will be randomly distributed about the centerline; absent of unusual patterns or trends, and within the ±3 sigma control limits. When the process departs from a state of control, due to a mean shift, then the data will plot outside of the control limits or might violate runs tests, such as a run of points above the centerline.

The assumption of independent errors in equation (1), which implies that the data must be independent, should be investigated when using Shewhart control charts, such as, XBar and R charts or Individual and Moving Range charts (XmR). In general, when the data are not independent and positively autocorrelated, the within subgroup variance used in the construction of the control limits, which is estimated by the average moving range or average of the subgroup ranges, will be underestimated. This in turn results in control limits that are too tight, which in turn result in an excess of unnecessary violations to runs tests, such as Nelson 1 and Nelson 2.

Statistics Note 8.1: In industrial applications, positive autocorrelation is the most common situation. For positive autocorrelation, the variance is underestimated. For negative autocorrelation, the variance is over estimated.

As Montgomery points out, the independence assumption is often violated when the sampling frequency out paces the inertial elements of the process. The phenomenon is referred to as autocorrelation. For example, on equipment using engineering process controls algorithms, such as, make-up air handlers, sensors measure and store data every second. Observations next to each other in time are more similar than observations that are spaced farther apart. In Section 10.4.1 in ISQC Chapter 10, this concept is illustrated using a tank with an incoming and outgoing concentration of a chemical. Therefore, the statistical monitoring strategy must be adapted for processes that generate autocorrelated outputs.

Montgomery discusses several approaches to accommodate an autocorrelated data structure, including model and model-free techniques. In a model-based approach the correlation structure of the data is modeled with a Time Series model, and the residuals from the fitted model are monitored using the control charts described in previous chapters, such as a CUSUM chart. Montgomery provides a simple model (ISQC equation 10.11) for this data, shown in equation 2.

$$x_t = \xi + \phi x_{t-1} + \varepsilon_t, \qquad\qquad t = 1, 2, \cdots \qquad\qquad (2)$$

where ξ and ϕ ($-1 < \phi < 1$) are unknown constants, and ε_t is normally and independently distributed with mean zero and standard deviation σ. This is a first-order autoregressive model where the observation at time t is predicted from the previous observation at time *t-1*.

A number of steps are needed to create residuals-based control charts. First, the specific form of the autocorrelation in a time series must be diagnosed and an initial model identified. The model is then fitted to the data and evaluated and adjusted as needed. In this book, ARIMA, or autoregressive integrated moving average, models will be used. The residuals, which are the actual value minus the predicted value from the ARIMA model, are approximately normally and independently distributed with mean 0 and constant variance. In other words, by using the residuals the autocorrelation is "removed" from the original time series data, and we can now chart the residuals with an XmR chart, for example, to retrospectively monitor the data.

Monitoring future data that is autocorrelated is accomplished by predicting future performance of the time series using the time series model with parameter estimates from the baseline data. In all cases the prediction equation will require data, or residuals, from previous time periods to predict the performance of the most recent data point. The residuals are calculated and stored in a data source and then run through the control charting application. The use of time series software is very helpful here.

As with any control charting technique, sometimes the control limits need to be updated to reflect the output from a desired process change. For time series-based control charts, this implies that the underlying autocorrelation structure has changed from the baseline process and the ARIMA model might need to be updated. A Cuscore control chart can be used to determine if it is time to re-estimate the time series model. A Cuscore chart is a CUmulative sum chart of SCORE statistics. Since the score statistics depend on the model being fit to the data, any changes in the model parameters can be detected with the Cuscore.

In this chapter we will show how to identify an appropriate time series model for a data stream, fit the specified model, obtain the residuals for the fitted model and monitor them using a XmR chart. We will also demonstrate how to use a Cuscore control chart to monitor the parameters from a time series model. Since the data for ISQC Chapter 10 was not available, these techniques will be illustrated using data from the Semiconductor industry.

JMP Platforms for Monitoring Autocorrelated Processes

The **Time Series** platform, within **Specialized Modeling,** is used to identify and fit a time series model, and the **Control Chart Builder** platform, under **Quality and Process,** is used to monitor the residuals. A general overview of these platforms was introduced in Chapter 2. Table 8.1 provides a summary of the features we find most useful in the **Time Series** platform.

Table 8.1 Overview of Features for JMP Time Series Platform

JMP Feature	Time Series
Series Diagnostic Tools	Autocorrelation plot, Partial Autocorrelation plot, AR Coefficients Chart, Variogram, Lag plot, Ljung-Box statistics, White noise test
Model Types	ARIMA, Seasonal ARIMA, smoothing models
Model Diagnostic Tools	Model summary statistics, residuals plots (autocorrelation and partial autocorrelation), Forecast plots
Saving Output	JMP table with actual values, predictions, standard errors, residuals, and prediction confidence intervals
Forecasting Capabilities	Select forecasting period from within platform, save prediction formula to new JMP table, create and submit SAS job

Examples from ISQC Chapter 10

As was noted previously, we were unable to obtain the data sets illustrated in ISQC Chapter 10. We will use a data set from the Semiconductor industry to illustrate the analysis shown in ISQC Section 10.4, but we will not be able to reproduce the exact ISQC Figures and Tables. Note, Cuscore charts are discussed in ISQC Section 10.7, but no examples are provided. The same JMP table will be used in all examples presented in this chapter.

Table 8.2 Summary of Examples from ISQC Chapter 10

ISQC Reference Number	JMP Table Name	JMP Platform	Key Points
Example 10.2 Residuals Control Chart	Chapter 8 – ISQC Example 10.2*	Time Series & Control Chart Builder	Diagnose and fit an ARIMA model and monitor Residuals. Add new data to model and monitor residuals.
Figure 10.17 Moving Centerline Forecast Control Chart	Chapter 8 – ISQC Example 10.2*	Time Series & Graph Builder	Create a moving center- line forecast control chart
Section 10.7 Cuscore Chart	Chapter 8 – ISQC Example 10.7	Formula Editor & Graph Builder	Monitor time series parameters using a Cuscore chart

ISQC Example 10.2 Residuals Control Chart

In this section, we will show how to monitor processes that have autocorrelated output by first fitting an ARIMA model using the **Time Series** platform, and then charting the residuals in the **Control Chart Builder**. As was described previously, we will be using a data set from the Semiconductor industry, in place of the data presented in ISQC Example 10.2.

The data used in this chapter is from J. Ramírez (1998), Monitoring Clean Room Air Using Cuscore Charts. In this paper, a parameter related to the pressure differential transmitter (PDT) of a make-up air handler, was analyzed using ARIMA models, residuals charts were used to monitor the process, and Cuscore charts were used to monitor the parameters of the fitted ARIMA model. All the analyses in the paper were done using SAS. We will reproduce the analyses in this paper using JMP. The JMP data table contains 1 pressure differential parameter with 1,000 baseline values, taken at a constant sampling interval of 5 minutes.

The following steps illustrate how to identify and fit an appropriate ARIMA model to the make-up air handler data. The JMP data table is called Chapter 8 – ISQC Example 10.2*.jmp to align with the example, but the * indicates the data is not the data shown in ISQC Chapter 10 Figure 10.12, since that data was not available. We begin with an XmR chart of the original data.

1. Open the JMP table Chapter 8 – ISQC Example 10.2*.jmp, which has two variables called *Sample ID* and *PDT1221*. In this table, Sample ID is the subgroup variable, and PDT1221 is the parameter.

2. Select **Analyze ▶ Quality Process ▶ Control Chart Builder** and drag **PDT1221** to the **Y** zone. Then click **Done**.

Figure 8.1 Individuals Control Chart for PDT1221

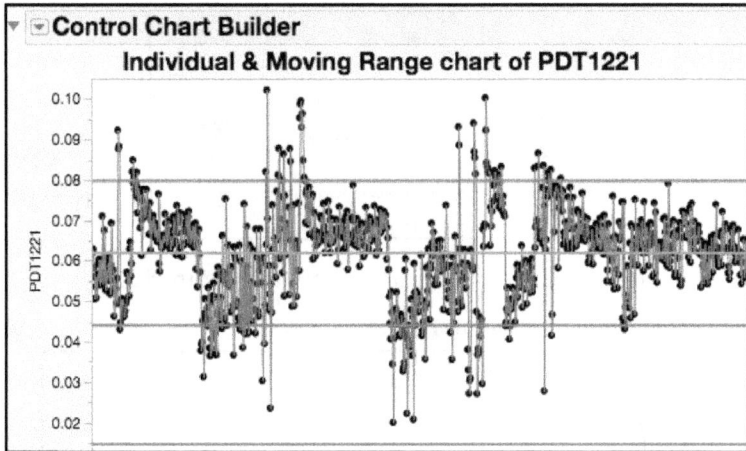

The Individuals control chart of PDT1221 is shown in Figure 8.1. The control limits for this chart were calculated in the usual way (see Chapter 3). It is interesting to note the significant number of points that exceed the upper and lower control limits. The short-term variation in this process, which is captured by the moving ranges, significantly underestimates the longer-term variation, due to the drift inherent in the process, resulting in groups of observations above or below the overall average (centerline). The violations of additional runs tests, such as 8-point on one side of the centerline, would also be significant. A similar scenario, with only 100 observations, is shown in ISQC Figure 10.12.

Statistics Note 8.2: A basic assumption of Shewhart charts is that the data are independent. Autocorrelated data do not meet this assumption and will most likely result in many false signals, like the ones in Figure 8.1.

This type of behavior is a tip off that the data might not meet the basic model assumptions for a Shewhart chart and it should be assessed for autocorrelation. This assessment is shown next.

1. Select **Analyze ▶ Specialized Modeling ▶ Time Series**.
2. A launch window will appear. From the window on the left-hand side, highlight **PDT1221** and then click **Y Time Series** (Figure 8.2).

Figure 8.2 Launch Window for Time Series Platform

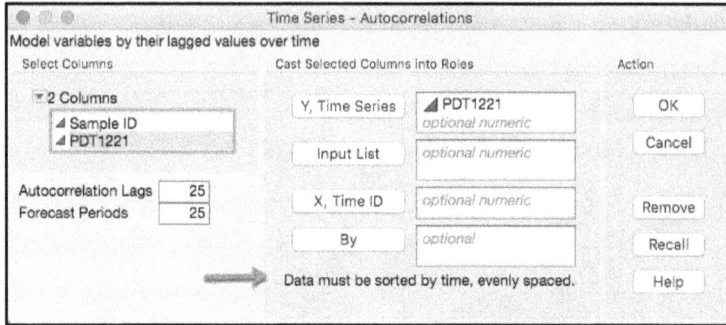

3. Click **OK** when finished. The default output is shown in Figure 8.3.

Figure 8.3 Default Time Series Output for PDT1221

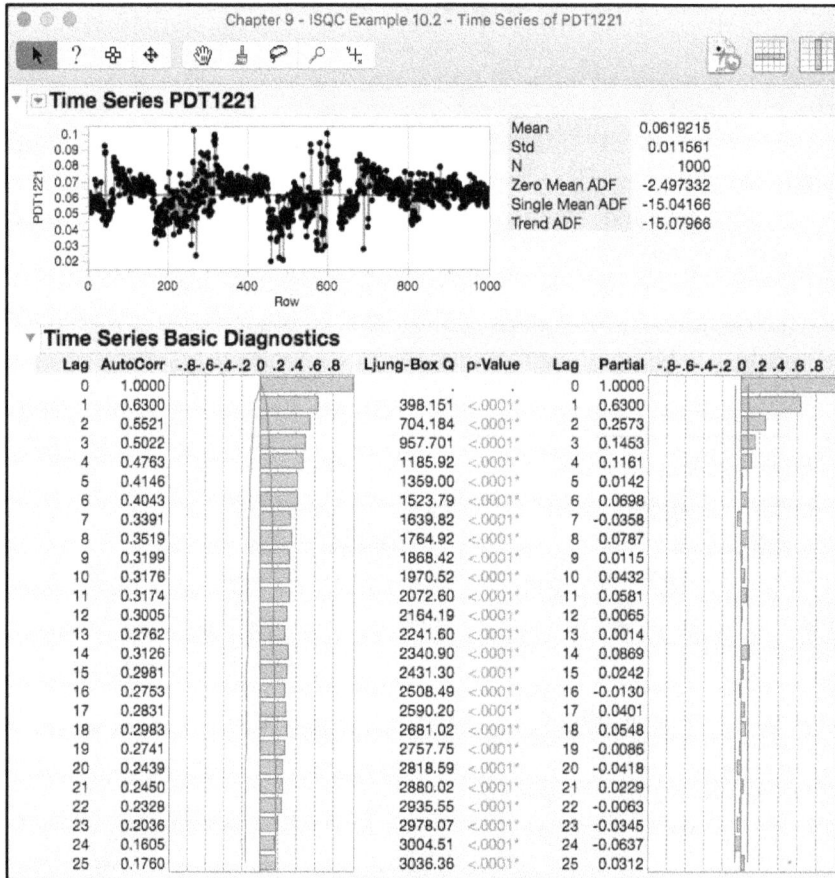

The default output shown in Figure 8.3 includes a **Time Series** plot of the data and underneath are the **Time Series Basic Diagnostics** plots. The plot of the raw data exhibits nonstationary behavior, where the data tends to trend upward for a period, followed by a downward trend. The

Autocorrelation plot, which is on the lower left-hand side of this output, shows the sample correlation between points at time t and t-p, where p is the lag of the data, $p = 0$ through 25. For example, for lag = 1, the sample correlation coefficient is 0.6300 shows the correlation between subsequent time points in the time series that is, (x_2, x_1), (x_3, x_2), (x_4, x_3), ..., (x_{1000}, x_{999}). The lag 1 Ljung-Box Q statistic of 398.151 and p-value <.0001, indicates that the correlation is statistically different from 0. Similarly, the correlation at lag 2 (0.5521) represents the correlation between every other point that is, (x_3, x_1), (x_4, x_2), (x_5, x_3), ..., (x_{1000}, x_{998}).

> • **JMP Note 8.1:** JMP has many helpful tools to diagnose the time series. The Autocorrelation and Partial Autocorrelation plots are used in conjunction to identify an initial model.

The Partial Autocorrelation plot, which is on the lower right-hand side of this output, shows the sample partial correlation between points at time t and t-p, where p is the lag of the data, $p = 0$ through 25. Note that the partial correlation at lag 1, 0.6300, is the same as the lag 1 correlation in the Autocorrelation plot, as expected. The partial correlation at lag 2 is 0.2573, which represents the correlation between every other point that is, (x_3, x_1), (x_4, x_2), (x_5, x_3), ..., (x_{1000}, x_{998}), after accounting for the correlation between previous lags, (x_2, x_1), (x_3, x_2), (x_4, x_3), ..., (x_{1000}, x_{999}). The partial correlations get weaker as the lags increase.

Statistics Note 8.3: For a given lag, time between observations, the autocorrelation function (ACF) is the correlation of the time series with itself. The partial autocorrelation function (PACF) is the partial autocorrelation of the time series with itself, not explained by autocorrelations at lower lags.

The Autocorrelation and Partial Autocorrelation plots are used together to diagnose the time series and arrive at an initial model. The Autocorrelation plot in Figure 8.3 is described as having a slow decay in correlations as the lags increase, while the Partial Autocorrelation plot is said to cut off after lag 3. The combination of these to behaviors is indicative of a nonstationary process, or a process that is drifting around. In order to further diagnose this series, the drift needs to be removed by differencing the series.

4. Click on the red triangle next to the **Time Series PDT1221** banner at the top of the window to bring up a list of options (Figure 8.4) A brief description for some of the options that we will illustrate in this chapter is shown below. To launch the online help, click on the **?** at the top of the window and then click anywhere in the default output shown in Figure 8.3.

Figure 8.4 Time Series Options

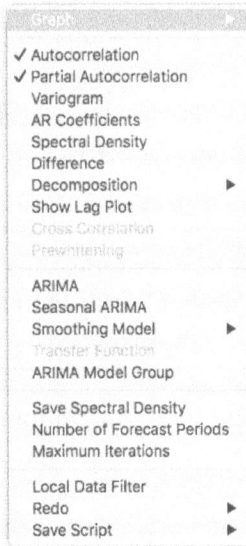

a. **Graph**: contains various options to alter the appearance of the Time Series plot of the data.

b. **Autocorrelation**: a plot that describes the correlation between all pairs of points for a given separation in time. This plot is used in conjunction with the Partial Autocorrelation plot to identify a preliminary model.

c. **Partial Autocorrelation**: a plot that describes the partial correlation between all pairs of point in the time series for a given a given separation in time. This plot is used in conjunction with the Autocorrelation plot to identify a preliminary model.

d. **Variogram**: a plot that measures the variance of the differences of points k lags apart and compares it to that for points one lag apart. It is computed from the autocorrelations and is used to identify if the time series is stationary. For a nonstationary series, the plot starts at 1 and increases until it reaches an asymptote. For a stationary process the plot, after lag 1, is relatively flat.

e. **AR Coefficients**: a graph of the coefficients approximated by those obtained from fitting a higher order auto-regressive model.

f. **Spectral Density**: a graph of the spectral density as a function of period and frequency. The plot is helpful to identify possible cycles (daily, monthly, quarterly, and so on) within the data.

g. **Difference**: a window appears to specify the non-seasonal and seasonal differences applied to the data and then graphed using the default output. This plot is used to help specify an initial model.

h. **Decomposition**: isolates and removes linear trends and seasonal cycles from the time series. This is used for better model estimation.

i.   **Show Lag Plot**: a plot that shows the data at time t on the Y axis and $t \pm p$ on the X axis. This plot is used to examine the correlation structure of the time series. It can be used to assess how strong the autocorrelation is at a given lag. The more elongated the ellipse of data, the higher the autocorrelation.

j.   **ARIMA**: shows a window that is used to specify the form of the ARIMA(p, d q) model to be fitted to the time series. The user enters the order for parameters, p (auto-regressive), d (difference) and q (moving average). The intercept can be excluded or included and the prediction level specified.

k.   **Seasonal ARIMA:**   shows a window that is used to specify the form of the seasonal ARIMA(p, d q)(P, D, Q)s model to be fitted to the time series. The user enters the order for parameters, p (auto-regressive), d (difference) and q (moving average) and P (seasonal autoregressive), D(seasonal difference) and Q (seasonal moving average). The intercept can be excluded or included and the prediction specified.

l.   **Smoothing Model**: shows a menu of different smoothing models to select from. These include models with simple smoothing and exponential techniques. Models are also included to account for linear trends and seasonality.

5.   From the red triangle next to the label **Time Series PDT1221** (Figure 8.3), select **Difference**. A dialog box will appear and in the left-hand side of the window labeled **Nonseasonal**, toggle the **Differencing Order** to **1**. Leave the **Seasonal Differencing Order** at **0** (Figure 8.5).

Figure 8.5 Differencing Dialog Box for PDT1221

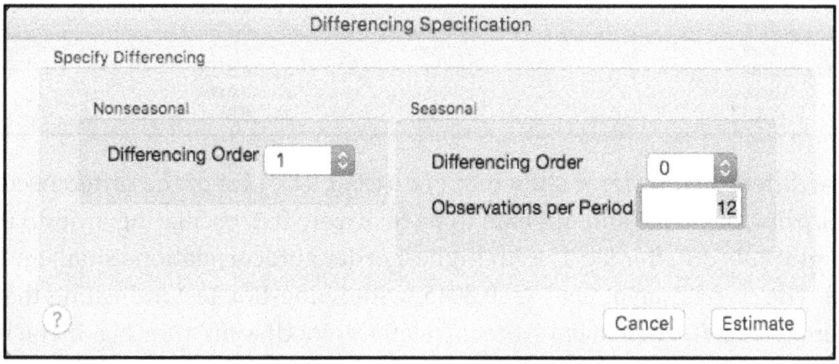

6. Click **Estimate** when finished. The default output is shown in Figure 8.6.

Figure 8.6 First Differencing Output of PDT1221

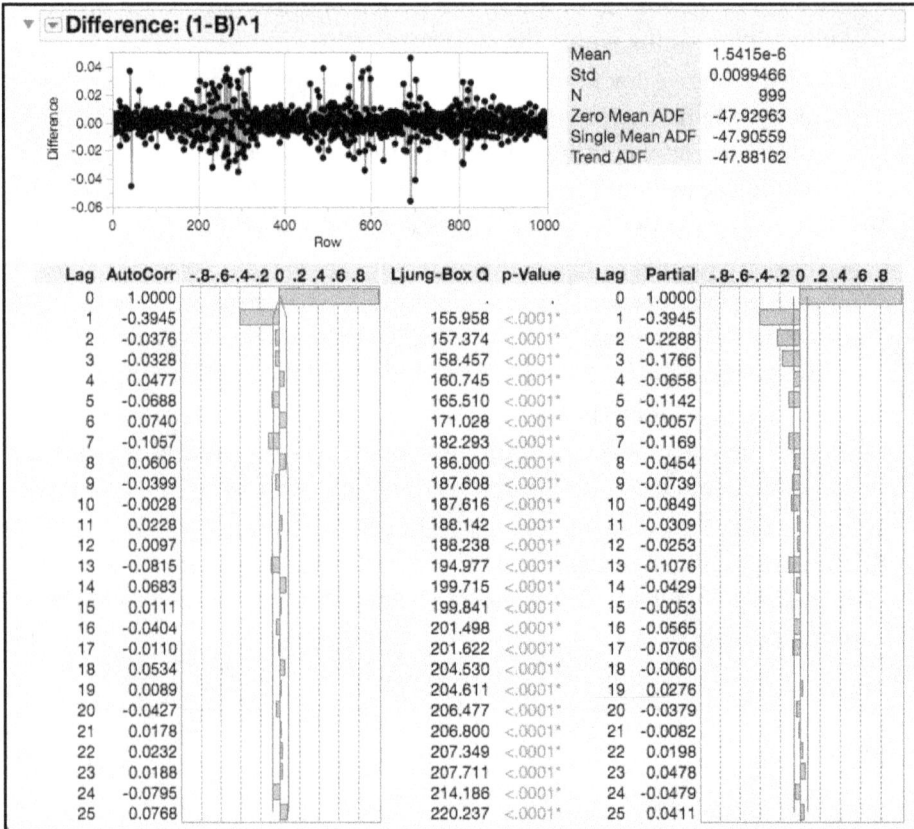

▼ ▾ **Difference: (1-B)^1**

Mean	1.5415e-6
Std	0.0099466
N	999
Zero Mean ADF	-47.92963
Single Mean ADF	-47.90559
Trend ADF	-47.88162

Lag	AutoCorr	-.8 -.6 -.4 -.2 0 .2 .4 .6 .8	Ljung-Box Q	p-Value	Lag	Partial	-.8 -.6 -.4 -.2 0 .2 .4 .6 .8
0	1.0000		.	.	0	1.0000	
1	-0.3945		155.958	<.0001*	1	-0.3945	
2	-0.0376		157.374	<.0001*	2	-0.2288	
3	-0.0328		158.457	<.0001*	3	-0.1766	
4	0.0477		160.745	<.0001*	4	-0.0658	
5	-0.0688		165.510	<.0001*	5	-0.1142	
6	0.0740		171.028	<.0001*	6	-0.0057	
7	-0.1057		182.293	<.0001*	7	-0.1169	
8	0.0606		186.000	<.0001*	8	-0.0454	
9	-0.0399		187.608	<.0001*	9	-0.0739	
10	-0.0028		187.616	<.0001*	10	-0.0849	
11	0.0228		188.142	<.0001*	11	-0.0309	
12	0.0097		188.238	<.0001*	12	-0.0253	
13	-0.0815		194.977	<.0001*	13	-0.1076	
14	0.0683		199.715	<.0001*	14	-0.0429	
15	0.0111		199.841	<.0001*	15	-0.0053	
16	-0.0404		201.498	<.0001*	16	-0.0565	
17	-0.0110		201.622	<.0001*	17	-0.0706	
18	0.0534		204.530	<.0001*	18	-0.0060	
19	0.0089		204.611	<.0001*	19	0.0276	
20	-0.0427		206.477	<.0001*	20	-0.0379	
21	0.0178		206.800	<.0001*	21	-0.0082	
22	0.0232		207.349	<.0001*	22	0.0198	
23	0.0188		207.711	<.0001*	23	0.0478	
24	-0.0795		214.186	<.0001*	24	-0.0479	
25	0.0768		220.237	<.0001*	25	0.0411	

The output of the differenced series is shown in Figure 8.6. The plot of the differenced data no longer exhibits drifting behavior and appears to be relatively flat, oscillating around 0. The Autocorrelation plot cuts off after lag 1, with highest order autocorrelations small and within the confidence band. The correlation at lag 1 is -0.3945, with Ljung-Box test indicating that is statistically different from 0. The Partial Autocorrelation plot dies off after lag 3. The behavior is these two plots suggests that a model with autoregressive (AR) order p=0, difference of order d=1, and a moving average (MA) of order q=1 may be appropriate. This model is the ARIMA(0,1,1) and can be written in a simplified way as IMA(1, 1), which stands for Integrated Moving Average model with order of differencing d = 1, and moving average order q = 1. Montgomery discusses the sample autocorrelation plot in ISQC Figure 10.13 for ISQC Example 10.2.

Statistics Note 8.4: A lot of industrial time series can be modeled using an IMA (1, 1) model. We call the IMA(1, 1) model the Swiss Army knife of time series models.

7. From the red triangle next to **Time Series PDT1221,** select **ARIMA.** A dialog box will appear. In the **d, Differencing Order** field enter the number **1,** in the **q, Moving Average Order** field enter the number **1** and leave **0** in the **p, Autoregressive Order** field. Uncheck the **Intercept** box (see Figure 8.7).

Figure 8.7 ARIMA Dialog Box

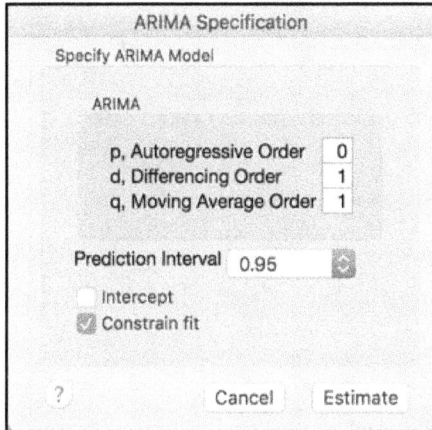

8. Click **Estimate** when finished. The default output is displayed in Figures 8.8 and 8.9.

Figure 8.8 Model Summary for IMA(1, 1)

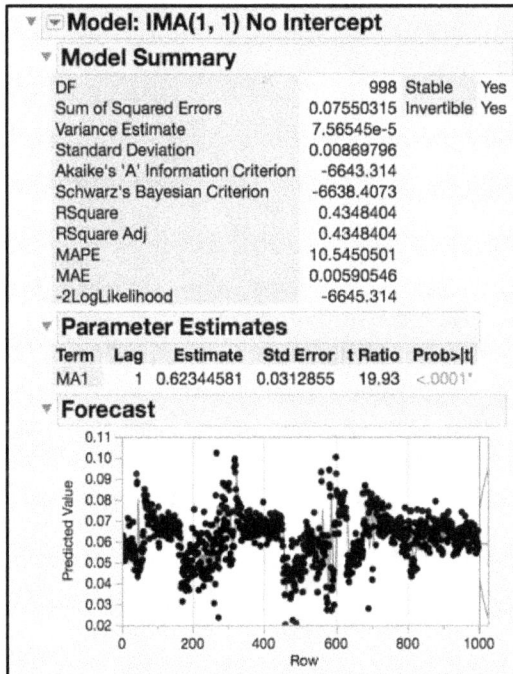

Figure 8.9 Residuals for IMA(1,1)

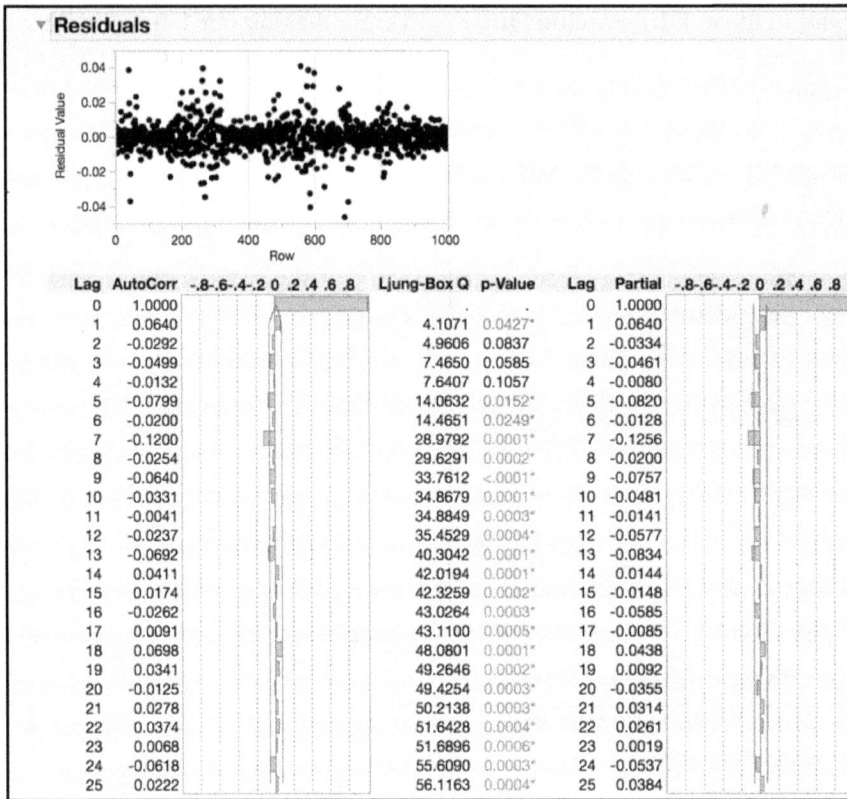

Lag	AutoCorr	-.8-.6-.4-.2 0 .2 .4 .6 .8	Ljung-Box Q	p-Value	Lag	Partial	-.8-.6-.4-.2 0 .2 .4 .6 .8
0	1.0000		.	.	0	1.0000	
1	0.0640		4.1071	0.0427*	1	0.0640	
2	-0.0292		4.9606	0.0837	2	-0.0334	
3	-0.0499		7.4650	0.0585	3	-0.0461	
4	-0.0132		7.6407	0.1057	4	-0.0080	
5	-0.0799		14.0632	0.0152*	5	-0.0820	
6	-0.0200		14.4651	0.0249*	6	-0.0128	
7	-0.1200		28.9792	0.0001*	7	-0.1256	
8	-0.0254		29.6291	0.0002*	8	-0.0200	
9	-0.0640		33.7612	<.0001*	9	-0.0757	
10	-0.0331		34.8679	0.0001*	10	-0.0481	
11	-0.0041		34.8849	0.0003*	11	-0.0141	
12	-0.0237		35.4529	0.0004*	12	-0.0577	
13	-0.0692		40.3042	0.0001*	13	-0.0834	
14	0.0411		42.0194	0.0001*	14	0.0144	
15	0.0174		42.3259	0.0002*	15	-0.0148	
16	-0.0262		43.0264	0.0003*	16	-0.0585	
17	0.0091		43.1100	0.0005*	17	-0.0085	
18	0.0698		48.0801	0.0001*	18	0.0438	
19	0.0341		49.2646	0.0002*	19	0.0092	
20	-0.0125		49.4254	0.0003*	20	-0.0355	
21	0.0278		50.2138	0.0003*	21	0.0314	
22	0.0374		51.6428	0.0004*	22	0.0261	
23	0.0068		51.6896	0.0006*	23	0.0019	
24	-0.0618		55.6090	0.0003*	24	-0.0537	
25	0.0222		56.1163	0.0004*	25	0.0384	

The **Model Summary** of the IMA(1, 1) fit is shown in Figure 8.8 and the **Residuals** of the fitted model are displayed in Figure 8.9. The **Model Summary** includes a number of statistics that quantify the quality of the fit of the model. For example, the Akaike's Information Criterion (AIC) is -6643 and the Schwarz's Bayesian Criterion is -6638, where smaller values indicate a better fit. The Standard Deviation (0.0087) and Variance (0.000076) are provided. The Mean Absolute Percentage Error (MAPE) is 10.55. This number is the sum of the ratio of the difference of the actual and predicted values to the actual value. The Mean Absolute Error (MAE) is 0.0059 and it is the average difference of the actual and predicted values. Finally, RSquare (0.4348) and Adjusted R-Square (0.4348) have the usual interpretation. Many of these statistics are more useful when comparing several models to each other, and we will review them in the Statistical Insights section.

The **Parameter Estimates** table provides the estimates, standard errors, t-ratios and p-values for the model parameters. The first order moving average parameter is 0.6234 and its standard error is 0.0313. The t Ratio for this parameter is 19.93 and its p-value is <.0001, indicating that it is statistically significant. Since the data were differenced, the intercept was omitted from the model

The **Residuals** output includes a plot of the model residuals and an Autocorrelation and Partial Autocorrelation plot of the residuals. These plots are used to determine if the autocorrelation has been removed from the original time series. Although, the trend plot of the residuals appears "flat", it can be difficult to visually determine if any autocorrelation remains. However, the Autocorrelation and Partial Autocorrelation plots indicate that an insignificant amount of correlation is left, with the lag 1 correlation estimate of 0.0640. Therefore, this time series can be adequately modeled using an IMA(1, 1).

Now that we have the fitted IMA(1, 1) model, we can monitor the residuals to determine if this process is in control, after removing the autocorrelation. The following steps illustrate how to chart the residuals in the **Control Chart Builder** platform.

9.   From the red triangle next to the banner **Model: IMA(1, 1) No Intercept** (Figure 8.8), select **Save Columns**. A JMP table is created that contains the actual data, predictions, prediction standard errors, residuals and the lower and upper confidence intervals (Figure 8.10)

Figure 8.10 JMP Table with IMA(1, 1) Residuals for PDT1221

	Actual PDT1221	Row	Predicted PDT1221	Std Err Pred PDT1221	Residual PDT1221	Upper CL (0.95) PDT1221	Lower CL (0.95) PDT1221
1	0.05586	1	•	•	•	•	•
2	0.06187	2	0.05586	0.0086979572	0.00601	0.0729076828	0.0388123172
3	0.06286	3	0.0591718286	0.0086979572	0.0036881714	0.0762195113	0.0421241458
4	0.05835	4	0.0607862314	0.0086979572	-0.002436231	0.0778339142	0.0437385487
5	0.05062	5	0.0598130624	0.0086979572	-0.009193062	0.0768607452	0.0427653797
6	0.05615	6	0.0562706929	0.0086979572	-0.000120693	0.0733183756	0.0392230101
7	0.05366	7	0.056224836	0.0086979572	-0.002564836	0.0732725187	0.0391771532
8	0.05084	8	0.0552556611	0.0086979572	-0.004415661	0.0723033439	0.0382079784
9	0.05667	9	0.0535906687	0.0086979572	0.0030793313	0.0706383515	0.036542986
10	0.05524	10	0.0547508153	0.0086979572	0.0004891847	0.0717984981	0.0377031326
11	0.05934	11	0.0549350576	0.0086979572	0.0044049424	0.0719827404	0.0378873749
12	0.06033	12	0.0565938893	0.0086979572	0.0037361107	0.073641572	0.0395462065
13	0.05403	13	0.058000781	0.0086979572	-0.003970781	0.0750484637	0.0409530982

10.   Click the JMP table to make it active. From the main menu, select **Analyze ▶ Quality and Process ▶ Control Chart Builder**. Highlight the **Residual PDT1221** variable in the Columns window in the left-hand side and drag it to the Y axis. When finished, click **Done**. The Chart is shown in Figure 8.11.

Figure 8.11 Individual and Moving Range Chart of IMA(1, 1) Residuals

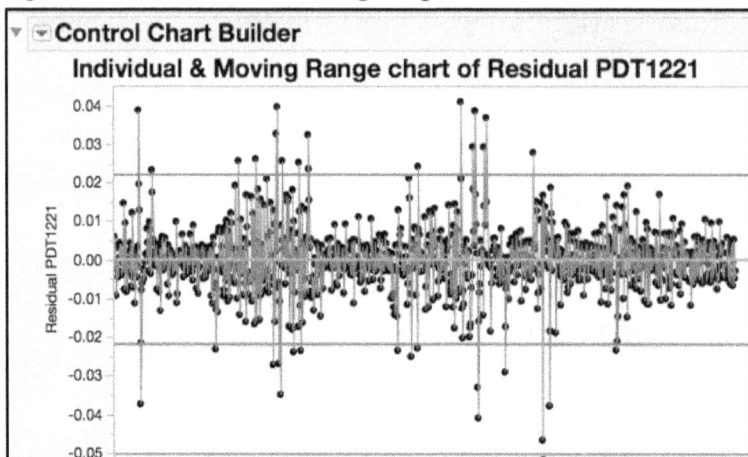

The Residuals control chart is shown in Figure 8.11. Since each time point represents a single value, sampled every 5 minutes, the subgroup size is n = 1. Therefore, an individual and moving range control chart is appropriate. The centerline of the Individuals chart is approximately 0 and the control limits are ±0.022. For such a large number of subgroups (N=999), it is expected that there will be a few points that exceed the control limits. However, there are 33 points that exceed the control limits in Figure 8.11, which seems excessive. This could be due to a couple of reasons. For one, we might have fit the wrong model to the time series and it needs to be re-specified. This can also indicate that the process is out-of-control and special cause variation should be investigated and removed. These options are investigated further in the Statistical Insights section of this chapter. Note Montgomery discusses the Residuals control chart in ISQC Figure 10.14 in ISQC Example 10.2.

ISQC Figure 10.17 Moving Centerline Centerline Forecast Control Chart

In the last section, we showed how to monitor an autocorrelated output by control charting the residuals from an IMA(1, 1) model. Since the residuals are not a direct plot of the data, additional analysis is required in order to investigate signals using the original data. While this approach is effective, it might be difficult for the practitioner to use. In ISQC Section 10.4, Montgomery discusses an approximate EWMA procedure for charting correlated data directly. This allows us to directly monitor the original data. An example of this approach is shown in ISQC Figure 10.17.

Another way to directly monitor autocorrelated output is to use the forecasts from the fitted ARIMA model and compare them to their $(1-\alpha) \times 100\%$ prediction interval. We prefer this approach because it is more robust to a variety of ARIMA models and is easier to implement in JMP. We will use the same data set described in the last example, which is a parameter related to the pressure differential transmitter (PDT) of a make-up air handler (J. Ramírez (1998)).

The following steps illustrate how use the forecasts from an IMA(1, 1) model to directly monitor the make-up air handler data. Recall, the JMP data table contains 1,000 baseline values, taken at a constant sampling interval of 5 minutes.

1. Open **Chapter 8 – ISQC Example 10.2*.jmp**, which has two variables called *Sample ID* and *PDT1221*. In this table, Sample ID is the subgroup variable, and PDT1221 is the parameter.

2. Select **Analyze ▶ Specialized Modeling ▶ Time Series.**

3. A dialog window will appear. From the window on the left-hand side, highlight **PDT1221** and then click **Y Time Series**. Click **OK** when finished.

4. From the red triangle next to **Time Series PDT1221,** select **ARIMA**. A dialog box will appear. In the **d, Differencing Order** field enter the number **1**, in the **q, Moving Average Order** field enter the number **1** and leave **0** in the **p, Autoregressive Order** field. Click on the toggle next to **Prediction Interval** and select **Other**. In the window that appears type **0.9973** and click **OK**. Uncheck the **Intercept** box. See Figure 8.12.

Figure 8.12 ARIMA Dialog Box for Prediction Interval

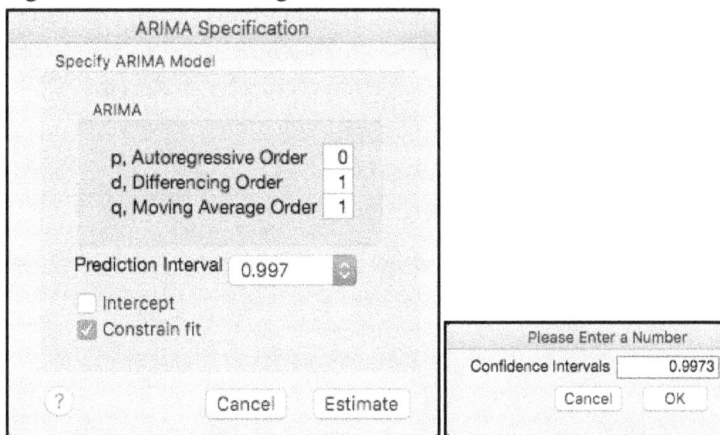

5. Click on the red triangle next to the banner **Model: IMA(1, 1) No Intercept** (see Figure 8.8), select **Save Columns.** A JMP table is created that contains the actual data, predictions, prediction standard errors, residuals and the lower and upper prediction intervals (see Figure 8.13). Note 25 additional rows are included in this table with forecasted results. These rows can be deleted before proceeding.

Figure 8.13 JMP Table with IMA(1, 1) Forecasts for PDT1221

	Actual PDT1221	Row	Predicted PDT1221	Std Err Pred PDT1221	Residual PDT1221	Upper CL (0.9973) PDT1221	Lower CL (0.9973) PDT1221
1	0.05586	1	•	•	•	•	•
2	0.06187	2	0.0558609272	0.0087023178	0.0060090728	0.0819676803	0.0297541741
3	0.06286	3	0.0591731707	0.0087023178	0.0036868293	0.0852799238	0.0330664176
4	0.05835	4	0.0607879117	0.0087023178	-0.002437912	0.0868946648	0.0346811586
5	0.05062	5	0.059815	0.0087023178	-0.009195	0.0859217531	0.0337082469
6	0.05615	6	0.0562728167	0.0087023178	-0.000122817	0.0823795698	0.0301660636
7	0.05366	7	0.0562270802	0.0087023178	-0.00256708	0.0823338332	0.0301203271
8	0.05084	8	0.0552579862	0.0087023178	-0.004417986	0.0813647393	0.0291512331
9	0.05667	9	0.0535930473	0.0087023178	0.0030769527	0.0796998004	0.0274862942
10	0.05524	10	0.0547532228	0.0087023178	0.0004867772	0.0808599759	0.0286464697
11	0.05934	11	0.0549374852	0.0087023178	0.0044025148	0.0810442383	0.0288307321
12	0.06033	12	0.0565963268	0.0087023178	0.0037336732	0.0827030799	0.0304895737
13	0.05403	13	0.0580032252	0.0087023178	-0.003973225	0.0841099783	0.0318964721
14	0.06033	14	0.0565080205	0.0087023178	0.0038219795	0.0826147736	0.0304012675

6. Click on the JMP table to activate it and select **Graph ▶ Graph Builder**. From the Columns window on the left-hand side, drag **Row** to the **X** zone (x axis). Next drag **Actual PDT1221** to the **Y** zone (y axis). Right-click in the graph window and select **Smoother ▶ Remove**.

7. Next, highlight and drag **Predicted PDT1221** to the **Y** zone (y axis).. In order to get an overlay plot, the variable needs to be dropped to the right of the **Y** axis. The two series will be plotted using two different colors, as is displayed in a legend on the right-hand side (Figure 8.14).

Figure 8.14 Actual and Predicted values for PDT122 Series

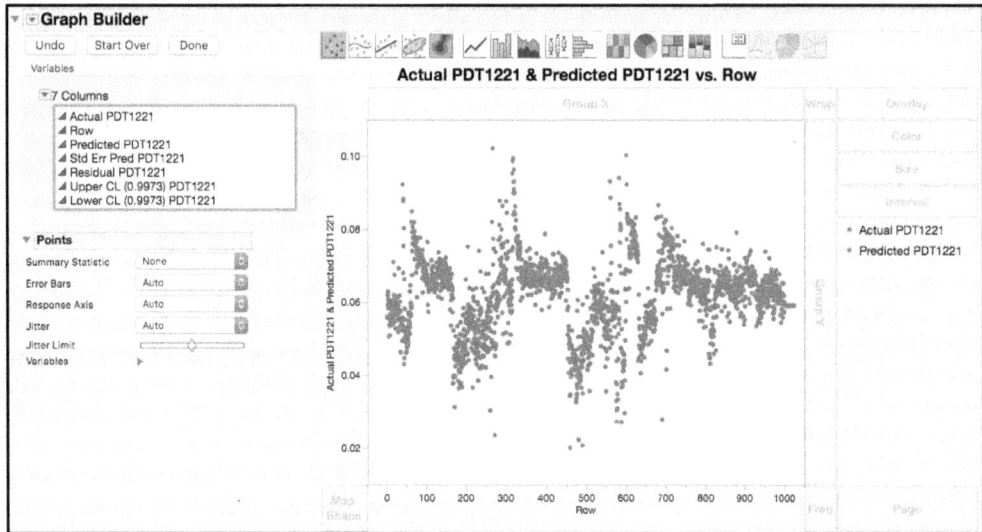

8. Right-click in the plot and select **Add ▶ Line**. This will add a line to both series. To remove the line for the actual points, right-click in the plot and select **Line ▶ Y ▶ Actual PDT1221**. To remove the points for the predicted values, select **Points ▶ Y ▶ Predicted PDT1221**. The plot is shown in Figure 8.15.

Figure 8.15 Actual values and Predicted Line for PDT122 Series

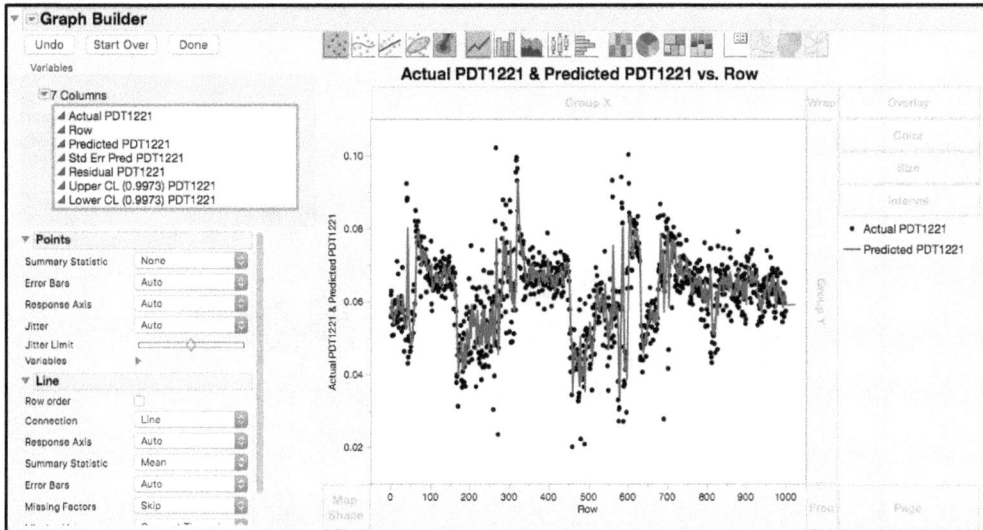

9. From the Columns window, drag **Upper CL (0.99) PDT1221** to the **Y** zone (y axis). To remove the points from the upper prediction limits, right-click in the plot and select **Y ▶ Points ▶ Upper CL (0.9973) PDT1221**.

10. Form the Columns window, drag **Lower CL (0.99) PDT1221** to the **Y** zone (y axis). To remove the points from the upper prediction limits, right-click in the plot and select **Y ▶ Points ▶ Upper CL (0.9973) PDT1221**. To remove the points from the lower prediction limits, right-click in the plot and select **Y ▶ Points ▶ Lower CL (0.99) PDT1221**.

11. To change point and line colors, right-click on the point (or line) legend in the right-hand side of the window and select the desired colors. Note marker types, sizes and line types can be changed in a similar manner. Click **Done** when finished. The plot is shown in Figure 8.16.

Figure 8.16 Forecast Control Chart for PDT1221

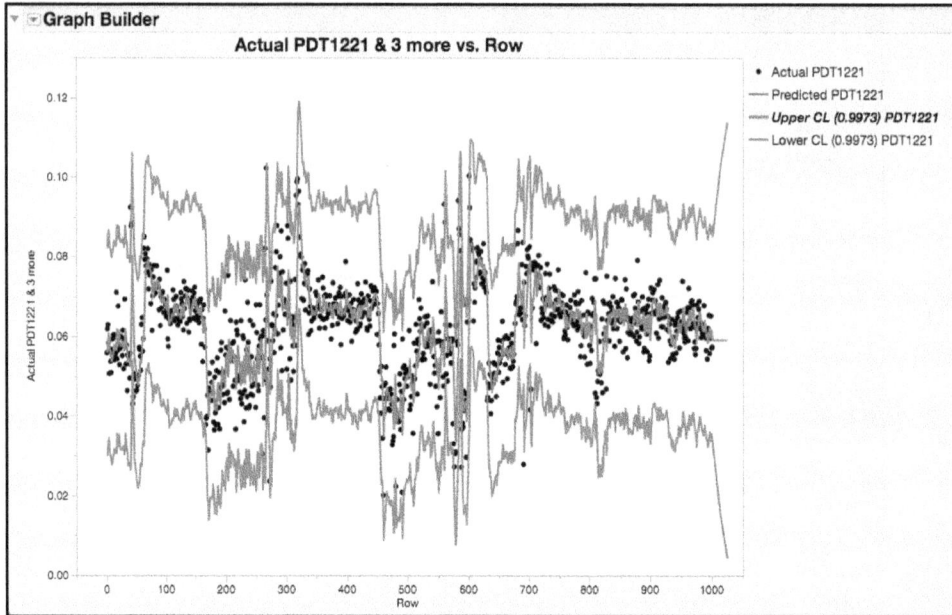

In order to create a control chart of the forecasts, the IMA(1, 1) model, discussed in the last example, must be fit to the baseline data. The parameter estimate shown in Figure 8.8 is used to develop a forecast equation, $\hat{y}_{t+1} = y_t - 0.62435\hat{a}_t$. In this equation, the next observation at time $t+1$ is predicted (or forecasted) from the sum of the current value at time t and a fraction of the current forecast error at time t. JMP calculates a $(1-\alpha)100\%$ prediction interval for each forecasted value. The prediction interval is an interval associated with an individual value. In this example, we selected a 99.73% confidence level in order to get interval that is more in line with the width of $\pm 3\sigma$ control limits.

Statistics Note 8.5: Under the normal distribution model, the probability that an observation falls between the usual $\pm 3\sigma$ control limits is 0.9973. See ISQC Section 5.3.2.

The chart in Figure 8.16 is a plot of the actual PDT1221 values, along with the forecasts and 99.73% prediction interval for each actual value. The forecasted results serve as the centerline of the control chart and the prediction interval bounds serve as the control limits. Unlike most control charts, the centerline and control limits are moving (or changing) with every subgroup.

Statistics Note 8.6: Under the normal distribution model, the probability that an observation at time t falls between the time series model prediction $\pm 3\sigma$ is 0.9973. This is shown in ISQC Equation (10.17).

Instead of using the 99.73% prediction intervals from the **Time Series** platform, we can generate limits as

$$LCL_t = \hat{y}_t - 3\sigma$$
$$UCL_t = \hat{y}_t + 3\sigma$$

$$t = 1, 2, \cdots \qquad\qquad (3)$$

These limits have a coverage probability of 0.9973 as shown in ISQC equation 10.17. The σ value is given by the **Standard Deviation** entry in the **Model Summary for IMA(1, 1)** shown in Figure 8.8. We leave it as an exercise for the reader to see that these limits are very close to the prediction limits produced by the **Time Series** platform.

For both type of limits, an out-of-control point is signaled by an actual value that plots outside of the prediction limits. For example, the PDT1221 value (0.0278) for observation n = 691 is below the lower prediction bound (0.0520). In fact, there are 20 values that are outside of the prediction interval in Figure 8.16. Twenty values are included in the 33 values that were outside of the control limits for the residuals control chart shown in Figure 8.11. This example is further discussed in the Statistical Insights section.

> **JMP Note 8.2:** To identify the points that fall outside the forecast control chart one can create two columns with indicator formulas Actual − Upper > 0 and Lower − Actual > 0. Any row with a value 1 indicates an out-of-control point. We can the use the Select Where tool to select those rows. Markers and colors can be then added to make it easier to see in the chart. If desired, one can create just one column with formula: (Actual − Upper > 0) | (Lower − Actual > 0). A value of 1 will indicate an out-of-control point regardless which bound.

Statistical Insights

In this section, we elaborate upon the modeling and interpretation of the parameter related to the pressure differential transmitter (PDT) of a make-up air handler. We further examine the signals obtained in the residuals and forecast charts presented earlier in this chapter to understand why there were so many signals. In the next section, we illustrate the powerful tools that JMP has in the **Time Series** platform to evaluate several models and select the best one. Finally, we show how to monitor future observations from an autocorrelated time series, using the ARIMA forecast model obtained from the baseline data, and we demonstrate how to use a Cuscore chart.

Evaluation of IMA(1,1) Chart

The Residuals control chart in Figure 8.11 and the Forecast control chart in Figure 8.16 both have 33 signals out of 999 subgroups. We would like to understand the cause for the excessive number of signals, which might be false alarms, or due to special cause variation or an ill-specified model.

We can use the Stability Ratio, discussed in Chapter 6, to gain insight into the type of variation (common cause or special cause) present in the IMA(1,1) residuals.

1. Open Chapter 8 – IMA Forecasts.jmp, which has variables called *Actual PDT1221, Predicted PDT1221, Residual PDT1221, Upper CL (0.9973) PDT1221* and *Lower CL (0.9973) PDT1221*.

2. From the main menu, select **Analyze ▶ Quality and Process ▶ Control Chart Builder**.

3. A launch window appears. Drag **Residual PDT1221** to the **Y** zone and click **Done**. Figure 8.17 shows partial output form the **Control Chart Builder**.

Figure 8.17 Stability Ratio of IMA(1, 1) Residuals

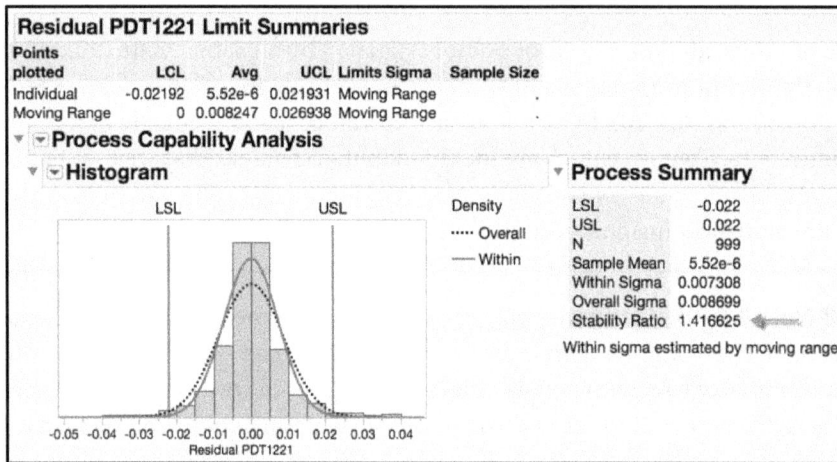

Recall from Chapter 6, the Stability Ratio is the ratio of the long-term variation and the short-term variation, estimated using the sample standard deviation and the average moving range, respectively. If the process is operating with common cause variation, the Stability Ratio will be close to 1; if special cause variation is present, then the Stability Ratio will be greater than 1. The **Process Screening** platform uses a default cut off of 1.5 to classify a parameter as unstable. However, since there is so much data, the Stability Ratio of 1.416 in Figure 8.17, based on the SR significance test for n = 999 observations, is indicative of an unstable process. Therefore, it is unlikely that the signals in Residuals and Forecast control charts are false alarms or false signals.

Statistics Note 8.7: To test for stability using the Stability Ratio, an approximate F test can be used. For an individual measurement and moving range chart the F-test numerator degrees-of-freedom are equal to (sample size – 1), and the denominator degrees-of-freedom are approximately 0.62×(sample size – 1).

For the Stability Ratio of 1.416 in Figure 8.17, an approximate F-test of the Stability Ratio, based on 999 observations, gives a p-value= 1.12E-6. This indicates that the residuals are not stable.

---

4. Click on Chapter 8 – IMA Forecasts.jmp to make it activate. In the JMP table, click in row 1 in the far left-hand column and scroll down until row 200 is visible and then hold the shift key and click in row 200. Rows 1 through 200 should be highlighted. To create a JMP table with the first 200 observations, select **Tables ▶ Subset**. Click **OK**.

5. Next select **Graph ▶ Graph Builder**.

6. From the Columns box, select **Row** and drag it to the **X** zone (x axis). Next, highlight **Actual PDT1221** and drag it to the **Y** zone (y axis).

7. Right-click in the plot and select **Smoother ▶ Remove**. Alternatively, hold the control key and click in the graph to bring up the menu.

8. Select **Residuals Signals** from the Columns box and drag it to **Color** box on the far right-hand side of the window.

9. Right-click in the plot and select **Add ▶ Line**.

10. Right-click on each symbol and line in the Residuals Signal legend and select appropriate symbols and line colors. Click **Done** when finished. The graph is shown in Figure 8.18.

Figure 8.18 Trend Plot of the First 200 PDT1221 with Annotated Signals

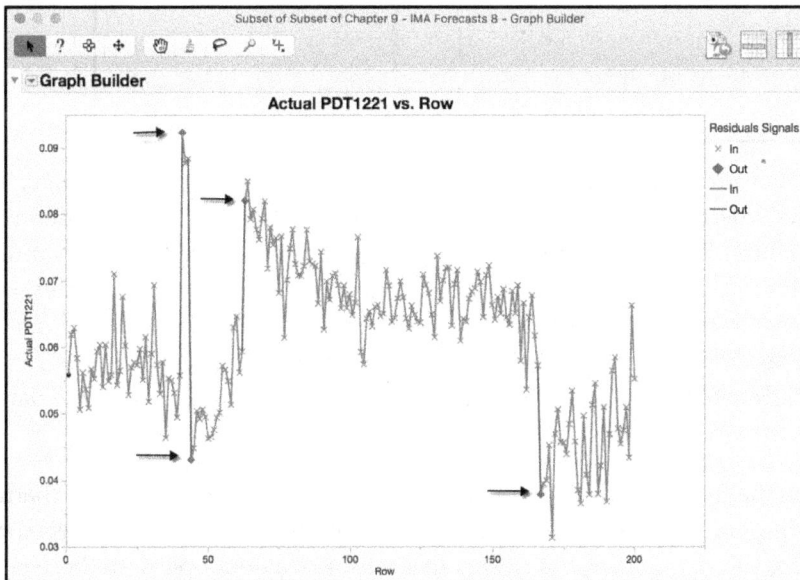

11. Repeat Step 5 through Step 11 using observations 200 to 300. The graph is shown in Figure 8.19.

Figure 8.19 Trend Plot of Observations 200-300 PDT1221 with Annotated Signals

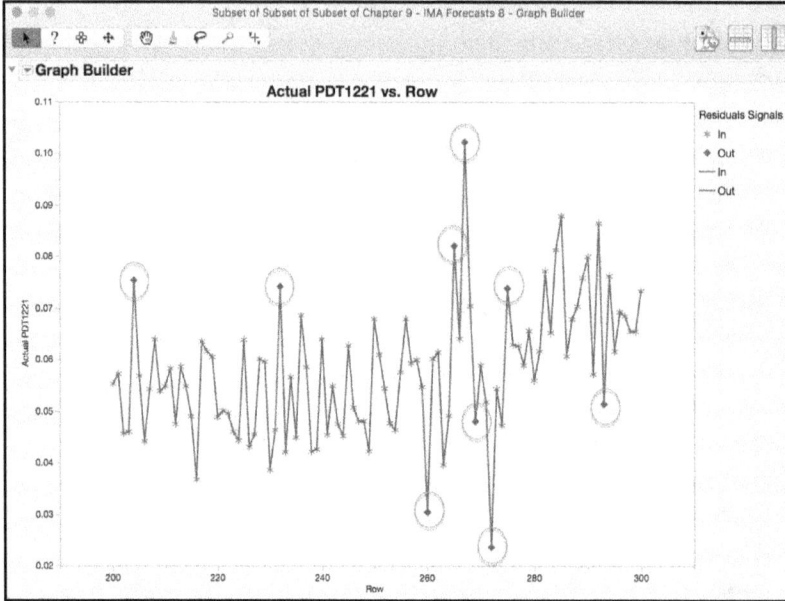

Subsets of the data were used in order to get a closer look at the patterns in the data. The graph in Figure 8.18 shows where the control limit excursions occurred in the first 200 observations (1,000 minutes or 16.7 hours). The first excursion shown in the graph occurred at observations 41 (205 minutes or 3.42 hours), after an increase of 0.037, from 0.056 to 0.092. The second excursion occurred at observation 44 (220 minutes or 3.67 hours), after a decrease of 0.045, from 0.088 to 0.43. Finally, the third excursion occurred at observation 63 (315 minutes or 5.25 hours), after an increase of 0.023, from 0.059 to 0.082. A similar assessment can be made using the data plotted in Figure 8.19.

The excursions in the Residual and Forecast control charts occur when there is a change in the consecutive readings of a certain magnitude. These situations can be due to a special cause in the clean room, where someone passes too closely to a sensor or a door opens. In these events, the sensor detects a change and the system needs time to catch up. There is also a chance that the model needs to be adjusted. These are explored further in the next example.

Time Series Model Comparisons

The diagnostic tools in the **Time Series** platform can be used to identify a preliminary model. For example, the following patterns in the ACF and PACF plots can be used to help specify the form and orders of an ARIMA model:

- Differencing: The ACF slowly declines, with many high correlations.
- AR(p): The PACF cuts off after lag *p* and the ACF decays in a sinusoidal manner.
- MA(q): The ACF cuts off after lag *q* and the PACF gradually decays toward zero in a sinusoidal manner.
- AR(p) & MA(q): The ACF and PACF do not have an abrupt cutoff, but both tail off slowly.

Even with the use of the diagnostic tools, the model fitting process can be an art and it is often necessary to fit several models to select the most appropriate one for the series. In the PDT1221 make-up air handler data, an IMA(1, 1) was identified to model the autocorrelation. In this section, we explore several other models and compare their performance to the IMA(1, 1) to determine if the fit can be improved. We begin, by refitting the IMA(1.1) model to PDT1221 and then fit several more forms of the ARIMA model, with parameter orders of 2 or less.

1. Open Chapter 8 – ISQC Example 10.2*.jmp.
2. From the main menu select **Analyze ▶ Specialized Modeling ▶ Time Series**.
3. In the launch window, select **PDT1221** and click **Y, Time Series** and click **OK**.
4. Next to the **Time Series PDT1221** banner, click on the red triangle and select **ARIMA**.
5. To fit an IMA(1, 1), in the dialog box, leave **0** in the **p, Autoregressive Order** field, enter the number **1** in the **d, Differencing Order** field, and enter the number **1** in the **q, Moving Average Order** field and. Click on the toggle next to **Prediction Interval** and select **0.99** from the list. Uncheck the **Intercept** box. Click **Estimate**.
6. To fit an IMA(1, 2), in the dialog box, leave **0** in the **p, Autoregressive Order** field, enter the number **1** in the **d, Differencing Order** field, and enter the number **2** in the **q, Moving Average Order** field and. Click on the toggle next to **Prediction Interval** and select **0.99** from the list. Uncheck the **Intercept** box. Click **Estimate.**
7. To fit an AR(1, 1), in the dialog box, enter the number **1** in the **p, Autoregressive Order** field, enter the number **1** in the **d, Differencing Order** field, and leave **0** in the **q, Moving Average Order** field. Click on the toggle next to **Prediction Interval** and select **0.99** from the list. Uncheck the **Intercept** box. Click **Estimate**.
8. To fit an AR(2, 1), in the dialog box, enter the number **2** in the **p, Autoregressive Order** field, enter the number **1** in the **d, Differencing Order** field, and leave **0** in the **q, Moving Average Order** field. Click on the toggle next to **Prediction Interval** and select **0.99** from the list. Uncheck the **Intercept** box. Click **Estimate.**
9. To fit an ARIMA(1, 1, 1), in the dialog box, enter the number **1** in the **p, Autoregressive Order** field, enter the number **1** in the **d, Differencing Order** field, and enter the number **1**

in the **q, Moving Average Order** field. Click on the toggle next to **Prediction Interval** and select **0.99** from the list. Uncheck the **Intercept** box. Click **Estimate.**

Figure 8.20 Model Comparisons for PDT1221

Report	Graph	Model	DF	Variance	AIC	SBC	RSquare	-2LogLH	Weights	.2 .4 .6 .8	MAPE	MAE
☑	☐	ARIMA(1, 1, 1) No Intercept	997	7.4322e-5	-6659.857	-6650.043	0.445	-6663.857	0.942100		10.543320	0.005911
☑	☐	IMA(1, 2) No Intercept	997	7.4746e-5	-6654.270	-6644.456	0.442	-6658.27	0.057659		10.522987	0.005898
☑	☐	IMA(1, 1) No Intercept	998	7.5654e-5	-6643.314	-6638.407	0.435	-6645.314	0.000241		10.545050	0.005905
☑	☐	ARI(2, 1) No Intercept	997	0.0000793	-6595.448	-6585.634	0.408	-6599.448	0.000000		10.726254	0.006039
☑	☐	ARI(1, 1) No Intercept	998	8.3614e-5	-6543.704	-6538.798	0.375	-6545.704	0.000000		11.059499	0.006237

Performance statistics for all the models fitted to the series are summarized in this platform under the **Model Comparison** banner, as is shown in Figure 8.20. Seven performance statistics are included: Variance, AIC (Akaike's Information Criterion), SBC (Schwarz's Bayesian Criterion), Rsquare, -2LogLH (Log Likelihood Function), MAPE (Mean Absolute Percentage Error) and MAE (Mean Absolute Error). The performance metrics were briefly described in the previous example and further details can be found in JMP Help. With the exception of Rsquare, smaller values indicate a better fitting model. For Rsquare, larger values indicate a better fitting model. Therefore, to compare these models, the statistics can be appropriately sorted in the **Time Series** platform, or saved to a JMP table and plotted, and interpreted for each model. By default, the models are sorted by the AIC value.

10. Right-click in the results of the **Model Comparison** output and select **Make into Data Table.** A JMP table will be created with these results.

11. Activate the JMP table by clicking on it. From the main menu, select **Graph Builder**.

12. From the Columns window, drag **Model** to the **X** zone (x axis). Next, drag **Variance** to the **Y** zone (y axis). To create a second plot below the Variance, drag **AIC** to the **Y** zone and drop it to the lower y axis tick marks and outside of the graph. Note, the symbol color will change when the cursor is in the right place on the **Y** zone.

13. To create a third plot below AIC, drag **MAPE** to the **Y** zone and drop it to the lower y axis tick marks and outside of the graph. Note, the symbol color will change when the cursor is in the right place on the **Y** zone.

14. Right-click in each plot and select **Add ▶ Line**. Click **Done** when finished.

Figure 8.21 Plot of Variance, AIC and MAPE for 5 Models

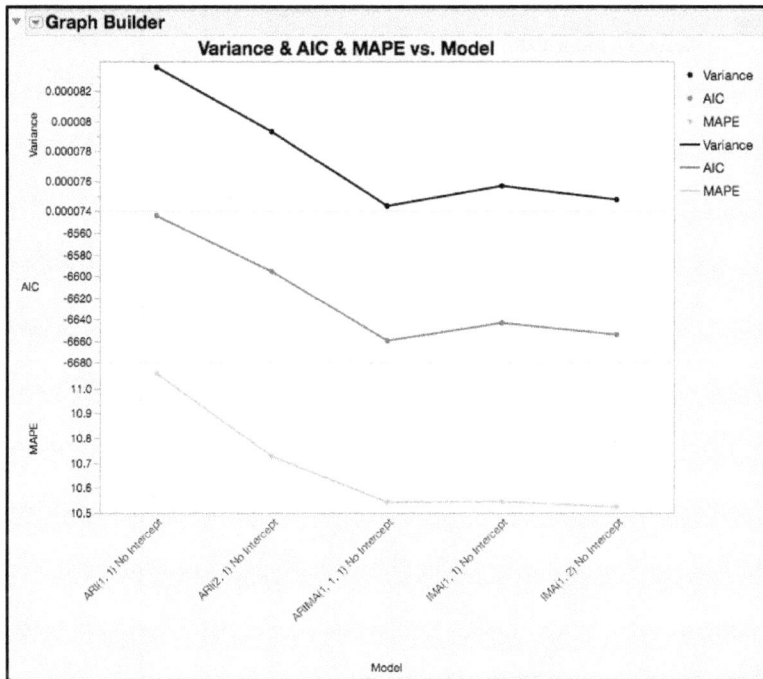

15. From the red triangle next to the Graph Builder banner, select **Save Script ▶ To Data Table...**.

16. Highlight the last two rows in the JMP table for the AR models and, from the main menu select **Rows ▶ Hide and Exclude**. Run the script that was saved to the JMP table in Step 15. The graph is shown in Figure 8.22.

Figure 8.22 Plot of Variance, AIC and MAPE for 3 Models

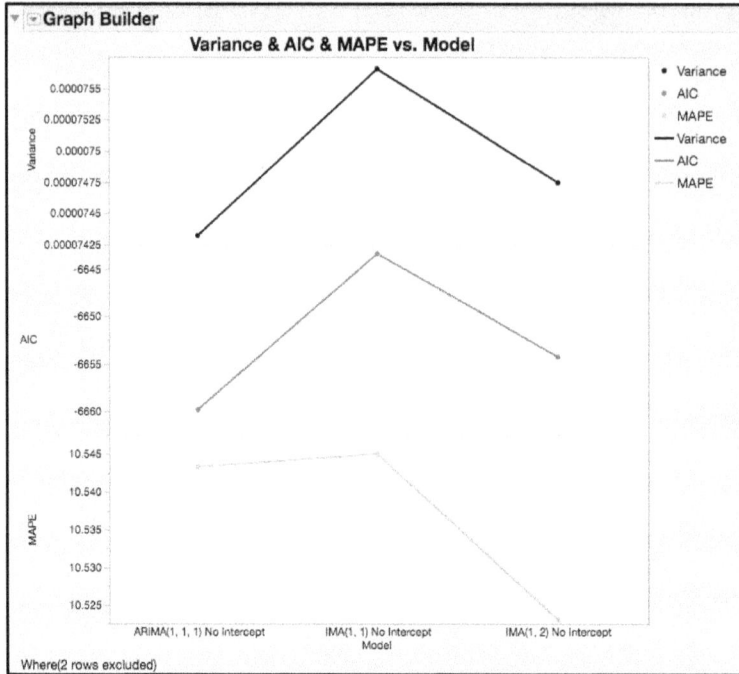

The plot in Figure 8.21 shows the Variance, AIC and MAPE for each of the five fitted models. For these three statistics, smaller values indicate a better fit. Although it is not shown graphically, the RSquare (see Figure 8.20) show a similar result, with the AR models having the lower RSquare values. Therefore, the first two AR(p, 1) models can be eliminated from the model choices, since they have the worst performing statistics; while the IMA(q, 1) models and ARIMA(1, 1, 1) are a better fit to the PDT1221 series.

Figure 8.22 shows the Variance, MAPE and AIC for the remaining three models. The ARIMA(1, 1, 1) model has the best Variance, AIC, and the IMA(1, 2) has the best MAPE. Our original model, IMA(1, 1), shouldn't be eliminated just yet, since it is the most parsimonious and has performance statistics with a similar magnitude as the other 2 models. The final decision will be made after observing the behavior of the residuals for each of the three models.

17. Go back to the **Time Series** output window.
18. Click on the red triangle next to the banner **Model: IMA(2, 1) No Intercept**, select **Save Columns.** A JMP table is created that contains the actual data, predictions, prediction standard errors, residuals and the lower and upper confidence intervals. Note 25 additional rows are included in this table with forecasted results. These rows can be deleted before proceeding.
19. Click on the red triangle next to the banner **Model: ARIMA(1, 1, 1) No Intercept**, select **Save Columns.** A JMP table is created that contains the actual data, predictions, prediction standard errors, residuals and the lower and upper confidence intervals. Note

25 additional rows are included in this table with forecasted results. These rows can be deleted before proceeding.

20. Open Chapter 8– All Forecasts.jmp. This table was created combining the forecast results from the IMA(1, 1), IMA(1, 2) and ARIMA(1, 1, 1). The columns include the actual results, forecasts, residuals and 99% prediction limits for each model. Columns have been added to identify points that are out-of-control using residuals and forecast charts. Finally, one column, **All Signals**, identifies rows where all charts produce a signal.

21. From the main menu, select **Analyze ▶ Screening ▶ Process Screening**. A launch window will appear. From the Columns window, select **Residual PDT1221**, these are the residuals for the IMA(1,1) model, and then click **Process Variables**. Next, select **Residual IMA(1, 2)** and click **Process Variable**. Finally, select **Residual ARIMA** and then click **Process Variable**. Click **OK** when done.

22. In the Process Screening table select all three rows and then, at the top of the output, click on the red triangle next to **Process Screening** and select **Quick Graph for Selected Items**.

Figure 8.23 Stability Ratios for Residuals

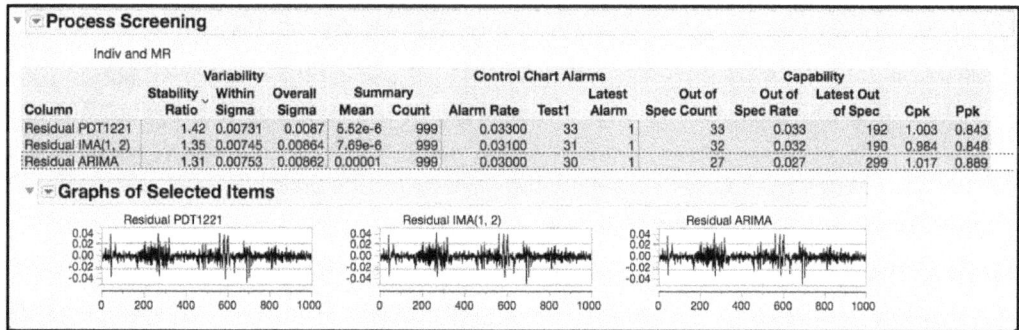

| | Variability | | | Summary | | Control Chart Alarms | | | Capability | | | | |
Column	Stability Ratio	Within Sigma	Overall Sigma	Mean	Count	Alarm Rate	Test1	Latest Alarm	Out of Spec Count	Out of Spec Rate	Latest Out of Spec	Cpk	Ppk
Residual PDT1221	1.42	0.00731	0.0087	5.52e-6	999	0.03300	33	1	33	0.033	192	1.003	0.843
Residual IMA(1, 2)	1.35	0.00745	0.00864	7.69e-6	999	0.03100	31	1	32	0.032	190	0.984	0.848
Residual ARIMA	1.31	0.00753	0.00862	0.00001	999	0.03000	30	1	27	0.027	299	1.017	0.889

23. Click on Chapter 8 – All Forecasts.jmp to activate it. From the main menu, select **Analyze ▶ Distribution**. Highlight **All Signals** in the column window and click **Y, Columns**. Click **OK** when finished. The output is shown in the left side of Figure 8.24.

24. Click the **Different** bar to highlight the rows in the JMP table. Click on Chapter 8 – All Forecasts.jmp and from the main menu, select **Tables ▶ Subset**. In the launch window, click **OK**. The output is shown in the right side of Figure 8.24.

Figure 8.24 Signals with Different Outcomes

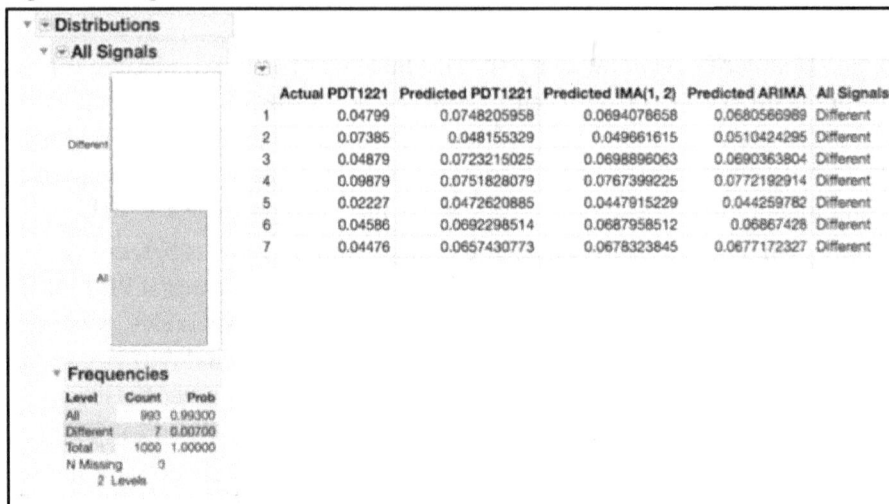

	Actual PDT1221	Predicted PDT1221	Predicted IMA(1, 2)	Predicted ARIMA	All Signals
1	0.04799	0.0748205958	0.0694078658	0.0680566989	Different
2	0.07385	0.048155329	0.049661615	0.0510424295	Different
3	0.04879	0.0723215025	0.0698896063	0.0690363804	Different
4	0.09879	0.0751828079	0.0767399225	0.0772192914	Different
5	0.02227	0.0472620885	0.0447915229	0.044259782	Different
6	0.04586	0.0692298514	0.0687958512	0.06867428	Different
7	0.04476	0.0657430773	0.0678323845	0.0677172327	Different

Frequencies

Level	Count	Prob
All	993	0.99300
Different	7	0.00700
Total	1000	1.00000
N Missing	0	
2 Levels		

We used the **Process Screening** platform to calculate the Stability Ratios for the residuals from the three different models, IMA(1, 1), IMA(1, 2) and ARIMA(1, 1, 1), shown in Figure 8.23. We already discussed this ratio for the IMA(1, 1) model. The three stability ratios are 1.42, 1.35 and 1.32. Although the Stability Ratio for the ARIMA(1, 1, 1) is the closest to 1, they are all similar. Even though they are all technically less than the JMP cutoff of 1.5, their corresponding p-values (see Statistics Note 8.6 for how to calculate the p-value for the Stability Ratio) all indicate that there is special cause variation in the residuals.

The spark graphs of the residuals, shown in Figure 8.23, show a similar pattern among the residuals from the IMA(1, 1), IMA(1, 2) and ARIMA(1, 1, 1). They all have a few larger residual values in the first 50 observations, with the next period of larger residuals occurring between observations 200 to 300, and finally, between observations 600 to 700. Note that it is difficult to carry out a detailed comparison using all the data. In order to take a closer look, the data needs to be plotted and compared in groups of approximately 200 observations.

The JMP table includes variables that flag a signal in the residuals chart using an XmR chart and forecast chart using the prediction limits. If the observation is within the limits, the cell is set to "In"; if the observation exceeds the limits, then the cell is set to "Out". For the IMA(1, 2) model, 31 observations had signals on both charts, while one observation had a signal on the residuals chart but not the forecast chart. For the ARIMA(1, 1, 1) model, 27 observations had signals on both charts, while 3 observations have a signal on the forecast chart but not the residuals chart. Note this information can be obtained using the **Fit Y by X** platform and placing Forecast Signals in the Y, Response and Residual Signals in the X, Factor.

The "All Signals" column in the JMP table has two values, "All" or "Different." A result of "All" means that the six chart types (2 per model type) all signaled in the same way; all signaled as 'in' or all signaled as "out." A result of "Different" means that the six chart types do not all signal in the same way for a particular observation. The output shown in Figure 8.24 provides the number of observations with each result in the "All Signals" column. There were 7 out of 999 observations, or 0.7%, that had different signal classifications among the 6 charts and 992 out of 999 observations, or 99.3% that had the same signal classification among the 6 charts.

Most of statistics used to evaluate the model performance, suggest that the ARIMA(1, 1, 1) model slightly edges out the performance of the IMA(1, 1) and IMA(1, 2) models. However, none of the statistics show a major improvement for the ARIMA(1, 1, 1) over the other two models. Most notably, 99.3% of the observations had the same signal identification for all three models. Since this model will be used to monitor future PDT1221 observations, and the model performance was similar, the most parsimonious model, the IMA(1, 1), is chosen as an appropriate model. It is also noted that, since all models perform similarly, the higher number of signals observed is most likely due to special cause variation, as we discussed previously.

Monitoring Future Autocorrelated Observations

Montgomery discusses Phase I and Phase II control chart applications in ISQC Chapter 5, Methods and Philosophy of Statistical Process Control. Retrospective monitoring is done in Phase I, where control limits are calculated and applied to the same data set and special cause variation is removed from the process. Once it is fairly stable, monitoring moves to Phase II, where control limits are locked down and applied to future process output. These phases can also be applied to monitoring autocorrelated data. In order to establish Phase II limits, we need a forecast model, obtained using the parameter estimates from an ARIMA model fit to the baseline data series, and control limits for the residuals, also calculated from the baseline data.

For the make-up air handler example, we selected the IMA(1, 1) model to represent the correlation in the PDT1221 results, that were sampled every 5 minutes. The forecast equation has the form, $\hat{y}_{t+1} = y_t - 0.62435a_t$. In this equation, the next observation at time $t+1$ is predicted (or forecasted) from the sum of the current value at time t and a fraction of the current forecast error at time t. Undoubtedly, it is more challenging to automate these algorithms, since the forecasts require previous observations and residuals. This task is simplified with the use of a statistical software application, such as JMP or SAS. In this example, we will use JMP to carry out Phase II monitoring of PDT1221.

1. Open Chapter 8 – ISQC Example 10.2*.jmp.
2. Select **Analyze ▶ Specialized Modeling ▶ Time Series**.
3. In the launch window, select **PDT1221** and click **Y, Time Series** and click **OK**.
4. Next to the **Time Series PDT1221** banner, click on the red triangle and select **ARIMA**.
5. In the dialog box, leave **0** in the **p, Autoregressive Order** field, enter the number **1** in the **d, Differencing Order** field, and enter the number **1** in the **q, Moving Average Order** field and. Click on the toggle next to **Prediction Interval** and select **0.99** from the list. Uncheck the **Intercept** box. Click **Estimate**.

6. From the red triangle next to the **Model: IMA(1, 1) No Intercept** banner, select **Save Prediction Formula**. This will create a new JMP table. To see the prediction equation, click on the plus next to **PDT1221 Prediction Formula+** in the Columns window. The formula is shown in Figure 8.25.

Figure 8.25 JMP Table with IMA(1, 1) Prediction Formula

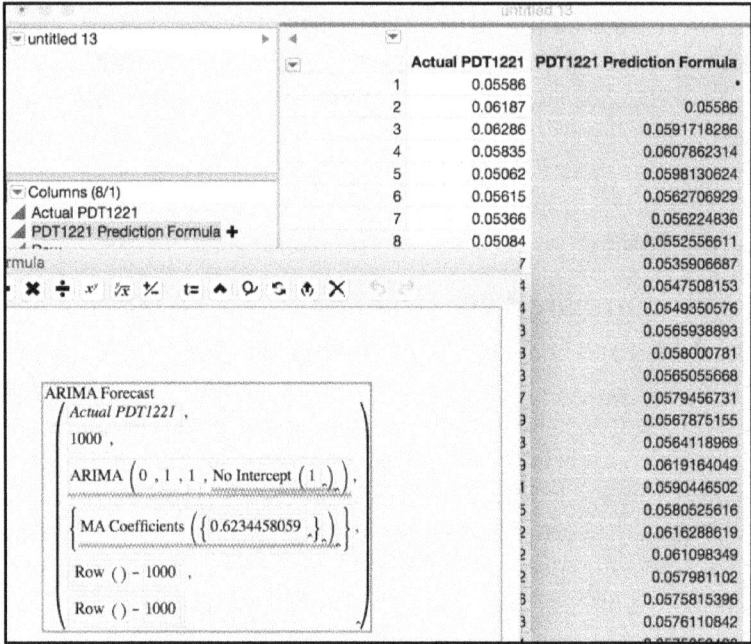

7. Select **Analyze ▶ Quality and Process ▶ Control Chart Builder**. Drag and drop **Residual PDT1221** from the Columns window on the left-hand side to the **Y** zone. From the red triangle select **Save Limits ▶ In Column**. Close the **Control Chart Builder**. The control limits for the original Residuals Chart will be added to the column properties. These can be viewed by clicking on the star next to **Residual PDT1221** in the Columns window in the new JMP table, as is shown in Figure 8.26.

Figure 8.26 PDT1221 Residuals Control Limits for New Observations

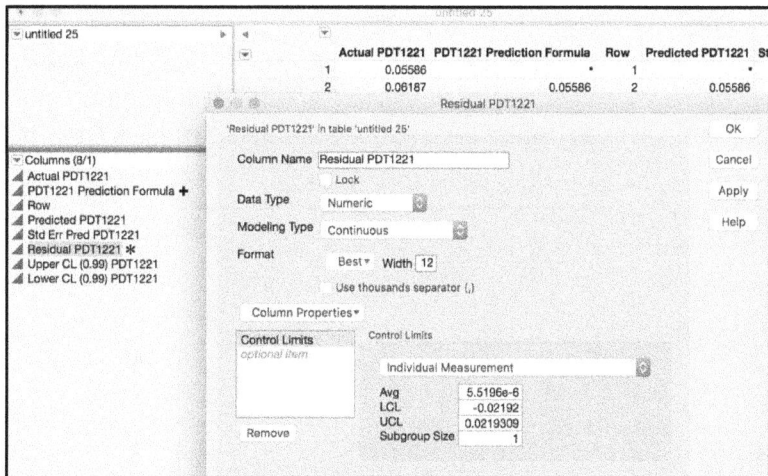

8. Highlight the **Residual PDT1221** column in the JMP table and select **Columns ▶ Formula….** In the Formula editor window, click **Actual PDT1221** in the Columns window. Next click on the minus '−' symbol at the top of the window. Then click on **PDT1221 Prediction Formula** in the Columns window (Figure 8.27). Click **OK** when finished.

Figure 8.27 Residuals Formula for New Observations

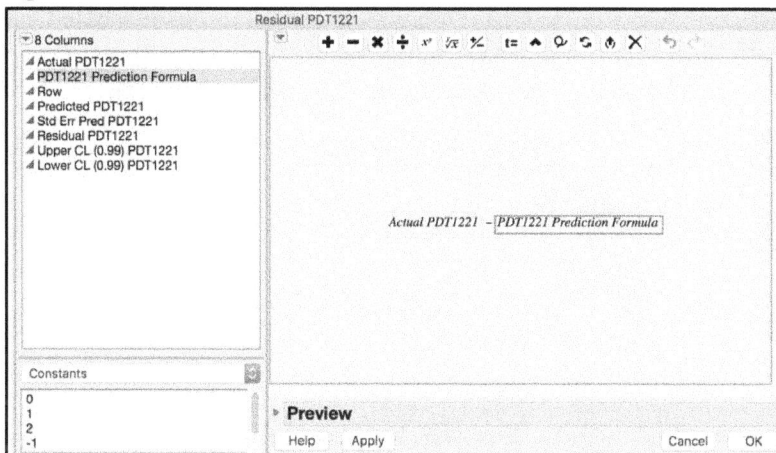

9. To add new observations to the JMP table, from the main menu select **Rows ▶ Add Rows….** In the dialog box enter **175** in the text entry box and click **OK**. Note we want to include 200 new observations. The JMP table already had 25 blank rows, to an additional 175 were added.

10. Type in or copy/paste the new observations into the **Actual PDT1221** column in rows 1001 through 1200. Note these observations are included in Chapter 8 – New PDT1221 Values.jmp. Note if you are going to copy the row values too, then first move the **Row**

column in the Columns window to the top of the list in the new table before copying the values from Chapter 8– New PDT1221 Values.jmp.

11. Click on the plus next to **PDT1221 Prediction Formula+** in the Columns window to launch the **Formula** window. Replace three values set to 1,000 with the value **1,200** by clicking each field and entering the new number (Figure 8.28). Click **OK** when finished.

Figure 8.28 ARIMA Forecast Formula to Predict Forward

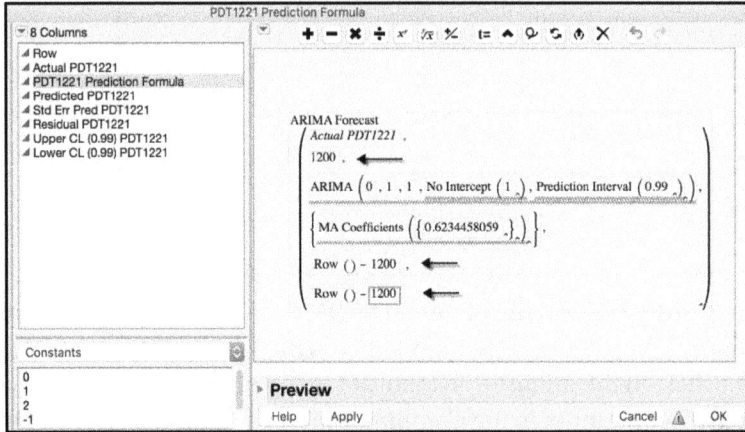

12. Add a new column to the JMP table and call it **Source**. Select **Character** for **Data Type**. Launch the **Formula** window and select **Conditional ▶ If**. Select **Row** in the Columns window to populate the expression. Then click **Comparison ▶ a <= b** and enter 1000 in the highlighted box. Click in the **then clause** box and type **"Baseline"**. Next click in the **else** clause and enter "**New**" (see Figure 8.29). Click **OK** when finished.

Figure 8.29 Plotting Symbol Variable for New Observations

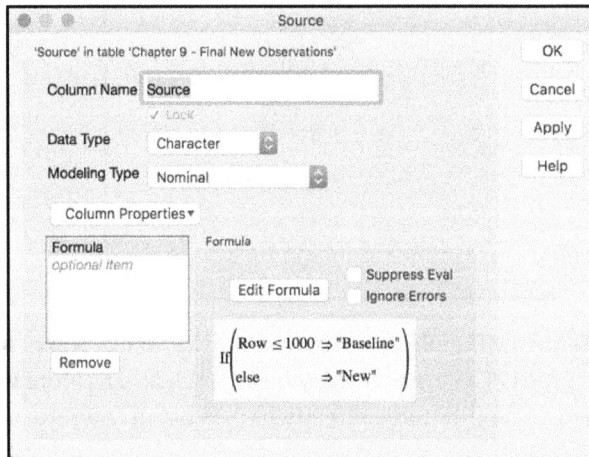

13. From the main menu select **Rows ▶ Color or Mark by Column…** and then select **Source** from the Columns window. In the **Markers** drop-down menu, select **Standard**. Click **OK** when finished.

14. From the main menu, select **Graph ▶ Graph Builder**. Drag **Actual PDT1221** to the **Y** zone and drag **Row** to the **X** zone. Drag **Source** to **Color**. Right-click in the plot and select **Smoother ▶ Remove.** Then right-click in the graph and select **Add ▶ Line**. The graph is shown in Figure 8.30.

Figure 8.30 New Observations for PDT1221

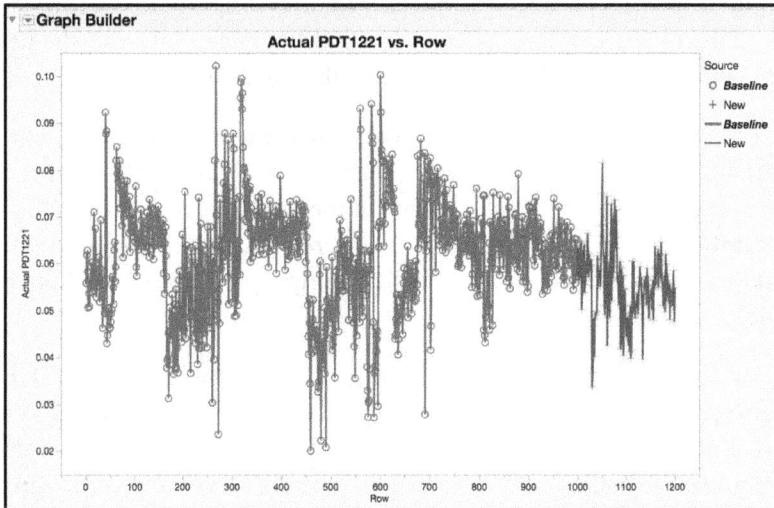

15. Click on the JMP table and select **Analyze ▶ Quality and Process ▶ Control Chart Builder**. Drag and drop the **Residuals PDT1221** column to **Y** zone and click **Done** when finished. The control chart is shown in Figure 8.31.

Figure 8.31 Residuals Control Chart for New Observations

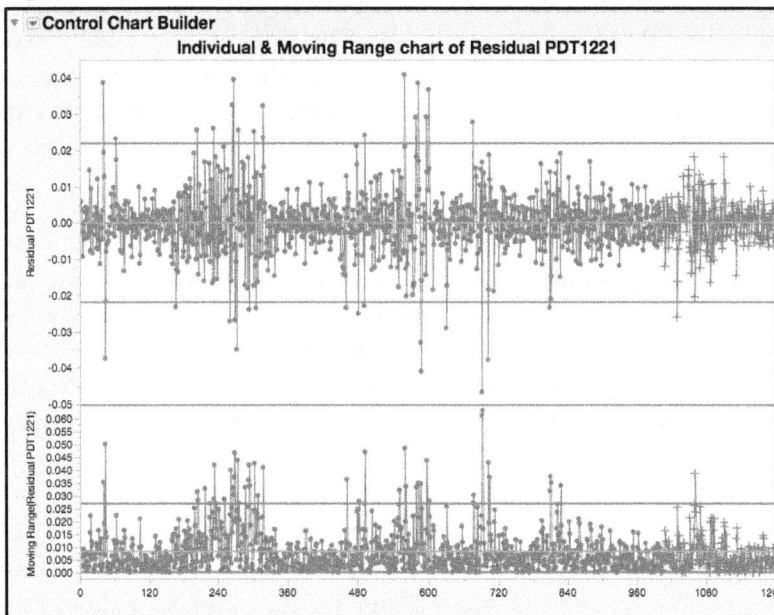

The previous steps illustrate an easier way to monitor future observations from an autocorrelated process using JMP. In Step 1 through Step 6, we refit the IMA(1, 1) model to the original 1,000 observations and created a new JMP table with the predicted values, standard error of prediction, residuals and 99% prediction limits. The prediction formula is also included in the table and the form of the equation is shown in Figure 8.25. In Step 7, an Individuals control chart is fit to the original residuals and the control limits are saved as a column property to the JMP table. Next, in Step 8 a formula is added to the Residuals column in the JMP table to calculate the new residuals using the difference between the actual and predicted values (see Figure 8.27).

Two hundred new observations are added to the JMP table in Step 9 and Step 10. For all 200 observations, the predictions are set to the same value. In Step 11, we update the prediction formula to provide unique predicted values for each of the new observations by changing the forecast range from n=1000 to n=1200, as is shown in Figure 8.28. Finally, in Step 12 through Step 15, a trend plot and Residuals control chart are created using a plotting symbol to more easily visualize the new observations.

The new observations are plotted using plus symbols in observations 1,001 to 1,200 in Figure 8.30. This time frame represents approximately 1,000 (=200 x 5) minutes, or 17 hours, of continuous operation of the make-up air handler. As compared with the previous 1,000 observations, or 83 hours, the new data appear to be less nonstationary. In fact, the behavior prior to observation 750 has more extreme behavior than appears in the series after observation 750.

The Residuals control chart in Figure 8.31 plots the old and new observations. For the new observations, one observation (n=1,030) exceeds the lower control limit, and one observation (n=1,061) exceeds the upper limit of the moving range chart. The point outside the upper control limit of the moving range chart indicates a large swing in observed values. For n=1,060 the PDT value is 0.07436, while for n=1,061 the value is 0.04253. This drop in differential pressure suggests that something has changed in the clean room or with the performance of the controller. As pointed out by J. Ramírez (1998), in those instances "the protocol for opening the doors of the fabrication areas was not followed. This resulted in sudden air flows in the fabrication area causing the observed changes in pressure."

The approach to Phase II monitoring illustrated in this section is appropriate for post-hoc monitoring of data. However, it would be difficult to use this approach for "real-time" monitoring. In order to perform "real-time" monitoring using Time Series models, a software solution would need to be developed that automatically uploads the new data into an appropriate data source, predicts the observation from the previous data, calculates the residual and generates the control chart. This can be accomplished using JMP scripting, but it is beyond the scope of this book.

Cuscore Chart

Montgomery discusses Cuscore Charts in ISQC Section 10.7, where he provides the general form of the Cuscore statistic and shows how to detect changes to the mean of residuals from a given model. Although no specific examples are provided in ISQC, a number of references are specified with more details and applications, including several by J. Ramírez. In this section, we review one of the Cuscore examples provided in the reference we have been using throughout this chapter, Monitoring Clean Room Air Using Cuscore Charts. More specifically, we will show how to use a Cuscore chart to determine if the IMA(1, 1) model needs to be updated, for Phase II monitoring. Recall, the forecast equation for the PDT1221 data series, $\hat{y}_{t+1} = y_t - 0.62435\hat{a}_t$. We will show how to use the formula editor in JMP to create the necessary formulas for monitor the moving average estimate, $\theta = 0.62437$, using the next 200 observations, 1001 to 1200, in the PDT1221 series.

Statistics Note 8.8: The Cuscore (Q) formulas for the IMA(1, 1) model are given in J. Ramírez (1998) and are shown below

$$\hat{a}_t = (1-\theta)\hat{a}_{t-1} + \theta \hat{a}_t$$

$$d_t = -\frac{\hat{a}_t}{(1-\theta)}$$

$$Q = \sum a_t \times d_t = \frac{-\sum a_t \times \hat{a}_t}{(1-\theta)}$$

---

1. Open Chapter 8 – ISQC Example 10.7.jmp. The table has a column for the observation number *t*, and the *PDT1221* observations. There are seven additional columns that are used to compute the Cuscore statistic, *IMA(1, 1) Theta, PDT1221 hat, at, at hat, dt, at×dt,* and *Cuscore.*

2. Click on the **plus** sign next to the **IMA(1, 1) Theta** variable in the **Columns** window in the left-hand side of the JMP table to display the formula (Figure 8.32).

Figure 8.32 Formula for IMA(1, 1) Theta

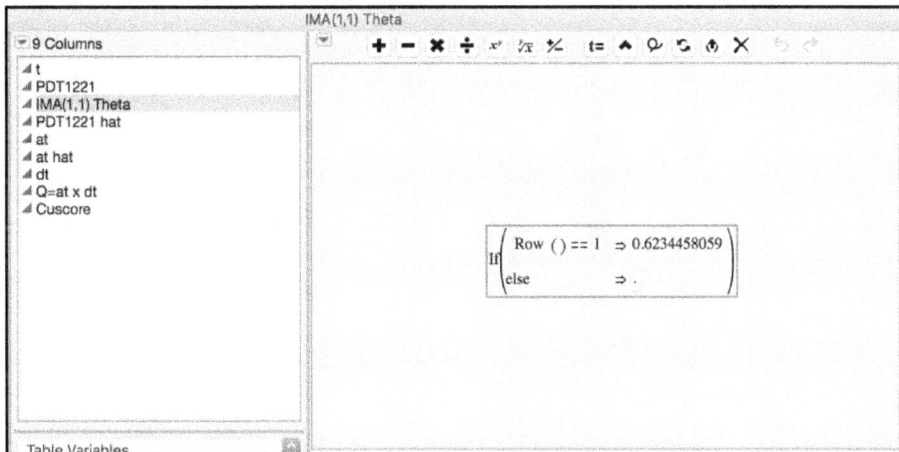

3. Click on the **plus** sign next to the **PDT1221 hat** variable in the **Columns** window in the left-hand side of the JMP table to display the formula (Figure 8.33).

Figure 8.33 Formula for PTD1221 hat

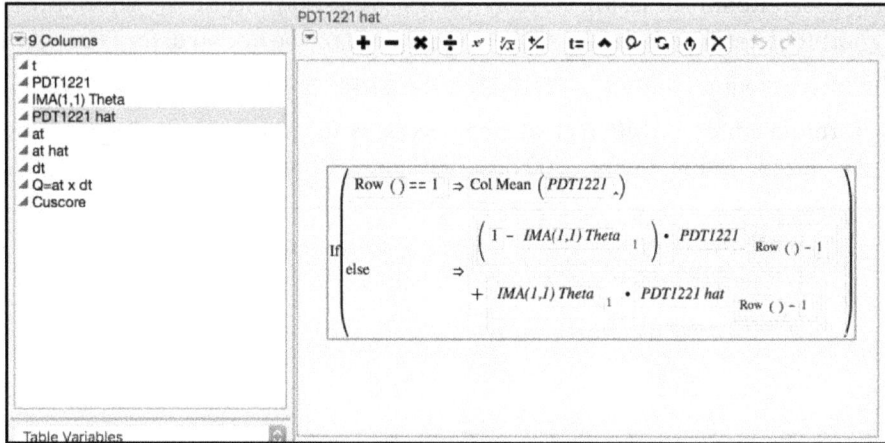

4. Click on the **plus** sign next to the **at** variable in the **Columns** window in the left-hand side of the JMP table to display the formula (Figure 8.34). The **at** is the residual, that is, the observed data **PDT1221** minus the predicted value **PDT1221 hat**.

Figure 8.34 Formula for at (residual)

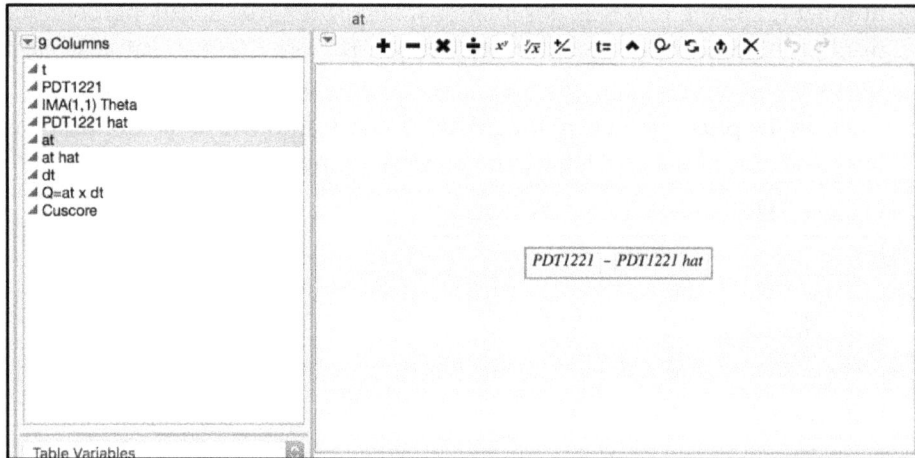

5. Click on the **plus** sign next to the **at hat** variable in the **Columns** window in the left-hand side of the JMP table to display the formula (Figure 8.35). **at hat** is the predicted residual.

Figure 8.35 Formula for at hat

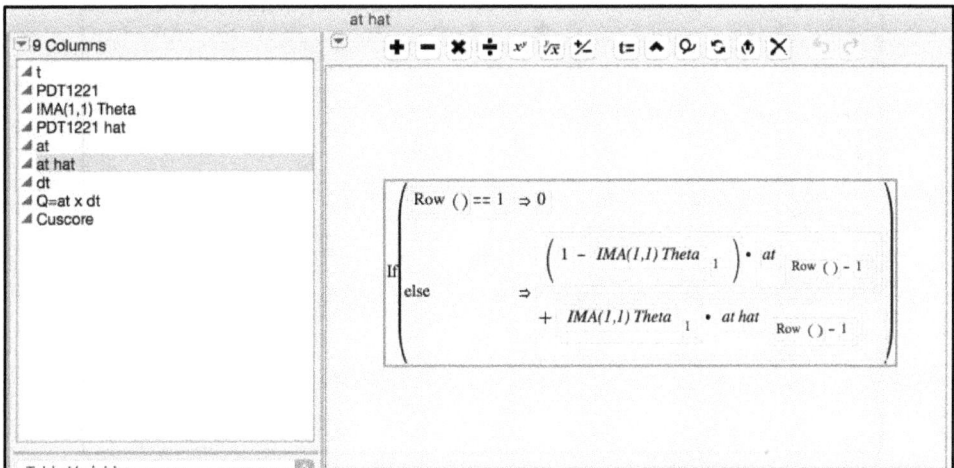

6. Click on the **plus** sign next to the **dt** variable in the **Columns** window in the left-hand side of the JMP table to display the formula (Figure 8.36). **dt** is derivative of the residual.

Figure 8.36 Formula for dt

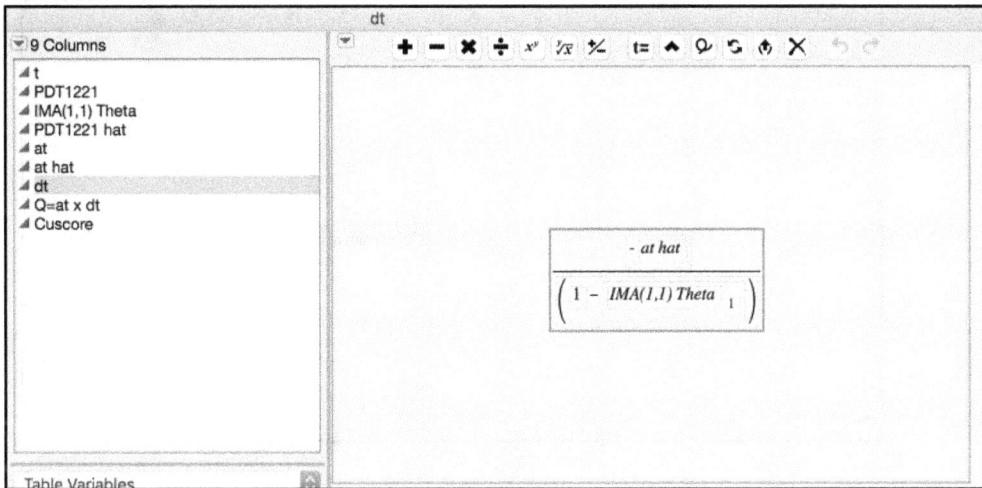

7. Click on the **plus** sign next to the **at × dt** variable in the **Columns** window in the left-hand side of the JMP table to display the formula (Figure 8.37).

Figure 8.37 Formula for at × dt

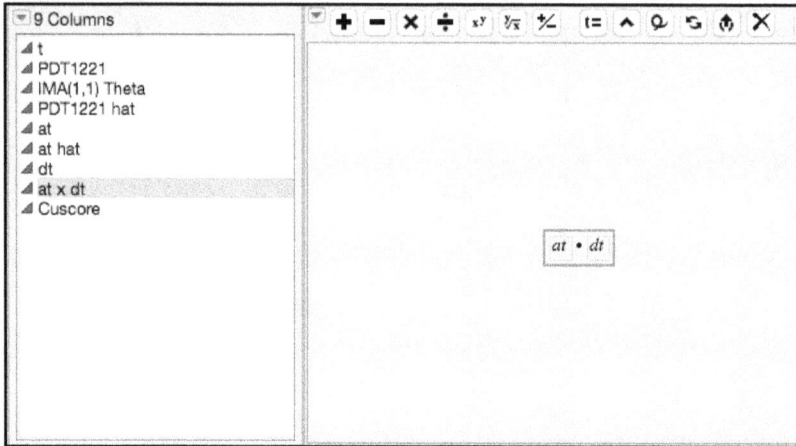

8. Click on the **plus** sign next to the **Cuscore** variable in the **Columns** window in the left-hand side of the JMP table to display the formula (Figure 8.38).

Figure 8.38 Formula for Cuscore

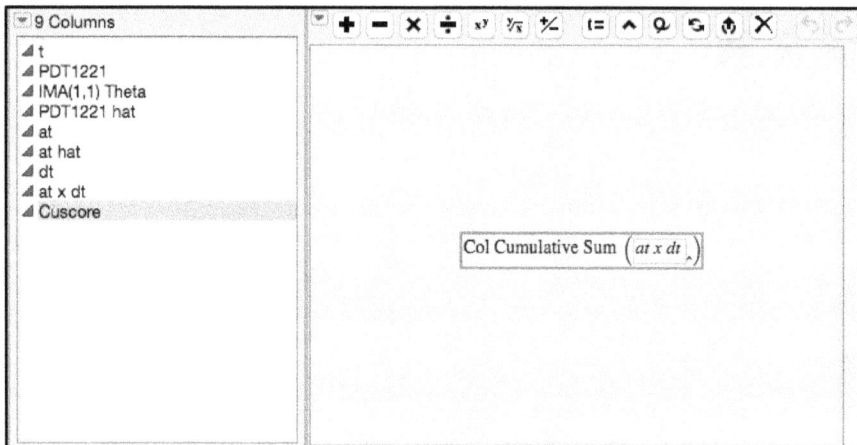

9. To create the Cuscore control chart we can use the **Graph Builder** but the table has a script to generate the graph. Click on the green arrow next to the JMP script in the upper left-hand window of the JMP table called **PDT1221 & Cuscore vs. t**. The chart is shown in Figure 8.39.

Figure 8.39 Cuscore Control Chart for Future Observations of PDT1221

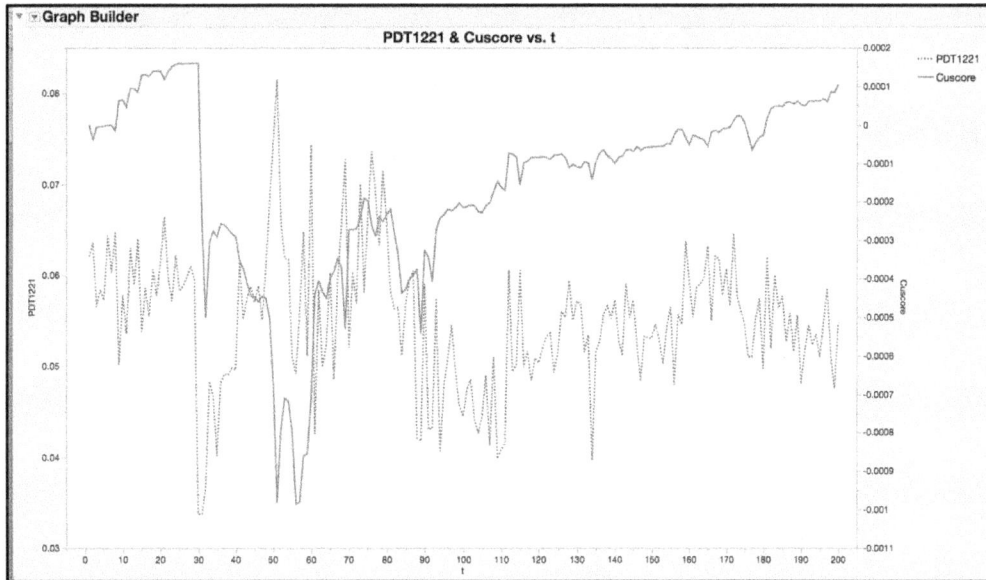

At the moment, a point-and-click platform is not available to create a Cuscore control chart in JMP. Therefore, a series of equations were manually created in a JMP table containing the next 200 observations for the PDT1221 series in order to generate the charting statistic. The estimate for the first order moving average parameter being monitored in the IMA(1, 1) forecast equation is entered in column IMA(1, 1) theta using the formula editor, as is shown in Figure 8.32. The next column, **PDT1221 hat**, contains the forecast equation for the IMA(1, 1). The formula shown in Figure 8.33 differs slightly from the one shown in the last example in Figure 8.28. Note the forecasts for both equations will converge quickly to the same value. The residuals are calculated in column **at** (see Figure 8.34). The next three columns (**at hat**, **dt** and **at×dt**) are used in the calculation of the Cuscore statistic, which is shown in equations 13, 14 and 15 of Ramírez (1998). Finally, the Cuscore statistic, in the last column is the cumulative sum of column **at×dt**.

The chart in Figure 8.39 plots the PDT1221 observations on the left-hand Y axis and the Cuscore on the right-hand Y axis. Abrupt changes in the Cuscore statistic signal a change in the IMA(1, 1) parameter. In this plot, the largest change occurs between observation 30 (n=1030) and observation and 31 (n=1031), where the Cuscore statistic drops from 0.00016 to -0.00027. This signal occurs in a similar region as where the Residuals control chart had a signal (n=1030), as is shown in Figure 8.31. The next largest jumps occur around observations 50 to (n=1050 to 1060), which agree with the large moving range value at observation 60 (n=1060). Note these drops can be selected in the JMP table by positioning your pointer over the appropriate point in the graph, clicking on it, and then finding the highlighted row in the JMP table. These drops can be located more easily by adding a column to the JMP table that contains the difference (**Row ▶ Dif**) of the Cuscore column and graphing it in time order.

> **JMP Note 8.3:** Selecting the points in a graph in the Graph Builder will automatically select the rows in the JMP table associated with those points.

The signals produced in the Cuscore control chart might be an indication that the moving average parameter has changed and the model needs to be updated. As shown in Ramírez (1998), a decision limit can be added to the Cuscore chart to signal when a change has occurred. We leave it as an exercise for the reader to fit an IMA(1, 1) model to the 200 new observations and compare the parameter estimate and forecast performance to the original 1,000 observations.

Chapter 9: Multivariate Process Monitoring and Control

Overview

This chapter illustrates how to construct multivariate control charts using examples from Chapter 11, Multivariate Process Monitoring and Control, of *Introduction to Statistical Quality Control* (ISQC), and includes discussions, tips and statistical insights on alternative ways to carry out these assessments. The techniques presented in this chapter are discussed for measurements using a continuous scale.

Two JMP platforms are highlighted in this Chapter, **Multivariate Control Chart** and **Multivariate**.

Multivariate Process Monitoring Review

In this chapter, we illustrate techniques to carry out process monitoring on multiple parameters simultaneously. These techniques are collectively found under the topic of Multivariate SPC. Recall the univariate techniques shown in Chapter 3, which focused on the construction of Shewhart's Xbar and Range or Individual Measurement and Moving Range charts, for example, to monitor a single parameter. The control chart types used and the calculation of the control limits were for a single parameter, and did not take into account other parameters being tracked. In contrast, multivariate techniques are designed to model the correlation among a group of parameters and signal when special cause variation impacts one or more of these relationships.

There are several key benefits of using multivariate process monitoring. First and foremost, is the ability to monitor the relationship among the parameters at a given process step or that are part of the same system. Very often we can measure several key attributes at the same manufacturing step, or we can measure the same attribute at several manufacturing steps throughout the process.

In both cases, these parameters might be related in some way. For example, in ISQC Chapter 11, Montgomery discusses the performance for inner and outer bearing diameters. Both diameters will determine the overall quality of the part. As is shown in ISQC Figure 11.2, the emphasis on the joint control region is of primary interest. The scatter plot shows one point that appear to deviate from the rest, while the univariate control charts demonstrate statistical control of each diameter independently.

The other important benefit to using multivariate SPC is the reduction of the overall type I error associated with multiple univariate charts. Recall, the type I error represents the probability that the chart signals in error. That is, a point outside of the limits will suggest the presence of special cause variation, when only common cause variation exists. In ISQC equation 11.1, the overall type I error rate is $\alpha' = 1-(1-\alpha)^p$, where $\alpha = 0.0027$. If five parameters are being charted for a particular process step, $\alpha' = 1-(1-0.0027)^5 = 1-0.9973^5 = 0.01343$; so instead of a false signal every 370 subgroups, we will get one every 74 subgroups (Jackson, 1956).

While there are several key benefits associated with using multivariate SPC techniques, there are also a few challenges. The algorithms are more complicated and hand calculations are not feasible; therefore, manual charting is not possible. Also, more data might be required to adequately reflect the baseline process and establish control chart limits, since more charting parameters must be estimated, including the covariance matrix and the means. Finally, when many parameters are jointly monitored, it is harder to visualize and interpret the behavior associated with an out-of-control signal, which can hinder our ability to identify a root cause and ultimately reduce the variation of the system.

Multivariate SPC includes a multifaceted collection of techniques that are designed to detect slightly different things, or accomplish a similar output in different ways. The Hotelling T^2 control chart is probably the most well-known multivariate chart. As is described in ISQC Section 11.3, the Hotelling T^2 chart plots a T^2 statistic, the multivariate generalization of Student's-t statistic, with control limits based on the F distribution. The limits, however, are adjusted depending on the control chart phase, Phase I or Phase II, and the number of future samples. Other multivariate charts include the multivariate EWMA (see ISQC Section 11.4), to detect smaller shifts in the mean vector, and the Generalized Variance (see ISQC Section 11.6), to simultaneously monitor the variation of several processes. Finally, latent structure methods (see ISQC Section 11.7), such as principal components or partial least squares can be used to modify the approach.

JMP Multivariate Monitoring Platforms

Three platforms, **Multivariate Control Chart, Multivariate** and **Principal Components**, are used to create multivariate control charts and explore multivariate relationships. These platforms were introduced in Chapter 2. In this chapter, we focus on the use of these platforms for continuous data. Table 9.1 provides a summary of the features we find most useful in each one.

Table 9.1 Overview of Features for JMP Multivariate SPC Platforms

JMP Feature	Multivariate Control Chart	Multivariate and Principal Components
Multivariate Statistical Techniques	Hotelling T^2 Control Chart Change Point Detection	Hotelling T^2 Test Outlier Detection
Latent Methods	Principal Components	Principal Components
Interpretation Aides	T Squared Partition	Principal components wheels Color maps Scatter plot matrix Ellipsoid 3-D plot Parallel Coord plot
Saving / Using Limits	Phase I limits use Beta distribution. Phase II limits use F distribution, sample size, and Saved Targets.	Principal component scores can be saved.

Examples from ISQC Chapter 11

The examples presented in this chapter from ISQC Chapter 11 are shown in Table 9.2. The examples will be reproduced using JMP, as are shown in ISQC. For some examples, additional output not provided in ISQC is shown to illustrate JMP functionality or elaborate on important points.

Table 9.2 Summary of Examples from ISQC Chapter 11

ISQC Example / Table Number	JMP Table Name	JMP Platform	Key Points
11.1 Textile Fiber	Chapter 9 – ISQC Table 11.1	Multivariate Control Chart	Create a Hotelling T^2 control chart for summarized data for 2 variables
Table 11.2 Grit Composition	Chapter 9 – ISQC Table 11.2	Multivariate Control Chart	Use a multivariate control limits based on sample size and/or phase for 2 variables.
Table 11.5 Cascade Process Data	Chapter 9 – ISQC Table 11.5	Fit Model Control Chart ▶ IR	Perform regression analysis as an alternative to multivariate methods for one output variable and 9 input variables.
11.2 Monitoring Variability	Chapter 9– ISQC Table 11.1	Graph Builder	Create a multivariate control chart for the variability of 2 variables.

ISQC Example / Table Number	JMP Table Name	JMP Platform	Key Points
Table 11.6 Chemical Process Data	Chapter 9 – ISQC Table 11.6	Multivariate Control Chart Multivariate Methods	Use principal components with four variables for multivariate monitoring.

ISQC Example 11.1 Textile Fibers

In this example, we will show how to construct a Hotelling T^2 chart in JMP. The data in ISQC Table 11.1 includes the Tensile Strength and Diameter for a textile fiber. The quality engineer has decided to jointly monitor these two parameters. These quality characteristics are presented for twenty samples, where each sample consists of n = 10 fiber specimens. In this example, the subgroup consists of a Sample of 10 fibers and subgroup size is 10. The Tensile Strengths and Diameters in ISQC Table 11.1 represent the averages for each subgroup.

As of JMP version 14, the **Multivariate Control Chart** platform does not handle summarized JMP tables. Therefore, to complete this example, the following steps illustrate how to construct a Hotelling T^2 chart using the formula editor to generate the equations for the T^2 statistic and the UCL, shown in the Solution section of Example 11.1, and the **Graph Builder** platform to generate the plot.

1. Open the JMP table **Chapter 9 – ISQC Table 11.1.jmp**, which has columns (variables) called *Sample Number, Subgroup Size, Tensile Strength (mean), Diameter (mean), Tensile Strength (variance), Diameter (variance)* and *Covariance*. In this table, Sample Number is the subgroup variable, Subgroup Size is the subgroup size for each mean and Tensile Strength (mean) and Diameter (mean) are the charting statistics. The remaining three variables are used to construct the Hotelling T^2 statistic.

2. Double click in the header of the JMP table to the right of the last column to add a new column. Then double click on the generic column label to open the column information window. Type **T2** in the **Column Name** field. Select **Fixed Decimal** in the drop-down menu next to **Format** and enter **2** in the **Dec** field. Select **Formula** from the **Column Properties** drop down menu (see Figure 9.1).

Figure 9.1 Column Information Window for T² Statistic

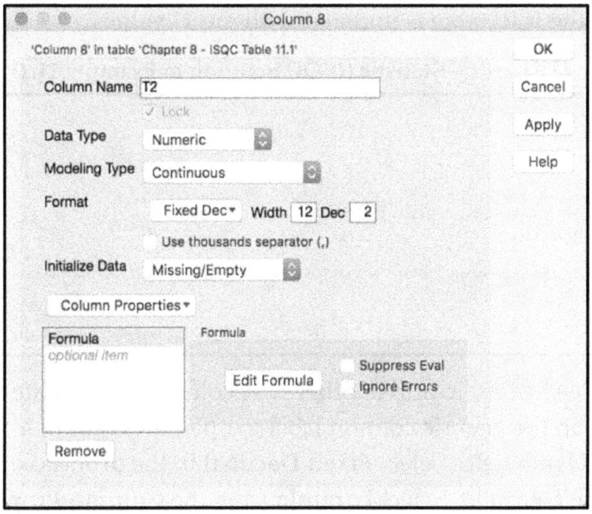

3. Click the **Edit Formula** button in Figure 9.1 to launch the Formula Editor. In order to reproduce ISQC equation 11.20, first build the structure of the equation, as is shown in Figure 9.2. To do this, hold down the shift key and press 9 to build the first set of parentheses. Next, click on the ÷ symbol at the top of the Formula window. With the bottom box highlighted, click on the × two times and then highlight the third box and click on the x^y.

Figure 9.2 Partial Formula Editor for T² Charting Statistic

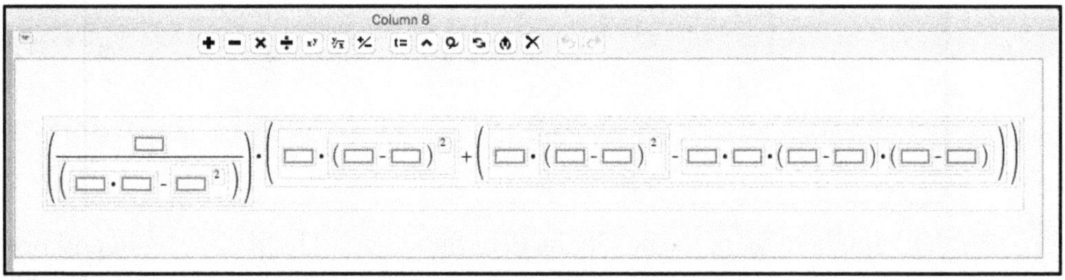

Highlight the outside of the box, click on the ×, hold down shift and press 9, and then click on the × again. Highlight the box that includes the product of the 2nd and 3rd boxes and click on the x^y. Highlight the box that is right inside of the parentheses and contains the three boxes and click on + at the top of the formula window.

4. Next, enter the information into the formula editor as is shown in Figure 9.3. Click **OK** when finished. Make sure the JMP table is updated with the T^2 values.

Figure 9.3 Formula Editor for T^2 Charting Statistic (ISQC Solution to Example 11.1)

$$\left(\frac{10}{\left(1.23 \cdot 0.83 - 0.79^{\ 2} \right)} \right) \cdot \left(\begin{array}{l} 0.83 \cdot \left(\textit{Tensile Strength (mean)} - 115.59 \right)^2 \\ + \left(1.23 \cdot \left(\textit{Diameter (mean)} - 1.06 \right)^2 - 2 \cdot 0.79 \cdot \left(\textit{Tensile Strength (mean)} - 115.59 \right) \cdot \left(\textit{Diameter (mean)} - 1.06 \right) \right) \end{array} \right)$$

5. Double click in the JMP table header to add another new column to calculate the upper control limit. Double click on the generic column label to open the column information. Type **UCL** in the **Column Name** field. Select **Fixed Decimal** in the drop-down menu next to **Format** and enter **2** in the **Dec** field. Select **Formula** from the **Column Properties** menu.

6. Click the **Edit Formula** button to launch the formula editor. First hold down the shift key and press 9 to build the first set of parentheses. Next, click on the ÷ symbol at the top of the Formula window. Click in the numerator and enter **2**. Next click on the × and enter **19**. Finally, click on the × and enter **9**. See Figure 9.4 for the full equation.

Figure 9.4 Formula Editor for UCL (ISQC Equation 11.20)

$$\left(\frac{\left(2 \cdot 19 \cdot 9 \right)}{\left(\left(20 \cdot 10 - 20 - 2 \right) + 1 \right)} \right) \cdot \text{F Quantile} \left(0.999 , 2 , 179 \right)$$

7. Click on the edge of the box in the denominator and hold down shift and press 9 to add a set of parentheses. Enter **20** in the first box and then click on the × and enter **10.** Click on the box that contains (20x10) and click on the – sign at the top of the window and enter **20** in the highlighted box. Click on the box that contains 20 and click on – and enter **2.** Highlight the box that contains the (20x10 – 10 –2) and click on the + at the top of the window and enter **1.** Highlight the outside parenthesis and click on the multiplication symbol, **x.** In the left-hand part of the window, select **Probability ▶ F Quantile.** Enter **0.999, 2** and **179** in the three consecutive boxes. The full equation is shown in Figure 9.4.

8. Click **OK** when finished. Check the table to ensure that the UCL is correctly entered.

Figure 9.5 Updated Table with T² and UCL

Sample Number	Subgroup Size	Tensile Strength (mean)	Diameter (mean)	Tensile Strength (variance)	Diameter (variance)	Covariance	T2	UCL
1	10	115.25	1.04	1.25	0.87	0.80	2.16	13.72
2	10	115.91	1.06	1.26	0.85	0.81	2.14	13.72
3	10	115.05	1.09	1.30	0.90	0.82	6.77	13.72
4	10	116.21	1.05	1.02	0.65	0.81	8.29	13.72
5	10	115.90	1.07	1.16	0.73	0.80	1.89	13.72
6	10	115.55	1.06	1.01	0.80	0.76	0.03	13.72
7	10	114.98	1.05	1.25	0.78	0.75	7.54	13.72
8	10	115.25	1.10	1.40	0.83	0.80	3.01	13.72
9	10	116.15	1.09	1.19	0.87	0.83	5.92	13.72
10	10	115.92	1.05	1.17	0.86	0.95	2.41	13.72
11	10	115.75	0.99	1.45	0.79	0.78	1.13	13.72
12	10	114.90	1.06	1.24	0.82	0.81	9.96	13.72
13	10	116.01	1.05	1.26	0.55	0.72	3.86	13.72
14	10	115.83	1.07	1.17	0.76	0.75	1.11	13.72
15	10	115.29	1.11	1.23	0.89	0.82	2.56	13.72
16	10	115.63	1.04	1.24	0.91	0.83	0.08	13.72
17	10	115.47	1.03	1.20	0.95	0.70	0.19	13.72
18	10	115.58	1.05	1.18	0.83	0.79	0.00	13.72
19	10	115.72	1.06	1.31	0.89	0.76	0.35	13.72
20	10	115.40	1.04	1.29	0.85	0.68	0.62	13.72

9. To create the control chart, select **Graph ▶ Graph Builder**. Select Sample Number and drag-and-drop into the **X** zone. Next, select T2 and drag-and-drop into the **Y** zone. Right-click on the plot area and add a line by selecting **Add ▶ Line**. Then select **UCL** drag-and-drop into the **Y** zone. Now right-click on the plot area and select **Points ▶ Y** to show T2 and UCL, and uncheck UCL to remove the points from the UCL line The control chart is shown in Figure 9.6.

Figure 9.6 Textile Fibers Hotelling T² Control Chart

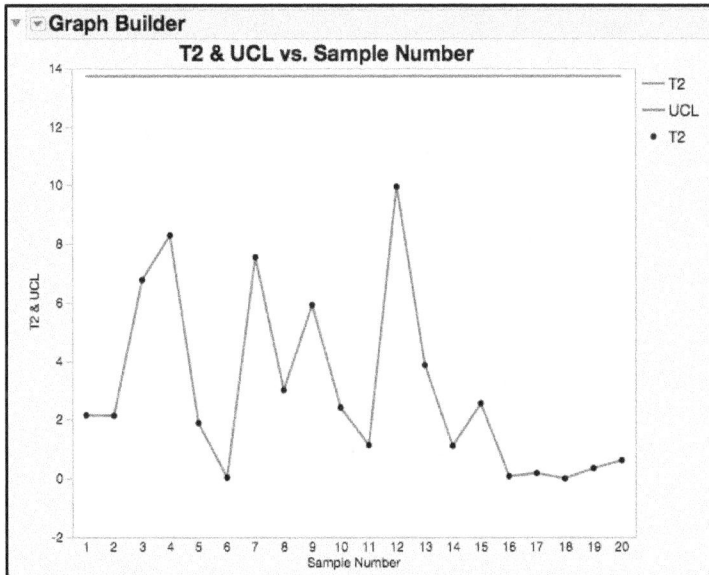

The control chart in Figure 9.6 corresponds to ISQC Figure 11.7. Since no points are out of control, the process is deemed stable. Obviously, this is not an ideal way to generate a multivariate control chart for pre-summarized data. While JMP is not able to work directly with summarized data, it is able to create multivariate charts when subgroup sizes exceed 1, where all individual results are included in the JMP table.

> ✦ **JMP Note 9.1**: For subgroup sizes greater than 1, the individual results must be available to produce the control chart using the Multivariate Control Chart platform.

ISQC Table 11.2 Grit Composition

In this table, data are presented from Holmes and Mergen (1993). There are 56 observations that represent the composition of "grit", using a percentage scale with three classifications, Large (L), Medium (M) and Small (S). It is desirable to monitor these variables using a multivariate approach. For each observation, since these three percentages add up to 100%, only the first two components, L and M, are used.

The following steps illustrate how to construct a Multivariate T^2 control chart using the **Multivariate Control Chart** platform.

1. Open **Chapter 9 - ISQC Table 11.2.jmp**, which has variables called *Sample Number, L (%), M (%)*, and *S (%)*. In this table, Sample Number is the subgroup variable, and L (%) and M (%) are the variables that will be charted.

2. Select **Analyze ▶ Quality and Process ▶ Control Chart ▶ Multivariate Control Chart**.

3. A launch window will appear (Figure 9.7) In the left-hand window, click on **L (%)** and **M (%)**, so they are both highlighted and then click **Y, Columns**. Note, for subgroup size = 1, there is no need to enter a **Subgroup** variable in this dialog box. Click **OK** when finished.

Figure 9.7 Launch Window for Multivariate Control Chart

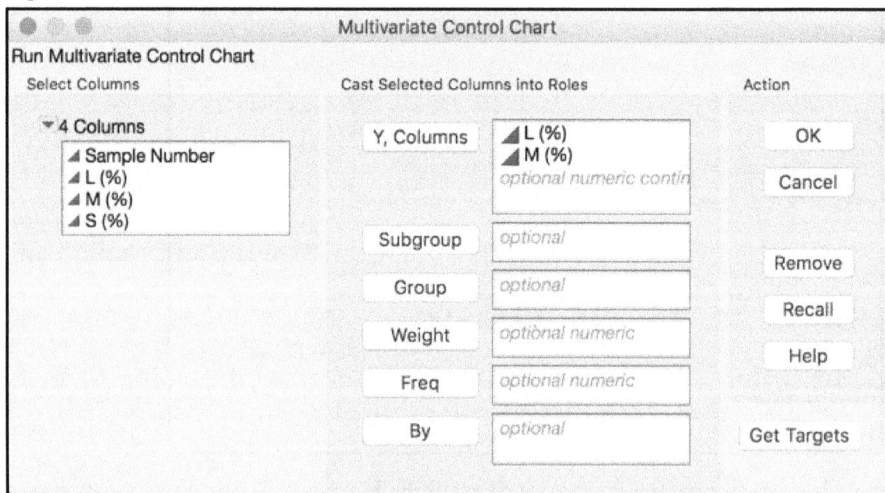

Figure 9.8 Default Multivariate Control Chart Output

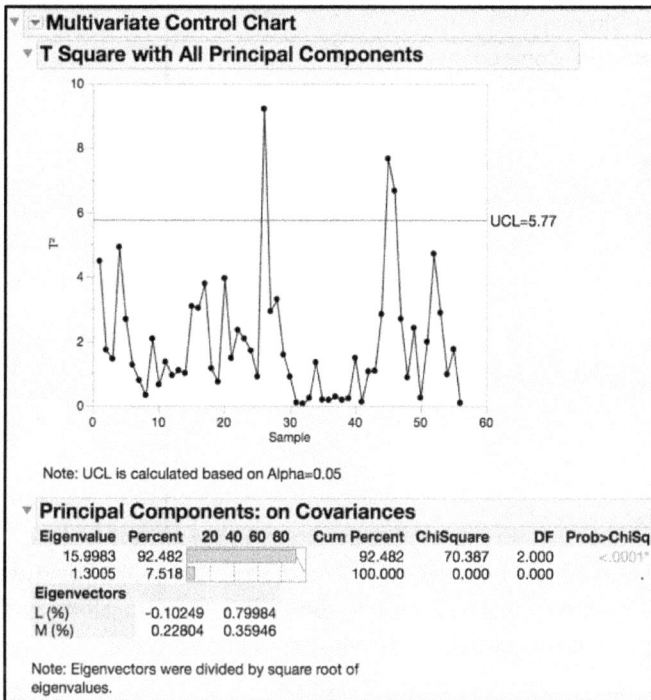

The output in Figure 9.8 shows the T² control chart along with the principal components and eigenvectors. The first eigenvalue of 15.99 corresponds to the first principal component, which is the linear combination of the standardized original variables that explains the largest amount of variation, 92.48% in this case. The eigenvectors show the coefficients of the principal components.

Statistics Note 9.1: The first principal component represents the direction, axis, of the component that explains the largest amount of variation. The eigenvector in the Grit Composition example has coefficients (-0.10249, 0. 22804), and can be written as -0.10249 × L (%) + 0.22804 × M (%).

4. The default output in Figure 9.8 uses alpha = 0.05 to compute the upper control limit for the T² chart. To change this value, from the red triangle next to the **Multivariate Control Chart** banner select **Set Alpha Level ▶ Other...**. Replace the default value of 0.05 with **0.00315** and click **OK**. The updated chart is shown in Figure 9.9.

Figure 9.9 Updated Multivariate T² Chart

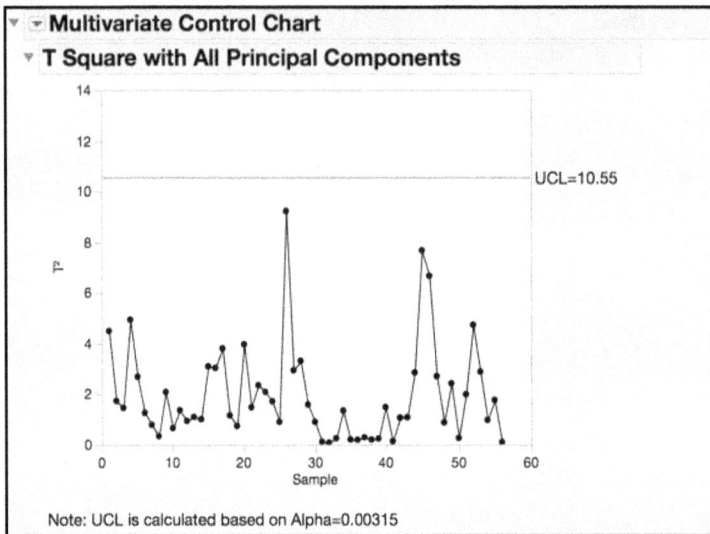

Note: UCL is calculated based on Alpha=0.00315

5. From the red triangle at the top of the window, select **Save T Square Formula**. This will save the T² charting statistics for each sample to the JMP data table, along with the formula used to compute these statistics, as is shown in Figure 9.10a.

Figure 9.10a T² Charting Statistics and Formula

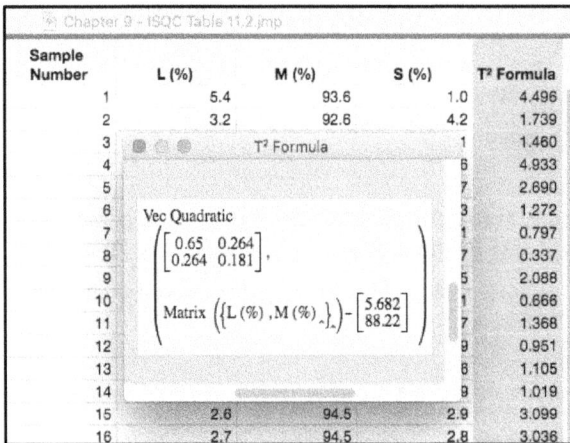

The control chart in Figure 9.9 is similar to the one shown in ISQC Figure 11.8 (a). Note the banner 'T Square with All Principal Components' indicates that this chart is created using all of the principal components, which is the same as using the raw data, as is done in this ISQC example. The upper control limit was adjusted using an unusual value for alpha = 0.00315, to reflect the fact that, for n=1, the Beta distribution is a better approximation to Phase I limits. This value was used in order to match the limit provided in ISQC, which is an exact control limit found using simulation methods for this data set.

The charting statistics were saved to the JMP table (see Figure 9.10a). These numbers exactly match the $T^2_{1,i}$ values found in ISQC Table 11.2. The formula editor shows the matrices that were used in the computations. The mean vector is shown in the lower right-hand side of the editor, with means of 5.682 and 88.22, for L and M, respectively. The Vec Quadratic 2x2 matrix, that is used in the matrix calculations for the charting statistic, is the inverse of the covariance matrix. The covariance matrix (or correlation matrix), as well as the inverse covariance (or correlation) matrices can be displayed from within the **Multivariate Control Chart** output by the selecting **Show Covariance** (or **Show Correlation**), **Show Inverse Covariance** (or **Show Inverse Correlation**) from the red triangle at the top of the window. These matrices are shown in Figure 9.10b. The covariance matrix is the same as the matrix S_1, used to create Figure 11.8 (a) in ISQC Section 11.3.

Figure 9.10b Covariance and Inverse Covariance for Grit Composition Example

Multivariate Control Chart		

▶ **T Square with All Principal Components**

▶ **Principal Components: on Covariances**

▼ **Covariance**

	L (%)	M (%)
L (%)	3.7702	-5.4955
M (%)	-5.4955	13.5285

▼ **Inverse Covariance**

	L (%)	M (%)
L (%)	0.6502	0.2641
M (%)	0.2641	0.1812

Statistics Note 9.2: The distribution used to calculate the UCL for a Multivariate T^2 control limit depends on the number of samples and the control chart phase (I or II). The beta distribution is used for phase I limits when n=1 (ISQC equation (11.27)), and the F-distribution is used for Phase II limits or larger data sets (ISQC equation (11.24)).

ISQC Table 11.5 Cascade Process Data

In Section 11.5 of ISQC, Hawkin's (1991) regression adjustment technique is introduced as a way to remove the effect of input variables form the output variables being charted. In this example, we show how to use the regression adjustment to remove the effect of deviations of input process variables, prior to generating control charts for the output variables. The approach obtains the residuals from a linear regression model, where the output variable of interest is regressed on the input variables, and plots them in a univariate manner using Shewhart charts discussed in ISQC Chapter 6 and Chapter 3 of this book.

The following steps illustrate how to construct an Individual Measurement and Moving Range control chart using the data in ISQC Table 11.5.

1. Open Chapter 9 - ISQC Table 11.5.jmp, which has variables called *x1* through *x9*, *y1* and *y2*. In this table, x1 – x9 are the input variables and y1 and y2 are the output variables.

2. From the main menu, select **Analyze ▶ Fit Model**.

3. Select **x1** through **x9** from the left-hand window and click the **Add** button under the **Construct Model Effects** label. Similarly, select **y1** from the left-hand window and click the **Y** button, under **Pick Role Variables** label (Figure 9.11). Click **Run** to fit the specified model.

Figure 9.11 Fit Model Dialog Window

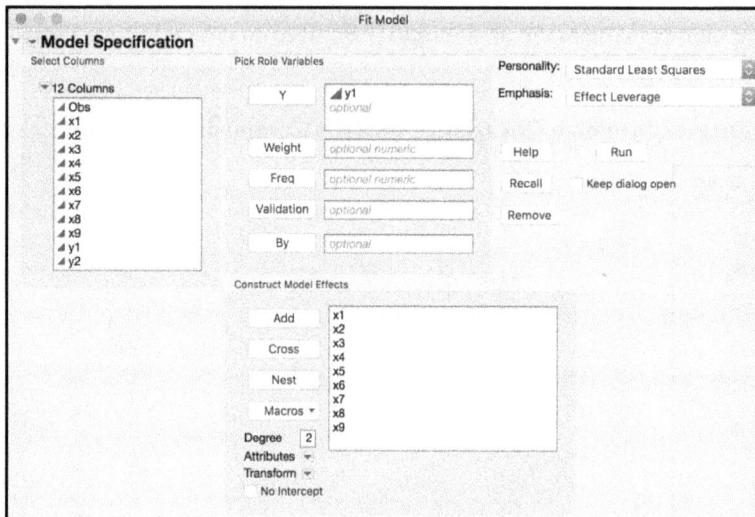

The partial regression output is shown in Figure 9.12a. The parameter estimates in the **Parameter Estimates** section of the report agree, up to rounding, to those shown in the fitted equation in ISQC Section 11.5.

Figure 9.12a Partial Regression Model Output for Cascade Data

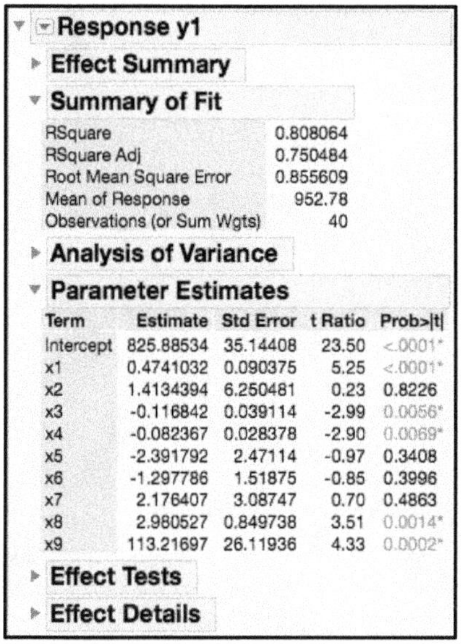

▼ ▾**Response y1**

▸ **Effect Summary**

▼ **Summary of Fit**

RSquare	0.808064
RSquare Adj	0.750484
Root Mean Square Error	0.855609
Mean of Response	952.78
Observations (or Sum Wgts)	40

▸ **Analysis of Variance**

▼ **Parameter Estimates**

Term	Estimate	Std Error	t Ratio	Prob>|t|
Intercept	825.88534	35.14408	23.50	<.0001*
x1	0.4741032	0.090375	5.25	<.0001*
x2	1.4134394	6.250481	0.23	0.8226
x3	-0.116842	0.039114	-2.99	0.0056*
x4	-0.082367	0.028378	-2.90	0.0069*
x5	-2.391792	2.47114	-0.97	0.3408
x6	-1.297786	1.51875	-0.85	0.3996
x7	2.176407	3.08747	0.70	0.4863
x8	2.980527	0.849738	3.51	0.0014*
x9	113.21697	26.11936	4.33	0.0002*

▸ **Effect Tests**

▸ **Effect Details**

As Montgomery points out in ISQC Section 11.5 "the residuals are found simply by subtracting the fitted value from this equation from each corresponding observation y1." These residuals are precomputed in JMP and can be saved easily from the regression output.

4. From the red triangle next to **Response y1** in the main banner, select **Save Columns ▸ Residuals**, as shown in Figure 9.12b.

Figure 9.12b Saving Residuals from Regression Model Output for Cascade Data

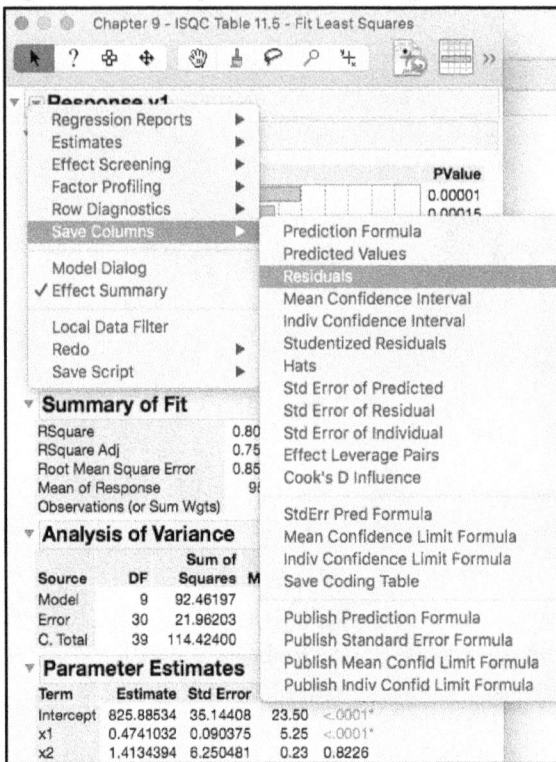

5. Click on the JMP table and confirm that a column for the residuals was added to the end of the table and then select **Analyze ▶ Quality and Process ▶ Control Chart ▶ IR**.

6. In the launch window select **y1** and **Residual y1** from the Columns list on the left-hand side of the window and then click the **Process** button to add them to the window. Click **OK** when done.

7. Turn on Test 1 in both Individuals charts by selecting **Tests ▶ Test 1** from the red triangle next to **Individual Measurement of y1** and **Individual Measurement of Residual y1**. The output is shown in Figure 9.13.

Figure 9.13 XmR Charts for y1 and Residuals of y1

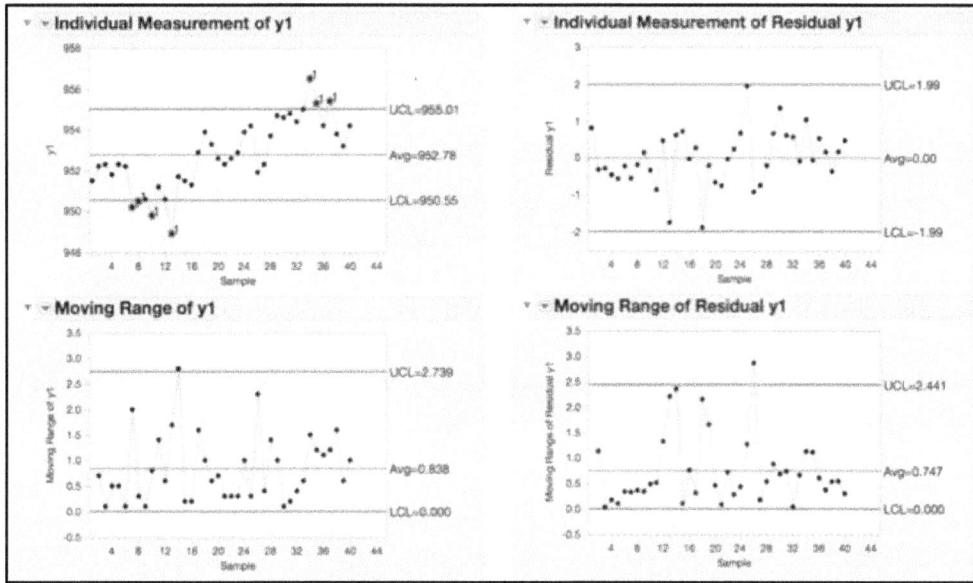

The XmR chart for y1, shown on the left-hand side of Figure 9.13, is similar to the one shown in ISQC Figure 11.9. The Individual Measurement control chart has a number of points that exceed the control limits, indicating a lack of statistical control. However, the Individual Measurement control chart of the residuals, shown on the right-hand side of Figure 9.13, and in ISQC Figure 11.10, has no signals. As Montgomery indicates, the state of process control has two very different interpretations in these two charts. The behavior of the original response, y1, includes the effects of changes in the process inputs, x1 through x9, which are most likely the cause of the signals. This can be studied by evaluating the regression fit output shown in Figure 9.12a. The parameter estimates are very similar to the ones presented in ISQC Section 11.5, with x1, x9 and x8 having the three largest **t Ratio**. The Adjusted R^2 is 0.75, indicating that 75% of the variation in the response is explained by the regression model. The residuals reflect the process behavior after the removal of the impact of the process inputs on y1 and, since there are no signals on the chart, confirm the process is in control when accounting for the variation in the process inputs.

The original response y1 exhibits a considerable amount of autocorrelation, as shown in the autocorrelation plot in ISQC Figure 11.1. In general, autocorrelation will result in a larger number of runs tests violations, due to short-term variation that is smaller than the longer-term variation and narrower control limits. If the autocorrelation is due to the influence of other factors, then the residuals from an appropriate model will be uncorrelated and more amenable to the assumptions of Shewhart charts, as shown in the autocorrelation plot for the residuals in ISQC Figure 11.12 ISQC Figures 11.1 and 11.12 are not reproduced here, but similar plots are discussed in detail in Chapter 8 in this book.

> **Statistics Note 9.3**: Regression adjustment is useful for removing the impact of input variables on an output variable. It can also be used in troubleshooting to help identify the root cause of a signal.

ISQC Example 11.2 Monitoring Variability

In this example, we will show how to construct a control chart for the generalized variance. As is the case with univariate approaches, like XBar and Range charts, multivariate approaches are available to monitor both the mean and variance of multiple parameters simultaneously. The approach uses the main diagonal elements of the $p \times p$ covariance matrix. This example uses the same data and scenario described in ISQC Example 11.1. Recall, there are 20 paired samples of tensile strength and diameter of a textile fiber, where each measurement represents the mean response for n = 10 fibers.

As of JMP 14, the **Multivariate Control Chart** platform in JMP does not have the capability of creating a chart to monitor the multivariate variances. The following steps illustrate how to construct a Multivariate Variance control chart using the formula editor, using the formulas in ISQC equation (11.36), and **Graph Builder**.

1. Open Chapter 9 - ISQC Example 11.2.jmp, which has *Sample Number, Subgroup Size, Tensile Strength (mean), Diameter (mean), Tensile Strength (variance), Diameter (variance)* and *Covariance*. In addition, *T2, UCL T2,* and *Sk* have been added to the JMP table, as is shown in ISQC Table 11.1. In this table, Sample Number is the subgroup variable, Subgroup Size is the subgroup size for each mean, and Sk is the charting statistic.

2. Double click in the header of the JMP table to add a new column.

3. Double click on the generic column label to open the column information window. Type **UCL Sk** in the **Column Name** field. Select **Fixed Decimal** in the drop-down menu next to **Format** and enter **2** in the **Dec** field. Select **Formula** from the **Column Properties** menu.

4. Click the **Edit Formula** button to launch the formula editor. Build the equation for the UCL, using ISQC Example 11.2 as a guide and Figure 9.14. First enter **0.4464** in the box and then click on the × at the top of the editor. Then hold down the shift key and press 9 to build the first set of parentheses. Next, enter **0.8889** in the first box then click on the + symbol at the top of the Formula window and enter **3** in the right-hand box. With this box highlighted, click on the × and enter **0.417**. Finally, with this box highlighted, click on the x^y and then click on ÷ and enter **1** and **2**. Before clicking **OK** to finish, click on the outer-most box and copy the entire formula.

Figure 9.14 Formula Editor for UCL Sk

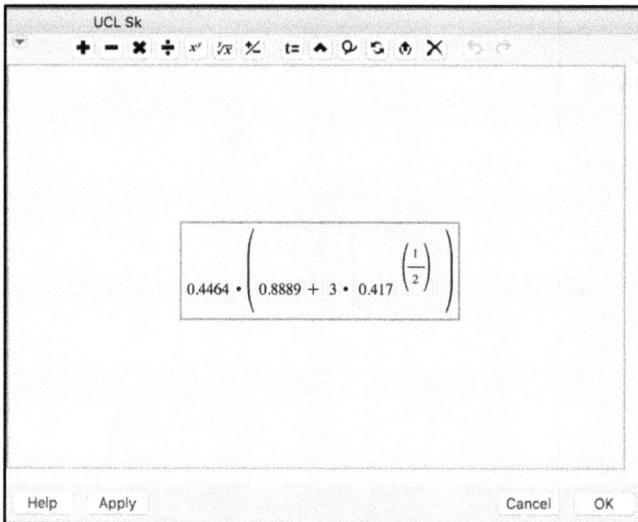

5. Double click on the generic column label to open the column information window. Type **LCL Sk** in the **Column Name** field. Select **Fixed Decimal** in the drop-down menu next to **Format** and enter **2** in the **Dec** field. Select **Formula** from the **Column Properties** menu.

6. Click the **Edit Formula** button to launch the formula editor. With the box highlighted, paste the previous formula into the box. Click on the + symbol and change it to a - symbol. Click **OK** when done.

7. Double click on the generic column label to open the column information window. Type **CL Sk** in the **Column Name** field. Select **Fixed Decimal** in the drop-down menu next to **Format** and enter **2** in the **Dec** field. Select **Formula** from the **Column Properties** menu.

8. Click the **Edit Formula** button to launch the formula editor. Enter **0.3968** in the box and click **OK** when done. The three control limits should be added to the JMP table (Figure 9.15).

Figure 9.15 Updated JMP Table with Variance Limits

Sk	UCL Sk	CL Sk	LCL Sk
0.45	1.26	0.3968	-0.47
0.41	1.26	0.3968	-0.47
0.5	1.26	0.3968	-0.47
0.21	1.26	0.3968	-0.47
0.21	1.26	0.3968	-0.47
0.23	1.26	0.3968	-0.47
0.41	1.26	0.3968	-0.47
0.52	1.26	0.3968	-0.47
0.35	1.26	0.3968	-0.47
0.1	1.26	0.3968	-0.47
0.54	1.26	0.3968	-0.47
0.36	1.26	0.3968	-0.47
0.17	1.26	0.3968	-0.47
0.33	1.26	0.3968	-0.47
0.42	1.26	0.3968	-0.47
0.44	1.26	0.3968	-0.47
0.65	1.26	0.3968	-0.47
0.36	1.26	0.3968	-0.47
0.59	1.26	0.3968	-0.47
0.63	1.26	0.3968	-0.47

9. From the main menu, select **Graph ▶ Graph Builder**. In the dialog window, select **Sk, UCL Sk**, and **CL Sk** from the left-hand area and drag and drop in the **Y** zone. Select **Sample Number** and drag-and-drop in the **X** zone. Right-click within the plot area and select **Add ▶ Line**. Right-click again within the plot area and select **Points ▶ Y**. From the menu, uncheck **UCL Sk** and **CL Sk**. Click **OK** when finished. The chart is shown in Figure 9.16.

Figure 9.16 Multivariate Control Chart for Monitoring Variability

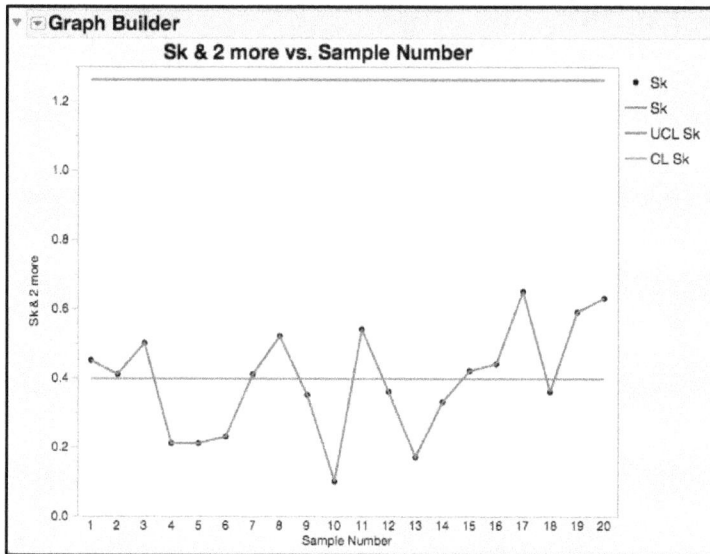

The limits shown in Figure 9.15 agree with those shown in ISQC Example 11.2. As with univariate charts, it is important to monitor both the means and variances in a multivariate scenario. For this approach to be used, there must be more than one observation ($n > 1$) for each variable and sampled point. Recall, the data in ISQC Example 11.1 consists of tensile strength and diameter measurements for $n = 10$ fibers for each sample. The chart in Figure 9.16 (also shown in ISQC Figure 11.13) is based on the sample *generalized* variance, where the charting statistic is the determinant of the sample covariance matrix $|S|$. For each sample number k, the sample covariance matrix is estimated using the sample variances and covariance. The determinant of a 2x2 matrix with elements a (variance y_1), b (covariance y_1 and y_2), c (covariance of y_2 and y_1) and d (variance y_2), is easily calculated as $a \times d - b \times c$. In other words, for a 2x2 covariance matrix the determinant is the product of the variances minus the squared covariance. For example, for the first row of ISQC Table 11.1, $|S_k| = |1.25 \times 0.87 - 0.8^2| = 0.45$. This formula is included in Chapter 9 - ISQC Example 11.2 All.jmp.

The variability chart for ISQC Example 11.2 does not show any signs of instability, since all points are within the control limits. Therefore, the process is considered to be stable and the limits can be used for Phase 2 monitoring.

ISQC Table 11.6 Chemical Process Data

In this example, we will show how to use principal components to gain further insight into multivariate data and how they can be used for process monitoring. Data from a chemical process are used to illustrate these techniques, where four potentially correlated variables, x_1, x_2, x_3, and x_4, are tracked.

The following steps illustrate how to fit the principal components using the **Multivariate** platform in **Multivariate Methods**.

1. Open Chapter 9 - ISQC Table 11.6.jmp, which has variables called *Observation, x1, x2, x3,* and *x4.* Observation is the subgroup variable and x1 through x4 are the responses of interest.

2. From the main menu, select **Analyze ▶ Multivariate Methods ▶ Multivariate.**

3. When the launch window appears, select **x1, x2, x3,** and **x4** in the **Columns** window on the left and click **Y, Columns** (Figure 9.17).

Figure 9.17 Dialog Window for Multivariate Platform

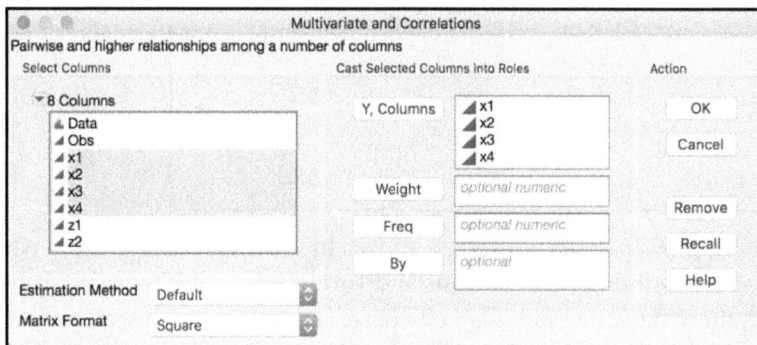

4. Click **OK** when finished. The default output includes a correlation matrix and scatterplot matrix, shown in Figure 9.18.

Figure 9.18 Default Output for Multivariate Analysis

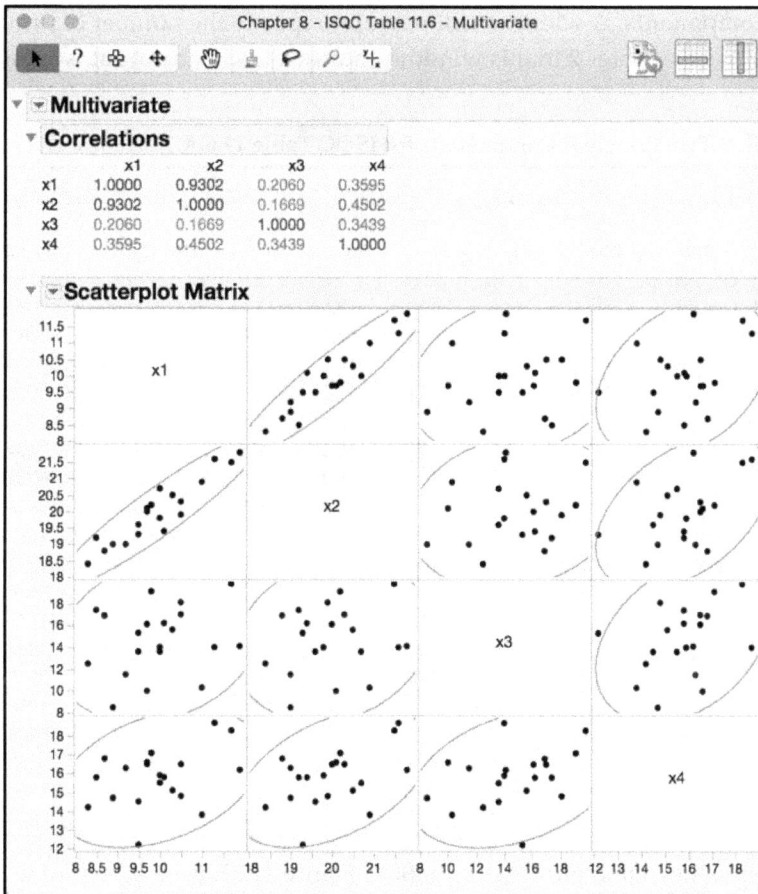

5. In order to create the principal components, click on the red triangle next to the **Multivariate** label and select **Principal Components ▶ On Correlations**. This will create a table of eigenvalues. From the red triangle next to **Principal Components / Factor Analysis** select **Eigenvectors**. The output is shown in Figure 9.19.

Figure 9.19 Principal Components for ISQC Table 11.6

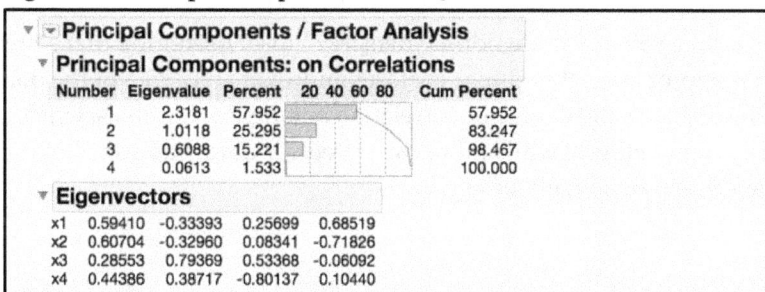

6. From the red triangle next to **Principal Components / Factor Analysis** select **Save Principal Components**. A window will pop up to specify the number of principal components to save. Enter **2** in this window and click **OK**. The output is shown in Figure 9.20.

Figure 9.20 First Two Principal Components for ISQC Table 11.6

	Data	Obs	x1	x2	x3	x4	Prin1	Prin2
1	Original	1	10	20.7	13.6	15.5	0.2916809057	-0.603401271
2	Original	2	10.5	19.9	18.1	14.8	0.2942812289	0.4915327967
3	Original	3	9.7	20	16.1	16.5	0.1973367668	0.6409365561
4	Original	4	9.8	20.2	19.1	17.1	0.839021999	1.4695791871
5	Original	5	11.7	21.5	19.8	18.3	3.2048763004	0.8791722746
6	Original	6	11	20.9	10.3	13.8	0.203271222	-2.29514172
7	Original	7	8.7	18.8	16.9	16.8	-0.992105021	1.6704642538
8	Original	8	9.5	19.3	15.3	12.2	-1.702412423	-0.360891326
9	Original	9	10.1	19.4	16.2	15.8	-0.142460551	0.5608080147
10	Original	10	9.5	19.6	13.6	14.5	-0.994981163	-0.314933594
11	Original	11	10.5	20.3	17	16.5	0.9446965931	0.5047110725
12	Original	12	9.2	19	11.5	16.3	-1.219502081	-0.091293581
13	Original	13	11.3	21.6	14	18.7	2.6086661091	-0.421763754
14	Original	14	10	19.8	14	15.9	-0.123784633	-0.087672826
15	Original	15	8.5	19.2	17.4	15.8	-1.104228758	1.4725928879
16	Original	16	9.7	20.1	10	16.6	-0.278251264	-0.947627123
17	Original	17	8.3	18.4	12.5	14.2	-2.656077942	0.1352883131
18	Original	18	11.9	21.8	14.1	16.2	2.3652800923	-1.304936966
19	Original	19	10.3	20.5	15.6	15.1	0.4113106837	-0.218930384
20	Original	20	8.9	19	8.5	14.7	-2.146618067	-1.178492813

The scatterplot matrix, shown in Figure 9.18 and ISQC Figure 11.15, is very helpful to visualize the correlations among the pairwise combinations of the four factors. For example, the scatter plot for x_1 and x_2 is shown in the cell in the first row and second column and is repeated in the cell in the second row and first column. The plot indicates a strong positive linear relationship between the two variables, which means that as x_1 increases, x_2 increases also. In contrast, the relationship between x_2 and x_3 (shown in cells 2, 3 and 3, 2) indicates a weak relationship between the two variables.

Each scatter plot includes a 95% bivariate normal density ellipse. Assuming that a bivariate normal distribution is appropriate, this ellipse encloses approximately 95% of the points. The shape of the ellipse also reflects the degree of correlation among the variables. Uncorrelated variables have a fairly round ellipse; while more correlated variables have an ellipse that is elongated, cigar shaped, and diagonally oriented.

Statistics Note 9.4: The bivariate density ellipse is a function of the correlation between the two variables. As the correlation between the two variables approaches either 1 or -1, the ellipse becomes elongated along its major axis. The direction of the axes is determined by the eigenvectors, and length proportional to the eigenvalues.

The sample correlation matrix is also included in the default output shown in Figure 9.18. This matrix confirms the degree and direction of the linear correlations shown in the matrix. The strongest correlation is 0.9302 between x_1 and x_2, and the smallest correlation is 0.1669 between x_2 and x_3. Additional information can be added to the correlation output, such as, confidence intervals for each correlation or, p-values associated with significance tests to determine if the sample correlation is different from 0.

The eigenvalues shown in Figure 9.19, and ISQC Table 11.7, indicate that the first two principal components explain over 83% of the variability in the original four variables. The eigenvectors are used to compute the principal components. For example, the first principal component score (z_1) is computed using $0.59410x_1 + 0.60704x_2 + 0.28553x_3 + 0.44386x_4$. By saving the principal scores to the JMP table, we can use the formula editor to view the equation for each principal component.

As Montgomery illustrates, the principal components can be used for multivariate monitoring in place of the original variables. In order to re-create the plot shown in ISQC Figure 11.16, the principal components must first be saved to the JMP table and then re-run in the **Multivariate** platform. New data can also be evaluated using the principal components computed from the baseline data, as is shown in ISQC Figure 11.17.

7.   After saving the first two principal components in Step 6, close or move the **Multivariate** output and click on Chapter 9 - ISQC Table 11.6.jmp to make it active.

8.   From the main menu, select **Rows ▶ Add Rows** and enter **10**, to add additional rows to the table. Enter the new results for x_1, x_2, x_3, and x_4 for observations 21 through 30 in Chapter 9 - ISQC Table 11.6. Note the principal component scores will automatically be calculated for the new rows. Also, enter '**New**' in the Data column for all ten rows. Note this data is already saved in Chapter 9 – ISQC Table 11.6 All.jmp.

9.   Select **Rows ▶ Color or Mark by Column...** and select **Data** in the left-hand column. Select **Standard** from the **Markers** drop-down menu and click **OK**.

10. Highlight rows 21 through 30 and from the main menu select **Rows ▶ Exclude /** **Unexclude**.

11. From the main menu, select **Analyze ▶ Fit Y by X**.

12. When the launch window appears, select **Prin2** in the **Columns** window on the left and click **Y, Response**. Click on **Prin1** and **X, Factor**. Click **OK** when finished.

13. Scale the Y axis by double-clicking on the numbers and enter **-4** as the **Minimum** and **4** as the **Maximum** with an **Increment** of 1. Click **OK**. Double-click on the X axis and enter **-4** as the **Minimum** and **7** as the **Maximum,** with an **Increment** of 0.5, and click **OK**.

14. To generate the bivariate ellipse, click the red triangle and select **Density Ellipse ▶ 0.95** as shown in Figure 9.21.

Figure 9.21 Selecting 95% Density Ellipse

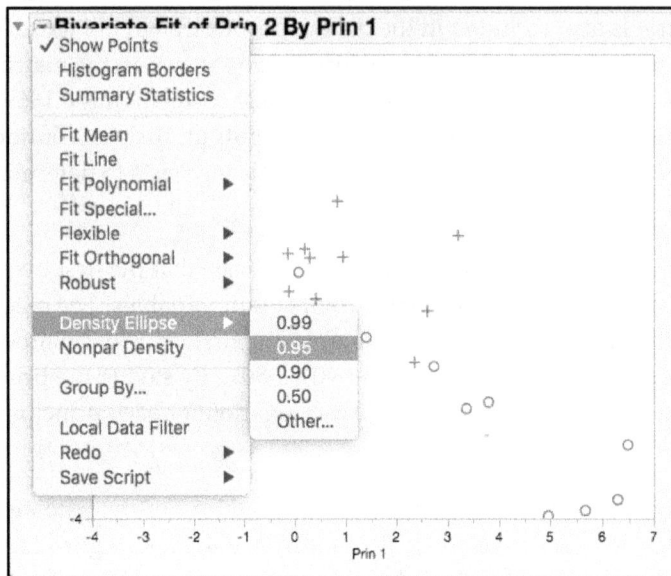

15. Now highlight rows 21 through 30 and from the main menu select **Rows ▶ Exclude / Unexclude** to display the new observations. The ellipse is shown in Figure 9.22.

Figure 9.22 Principal Components Normal Density Ellipse

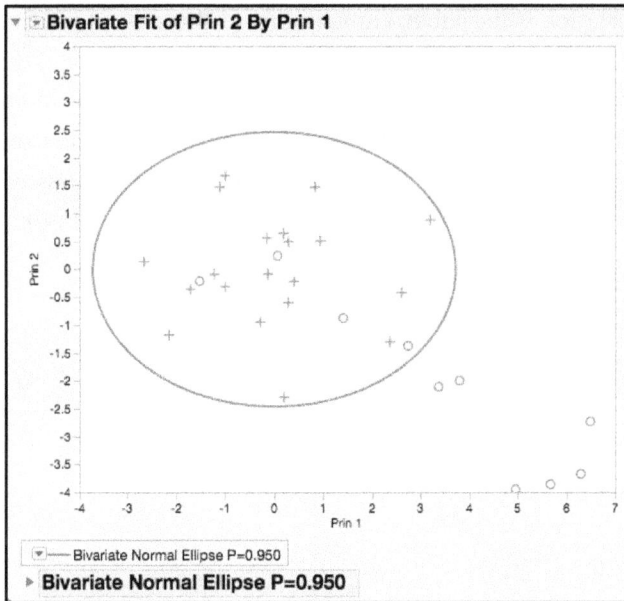

The bivariate normal density ellipse for the first two principal components is similar to the one shown in ISQC Figure 11.17. The ellipse shape is determined by the original data, since the new data was excluded from calculations in the JMP table. The observation number will show up in the plot if you position your pointer over a point. The new observations beginning with number 24 are outside of the density ellipse, indicating a shift in the process. Additional analyses must be done in order to investigate the out of control points, which is discussed in the next section of this chapter.

Statistical Insights

In this section, we elaborate upon some of the examples provided in Chapter 11 of ISQC. The examples highlighted in this section include several important concepts we have encountered over our many years of applying SPC successfully to a variety of industries. For most of these examples, additional output not provided in ISQC is included to illustrate JMP functionality or further elaborate on important points considered by the authors.

Additional Tools in the Multivariate Control Chart Platform

The examples presented throughout the chapter did not fully explore the functionality in the **Multivariate Control Chart** platform in JMP. In this section, we provide additional tools that can be used to monitor multiple variables using Multivariate SPC and the data in ISQC Table 11.6. First, we save the phase I multivariate limits using the baseline data and the original variables, x1, x2, x3 and x4, and then we apply it to the new data using phase II limits and interpret the output.

1. Open Chapter 9 – ISQC Table 11.6 All.jmp, which has variables called *Data, Observation, x1, x2, x3,* and *x4.* Observation is the subgroup variable, x1 through x4 are the charting variables.

2. Highlight rows 21 through 30 and select **Rows ▶ Delete Rows** from the main menu.

3. Select **Analyze ▶ Quality and Process ▶ Control Chart ▶ Multivariate Control Chart**.

4. When the launch window appears, select **x1, x2, x3**, and **x4** in the **Columns** window on the left and click **Y, Columns**. Click **OK** when finished. See output in Figure 9.23.

Figure 9.23 T² Chart for x1, x2, x3, and x4

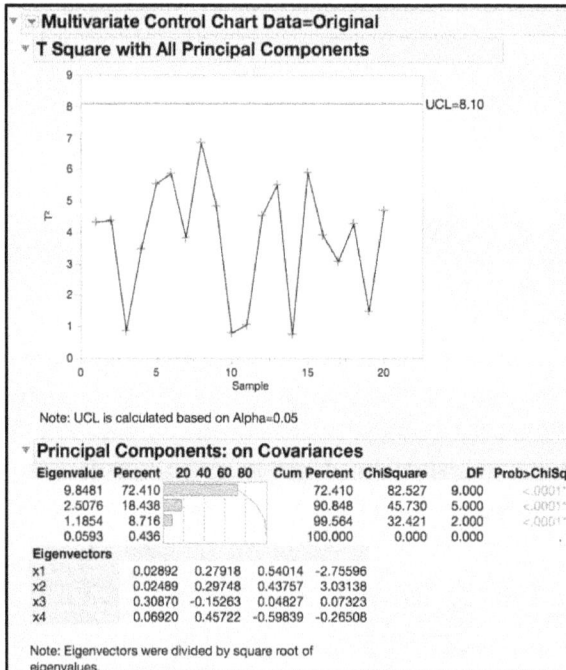

The **T Square Partitioned** option, available in this output window, produces control charts using the principal components by splitting them into two groups, Big and Small. The Big group consists of the first *r* principal components and the remaining components go to the Small group. JMP suggests a default number of components, but the user can select *r* based on the principal components that account for the most variation. Note that **T Square with All Principal Components**, shown in the default output, Figure 9.23, is the sum of the T² values in the Big and Small Principal Components charts.

5. From the red triangle next to **Multivariate Control Chart** select **T Square Partitioned**. A window will appear to enter the number of major principal components for the T Square. The JMP suggested default is 2 principal components, as shown in Figure 9.24, which, in this case, explain about 91% of the variation.

Figure 9.24 JMP Suggested Number of Principal Components

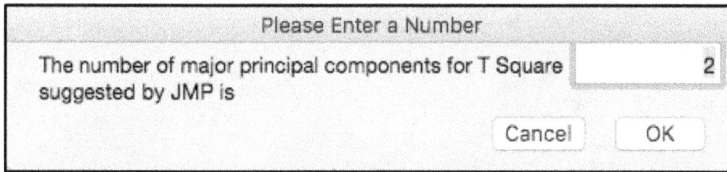

Please Enter a Number
The number of major principal components for T Square suggested by JMP is	2
	Cancel · OK

6. Click **OK**. The output is shown in Figure 9.25, where the **T Square with Big Principal Components** is based on X1 and X2, while the **T Square with Small Principal Components** is based on variables X3 and X4.

Figure 9.25 T Square Partitioned Output into Major and Minor Components

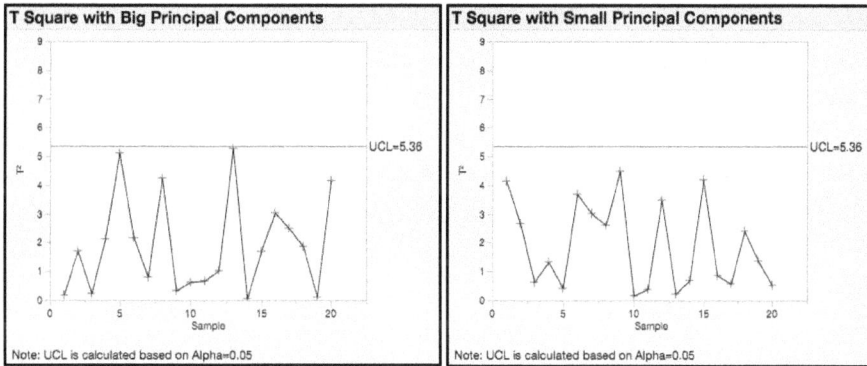

Statistics Note 9.5: The UCL for the T Square with Big Principal Components is calculated using ISQC Equation (11.27), which depends on the beta distribution. Since the number of major principal components suggested by JMP was 2, we set p=2 in Equation (11.27)

$$\text{UCL} = \frac{(20-1)^2}{20}\text{Beta}_{0.05,\frac{2}{2},\frac{20-2-1}{2}} = \frac{19^2}{20}\text{Beta}_{0.05,1,\frac{17}{2}} = 5.36$$

The lower panel of ISQC Table 11.6 contains 10 new observations. We can use the limits calculated using the first 20 observations to monitor the new data. We can save the upper control limit and associated statistics directly from the **Multivariate Control Chart** output.

7. From the red triangle next to Multivariate Control Chart select **Save Targets Statistics** and then close this window. The target statistics will be saved to a JMP table. Activate this table by clicking on and then select **File ▶ Save As ...** and type in **Chapter 9 – ISQC Table 11.6 Targets** in the name field and click **Save**. Leave this table open.

Figure 9.26 Target Statistics for Chapter 9 - ISQC Table 11.6 Example

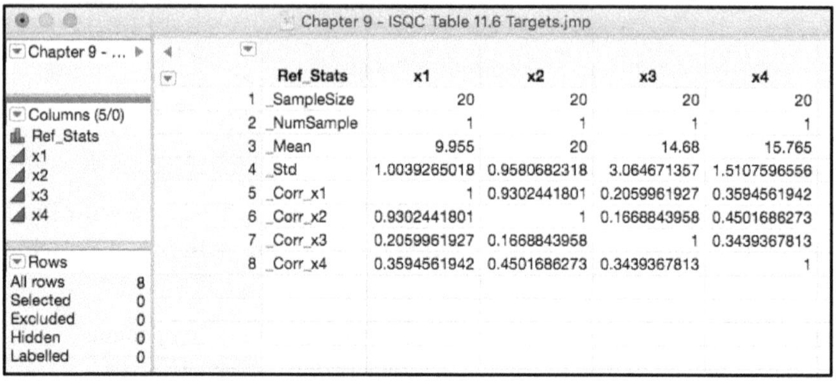

	Ref_Stats	x1	x2	x3	x4
1	_SampleSize	20	20	20	20
2	_NumSample	1	1	1	1
3	_Mean	9.955	20	14.68	15.765
4	_Std	1.0039265018	0.9580682318	3.064671357	1.5107596556
5	_Corr_x1	1	0.9302441801	0.2059961927	0.3594561942
6	_Corr_x2	0.9302441801	1	0.1668843958	0.4501686273
7	_Corr_x3	0.2059961927	0.1668843958	1	0.3439367813
8	_Corr_x4	0.3594561942	0.4501686273	0.3439367813	1

(Window: Chapter 9 - ISQC Table 11.6 Targets.jmp; Columns (5/0): Ref_Stats, x1, x2, x3, x4; Rows — All rows 8, Selected 0, Excluded 0, Hidden 0, Labelled 0)

8. Make Chapter 9 – ISQC Table 11.6 All.jmp active by clicking on it. Highlight rows 21 through 30 and then, from the main menu, select **Rows ▶ Exclude/Unexclude** and then select **Rows ▶ Hide/Unhide.** This will unexclude and unhide the rows in the table.

9. Select **Analyze ▶ Quality and Process ▶ Control Chart ▶ Multivariate Control Chart.**

10. When the launch window appears, select **x1, x2, x3,** and **x4** in the **Columns** window on the left and click **Y, Columns.** Select **Get Targets** and when the dialog window appears, highlight Chapter 9 – ISQC Table 11.6 Targets.jmp and click **Open.** Click **OK** when finished. The output is shown in Figure 9.27.

Figure 9.27 Default T² Output Using Target Statistics

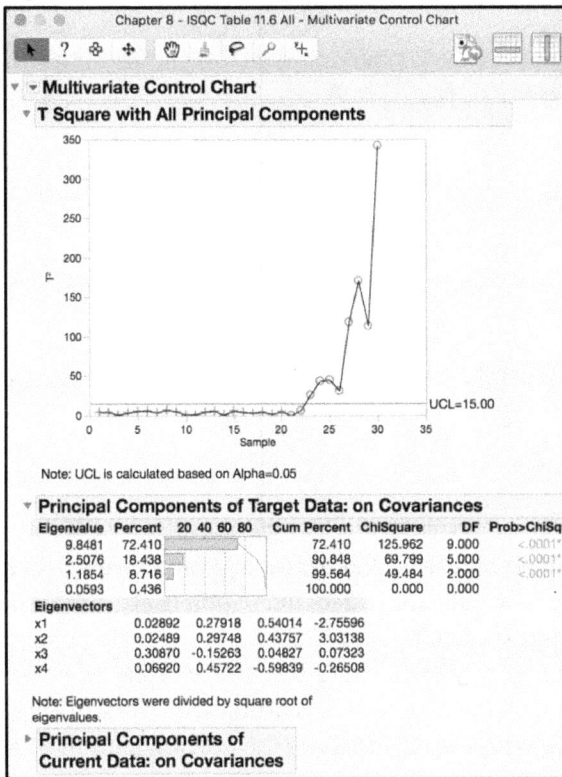

It is easy to get lost in multivariate output, so before interpreting the results shown above, we should explore the data using simple plots. The overlay plot in Figure 9.28 was created using the **Graph Builder** platform in JMP. From this plot it is easy to see that the behavior of x1 and x2 changes, starting around observation 24. At this observation, the mean of x1 appears to be increasing while the variance of x2 appears to increase. There are no observable changes to x3 and x4. In Figure 9.18, we saw that x1 and x2 had the highest linear correlation, which is most likely impacted by these changes as well. Now let's see how these shifts impact the multivariate techniques explored in this section.

Figure 9.28 Overlay Plot of x1, x2, x3 and x4

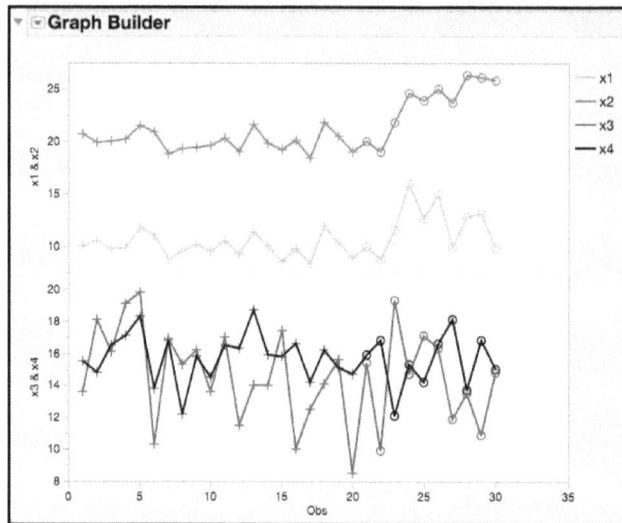

The T[2] chart of the baseline data shown in Figure 9.23 indicates that the process is in a state of statistical control. However, when the new data are added, many points are out of control, as is shown in Figure 9.27. Recall, the T[2] chart shown in Figure 9.27 was constructed using the statistics calculated from the baseline data (see Figure 9.26), so the new chart is designed to detect departures from this baseline data. In fact, the new results, starting with the 23rd observation, are all out of control. This seems to line up with the beginning of the upward shift in x1 and the variance increase in x2 that we observed in Figure 9.28.

Partial Contribution in the Principal Components Platform

In the previous section we explored some of the additional tools in the **Multivariate Control Chart** platform. From Figure 9.23 we saw that the first two principal components explained over 90% of the variation. What if we want to know the contribution of each of the variables to a given principal component? The **Partial Contribution of Variables** in the **Principal Components** platform produces a table and a plot with the individual contributions of each variable to a principal component. Here we continue exploring the Chemical Process data.

1. Open Chapter 9 – ISQC Table 11.6 All.jmp, which has variables called *Data, Observation, x1, x2, x3,* and *x4.* Observation is the subgroup variable, x1 through x4 are the charting variables.
2. Highlight rows 21 through 30 and select **Rows ▶ Hide and Exclude** from the main menu.
3. Select **Analyze ▶ Multivariate Methods ▶ Principal Components**.
4. In the launch window, select columns **x1, x2, x3,** and **x4** and placed in the **Y, Columns** as shown in Figure 9.29. Then click **OK**.

Figure 9.29 Principal Components Launch Window

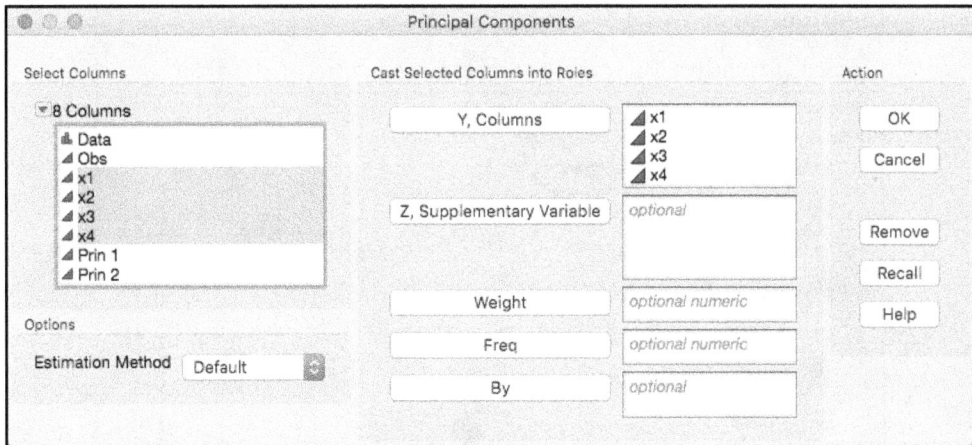

5. By default, the **Principal Components** output shows the eigenvalues, from largest to smallest, with a bar chart of the percent of the variation accounted for by each principal component. It also shows a Score plot and a Loadings plot for the first 2 principal components (Figure 9.30).

Figure 9.30 Principal Components Default Output

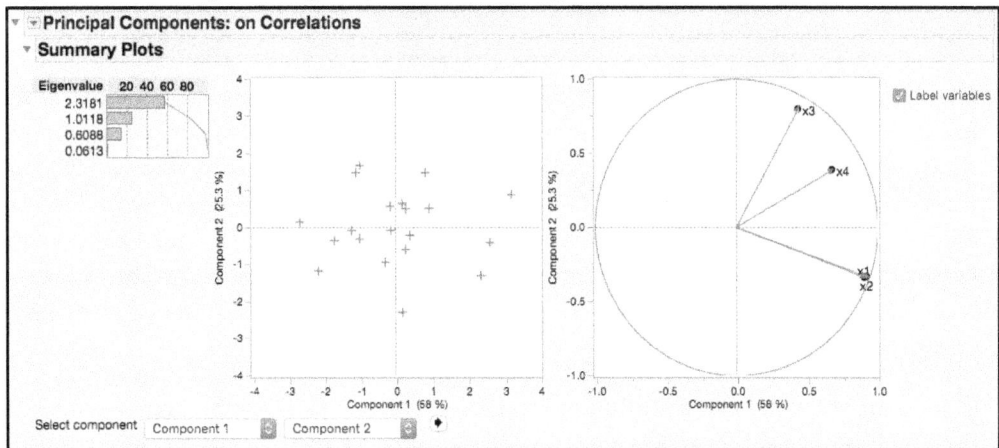

JMP Note 9.2: The Score plot plots each component against each other. The Loadings plot shows the unrotated loadings between the variables and principal components. The closer the value is to 1, the greater the effect of the component on the variable.

6. Click the red triangle next to **Principal Components: on Correlations** and select **Partial Contribution of Variables.** The output is shown in Figure 9.31.

Figure 9.31 Principal Components Partial Contribution of Variables

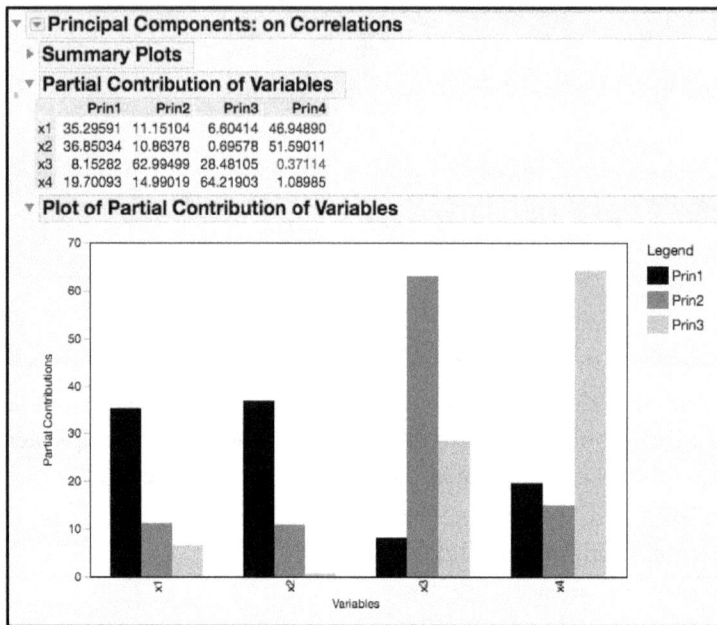

The **Partial Contribution of Variables** report in Figure 9.31 shows the principal components, **Prin1 – Prin4** as columns, and the variables, **x1-x4**, as rows. From the **Prin1** column we see that the contribution of x1 to Prin1 is 35.3%, and that x2 contributes 36.9%. In other words, the first principal component is being driven primarily by x1 and x2. This agrees with the Loadings plot, Figure 9.30, where the values of the vectors x1 and x2 are close to 1 in the Component 1 axis. The largest contributor to the **Prin2** column is variable x3, with a contribution of 63%. In agreement with the Loadings plot in Figure 9.30, where the vector x3 is close to the value of 1 in the Component 2 axis.

In the **Plot of Partial Contribution of Variables,**in Figure 9.31 the first bar, representing the first principal component (Prin1), is the tallest bar for both x1 and x2, with the bar for x2 slightly higher than the one for x1. This means that the first principal component is being driven primarily by x1 and x2 together. This agrees with the high positive correlation between x1 and x2, 0.9302, shown in Figure 9.18. For variable x3, the tallest bar is the one corresponding to the second principal component (Prin2), with a contribution of 63%. Finally, for variable x4, the tallest bar is the one corresponding to the third principal component (Prin3), with a contribution of 64%.

References

ASTM International. 2010. Manual on Presentation of Data and Control Chart Analysis, (STP) 15D, 8th edition, Newburyport, MA.

Automotive Industry Action Group. 2010. "Measurement Systems Analysis: Reference Manual, 4th edition." Automotive Division, American Society for Quality, Milwaukee, WI. http://www.rubymetrology.com/add_help_doc/MSA_Reference_Manual_4th_Edition.pdf

Britt, K. A., B. Ramírez, and T. Mistretta. 2016. "Process monitoring using statistical stability metrics: Applications to biopharmaceutical processes." *Quality Engineering* 28(2): 193–211.

Box, G., and J. Ramírez. 1991. "Sequential methods in statistical process monitoring." Report No. 65, Center for Quality and Productivity Improvement, University of Wisconsin-Madison, Madison, WI.

Box, G., and J. Ramírez. 1992. "Cumulative Score Charts." *Quality and Reliability Engineering International* 8(1):17-27.

Cantell, B. S. 1993. "Overdispersion in Binomial Models: An Application to Yield Data." MS thesis, Worcester Polytechnic Institute, Worcester, MA. (unpublished)

Cantell, B., R. Collica, and J. Ramirez. 1998. "Statistical Process Monitoring of Correlated Binary and Count Data Using Mixture Distributions." Proceedings of the Twenty-Third Annual SAS Users Group International Conference, Paper 237.

Hawkins, D. M. 1981. "A CUSUM for a Scale Parameter." *Journal of Quality Technology* 13(4): 228-235.

Hawkins, D. 1991. "Multivariate Quality Control Based on Regression-Adjusted Variables." *Technometrics* 33(1): 61-75.

Howe, W. G. 1969. "Two-sided Tolerance Limits for Normal Populations - Some Improvements." *Journal of the American Statistical Association* 64(326): 610-20.

Jackson, J. E. 1956. "Quality Control Methods for Two Related Variables." *Industrial Quality Control* 12 (7): 4-8.

Montgomery, D. C., and C. M. Mastrangelo. 1991. "Some statistical process control methods for autocorrelated data (with discussion)." *J. Q. Technol.* 23(3): 179–204.

Montgomery, D. C., C. L. Jennings, and M. Kulahci. 2011. *Introduction to time Series Analysis and Forecasting*. Hoboken, NJ: John Wiley & Sons, Inc.

Montgomery, D. C. 2013. *Introduction to Statistical Quality Control*. 7th Edition. Hoboken, NJ: John Wiley & Sons, Inc.

Ramírez, B., and G. Runger. 2006. "Quantitative techniques to evaluate process stability." *Quality Engineering* 18 (1): 53–68.

Ramírez, J. G., and J. Juan. 1989. "A contour nomogram for designing cusum charts for variance." Report No. 33, Center for Quality and Productivity Improvement, University of Wisconsin-Madison, Madison, WI. (unpublished)

Ramírez, J. G. 1989. "Sequential methods in statistical process monitoring." Ph. D. thesis, University of Wisconsin-Madison, Madison, WI. (unpublished)

Ramírez, J. G. 1998. "Monitoring Clean Room Air Using Cuscore Charts." *Quality and Reliability Engineering International* 14(4): 281-89.

Ramírez, J. G., and B.S. Ramírez. 2009. *Analyzing and Interpreting Continuous Data Using JMP: A Step-by-Step Guide*. Cary, NC: SAS Institute Inc.

Ramírez, J. G., and A. D. Zangi. 2014. "Designing Insightful Process Behavior Charts." JMPer Cable, Issue 29, Summer 2014. Cary, NC: SAS Institute Inc. https://www.jmp.com/content/dam/jmp/documents/en/newsletters/jmper-cable/29_summer_2014.pdf

Ramírez, J. G., 2016. "A Health Screening for Your Processes." JMP Foreword: 11-12. September 2016.

Ramírez, J. G. 2018. "The Cusum chart: A targeting radar." JMP Foreword: 17-118. March 2018.

Ramírez, J. G. 2018. "Discussion of "Scaling-up process characterization." *Quality Engineering* 30(1): 79-87.

Sall, J., 2018. "Scaling-up process characterization." *Quality Engineering* 30(1): 62-78.

Wheeler, D. J., and D. S. Chambers. 1992. *Understanding Statistical Process Control*, 2nd ed. Knoxville, TN: SPC Press.

Wheeler, D. J., 2006. *EMP III: Evaluating the Measurement Process & Using Imperfect Data*. Knoxville, TN: SPC Press.

Wheeler, D. J., 2009. "An honest gauge R&R study." Paper presented at the 2006 ASQ/ASA Fall Technical Conference. Knoxville, TN: SPC Press. http://www.spcpress.com/pdf/DJW189.pdf

JMP 14 Online Documentation. Quality and Process Methods online help. http://www.jmp.com/support/help/14/quality-and-process-methods.shtml

www.ingramcontent.com/pod-product-compliance
Lightning Source LLC
Chambersburg PA
CBHW081048220326
41598CB00038B/7019